1007401761

The Nature of Scientific Thinking

Also by Jan Faye

AFTER POSTMODERNISM
NIELS BOHR: HIS HERITAGE AND LEGACY
RETHINKING SCIENCE
THE REALITY OF THE FUTURE

The Nature of Scientific Thinking

On Interpretation, Explanation, and Understanding

Jan Faye
*Department of Media, Cognition and Communication,
University of Copenhagen, Denmark*

© Jan Faye 2014

All rights reserved. No reproduction, copy or transmission of this publication may be made without written permission.

No portion of this publication may be reproduced, copied or transmitted save with written permission or in accordance with the provisions of the Copyright, Designs and Patents Act 1988, or under the terms of any licence permitting limited copying issued by the Copyright Licensing Agency, Saffron House, 6–10 Kirby Street, London EC1N 8TS.

Any person who does any unauthorized act in relation to this publication may be liable to criminal prosecution and civil claims for damages.

The author has asserted his right to be identified as the author of this work in accordance with the Copyright, Designs and Patents Act 1988.

First published 2014 by
PALGRAVE MACMILLAN

Palgrave Macmillan in the UK is an imprint of Macmillan Publishers Limited, registered in England, company number 785998, of Houndmills, Basingstoke, Hampshire RG21 6XS.

Palgrave Macmillan in the US is a division of St Martin's Press LLC, 175 Fifth Avenue, New York, NY 10010.

Palgrave Macmillan is the global academic imprint of the above companies and has companies and representatives throughout the world.

Palgrave® and Macmillan® are registered trademarks in the United States, the United Kingdom, Europe and other countries

ISBN: 978-1-137-38982-4

This book is printed on paper suitable for recycling and made from fully managed and sustained forest sources. Logging, pulping and manufacturing processes are expected to conform to the environmental regulations of the country of origin.

A catalogue record for this book is available from the British Library.

A catalog record for this book is available from the Library of Congress.

Transferred to Digital Printing in 2014

Contents

Preface and Acknowledgements   vii

1 **Forms of Understanding**   1
    1 A subjective feature of explanation   2
    2 The unification view   4
    3 The causal-mechanical view   12
    4 The visualization view   15

2 **Understanding as Organized Beliefs**   24
    1 The content view   26
    2 Skills and intelligibility   34
    3 The organization view   37
    4 Embodied and embedded understanding   41
    5 Levels of understanding   48
    6 Norms of understanding?   51
    7 Cognitive schemata   56

3 **On Interpretation**   60
    1 Two notions of interpretation   60
    2 Interpretations in the natural sciences   64
    3 Roads to interpretation   68
    4 Interpretation and the discovery of scientific hypothesis   75
    5 Interpretation and the construction of scientific concepts   79

4 **Representations**   85
    1 What is a representation?   87
    2 Representing and understanding   91
    3 What do laws represent?   99
    4 The linguistic approach to theories   103
    5 Models as focal points of scientific explanation   108

5 **Scientific Explanation**   114
    1 Nomic explanations   117
    2 The Hempelian view   122
    3 The received criticism   125

6 **Causal Explanations**   136
    1 From embodiment to modal reflections   137
    2 Causal explanation as reflective understanding   142

3	Causation in physics	147
4	Causation in biology	149
5	Causal beliefs about causal facts	154
6	Context-dependent relations and context-dependent descriptions	157

7 Other Types of Explanations — 162
1. Types of explanations — 163
2. Structural explanations — 168
3. Functional explanations or evolutionary explanations? — 171

8 The Pragmatics of Explanation — 183
1. Why is explanation a matter of pragmatics? — 186
2. The contrastive dimension of why-questions — 193
3. van Fraassen on explanation — 195
4. Critiques of van Fraassen — 203

9 Not Just Why-questions — 210
1. Why-questions in science — 213
2. The paradigm argument — 216
3. The relation argument — 220
4. The reason argument — 225
5. The translation argument — 231
6. The contrast-class argument — 234
7. The relevance argument — 237

10 A Rhetorical Approach to Explanation — 241
1. The problem context — 241
2. Individual cognitive interests — 245
3. Explanation as speech act — 248
4. Explanation as a rhetorical means of communication — 251
5. The rhetorical situation — 255
6. Explanatory relevance — 259
7. Explanatory force — 262

11 Pluralism and the Unity of Science — 270
1. The epistemological issue — 272
2. The methodological issue — 276
3. The ontological issue — 279
4. Conclusion — 284

Notes — 289

Literature — 316

Index — 327

Preface and Acknowledgements

This is a book about the epistemology of science. It attempts to account for the scientific aim of understanding nature on the basis of a naturalized and pragmatic approach to the philosophy of science. My main purpose is to describe how our human cognitive capacities allow scientists to acquire an understanding of nature by means of representation, interpretation, and explanation. This means that this is not a book about the methodology of science focusing on certainty and the justification of scientific belief. These issues have dominated philosophy of science for a very long period of time since the logical positivists set up theories of confirmation and Popper concocted theories of verisimilitude or corroboration. I agree with the American philosopher Linda Zagzebski who remarks: "the concept of justification is associated both historically and conceptually with the perceived danger of scepticism."[1] The success of science proves more than anything that scepticism is a delusion of our philosophical imagination, which Descartes forcefully persuaded us to accept as a permanent challenge for any epistemic inquiry. It's time to leave this delusion behind. After Darwin I think it is correct to adopt the naturalistic standpoint that the capacity of humans to present the world perceptually is a product of evolution and due to our adapting to our external environment.[2]

Even though countering scepticism no longer rules the game within the profession of philosophy of science, there are many unsolved problems associated with the exact relation between nature, on the one hand, and the scientists' representing, understanding, and explaining nature, on the other. My purpose is to show not only that understanding shapes explanation but also that explanation provides understanding. I regard 'intention' and 'understanding' as central categories in a naturalized account of explanation, interpretation, and representation. A concept like 'intention' gives us the opportunity to see explanation, interpretation, and representation as having an empirical content in virtue of being the result of human intention and constituting important parts of scientific communication. In this light, an explanation is an intended report of the explainer's understanding.

First and foremost I argue that *understanding is distinct from knowledge*. In some situations it seems right to say that a person understands something without possessing knowledge. For example, there are situations where an explanation arises from understanding, but where this pre-existing understanding later turns out to be based on factually false beliefs, or those situations where it is impossible to prove that the pre-existing understanding is correct or incorrect. Think of explanations provided by past, now discarded scientific theories, or think of religious or philosophical explanations that are impossible to test empirically. In those cases, what justifies the explanation seems to be the understanding which shapes the explanation. But just as an explanation may be false, the understanding on which it builds may be inadequate.

This brings me to another long-lived philosophical dispute between the Platonist and the Aristotelian views of human understanding. Reviewing the huge literature on explanation, one realizes that most is inspired by the Platonist dream of a mind-independent understanding of the world. According to this epistemological orientation, real understanding requires apprehending true propositions; it does not concern people's beliefs in them. Explanation yielding such an understanding is not man-made, but the objective result of scientific activity. Platonists consider 'understanding' to be the product of relating true propositions together such that ideally they fit into a unified deductive system. Although Platonists may admit that there is indeed a worldly pragmatic side of explanation, they insist that it can be ignored because it does not contribute to the God's eye view. In contrast, the Aristotelian view sees human comprehension as internal rather than external to the natural world. Understanding is mind-dependent, even though it may concern objective matters. It is mind-dependent because it involves people's actual beliefs and their relations and organization. Therefore explanation that expresses people's understanding depends both on their mental resources as well as on the object of their understanding. So the explanatory product is as much a result of human explanatory activity as of the facts themselves. According to the Aristotelians, pragmatics of explanation should not be disregarded because it is essential for understanding of explanations in everyday life as well as in scientific practice. This book stands by the Aristotelian view.

I propose that in its origin, understanding is a neuro-biological way of organizing our everyday beliefs. But when we are dealing with cognitively advanced levels of reflection, as we are in science, the organization of our beliefs is enhanced by the use of theories and models. However, theories and models help us produce explanations that can be justified by observation and experiment. My thesis is that an explanation is an answer to an explanation-seeking question, and as such I see explanation as part of a *rhetorical practice of communication*, the function of which is to convey information and understanding. An interpretation helps the interpreter to gain understanding, whereas an explanation expresses the explainer's understanding and communicates it to the explainee, thereby providing understanding to her. In short, this book proposes a theory in which explanation, interpretation, representation, and understanding, are connected by considering "interpretation" in one sense of the term as the construction of a representation and by regarding it in another sense as a kind of explanation.

From the perspective of interpersonal communication, a theory of explanation should not be normative in the sense of holding that an answer must fulfil certain logical or ontological constraints to be considered an explanation. Thus, I neither stipulate the sorts of formal conditions associated with the approach of Carl Hempel and those who have followed his lead, nor demand causal connections as necessary for acceptable scientific explanations as has been maintained since Aristotle. Whether an explanation must fulfil certain norms for being considered 'satisfactory' in science (or in any other areas where people produce explanation) is contingent on the situation in which the utterance of the explanation acts as a response to an explanation-seeking question. As long as an answer is a response to an explanation-seeking question and provides the questioner (and others) with understanding, it *necessarily – ipso facto –* functions as an explanation. I shall argue that what characterizes a pragmatic view of explanation, in contrast to other views, is that a pragmatic approach primarily focuses on *the context in which the explanation-seeking question is asked* and the *functional role of the answer in terms of providing understanding for the questioner*. In short, a pragmatic approach attempts to give a descriptive analysis of explanation as a human, linguistic activity rather than a prescriptive definition of the sort given by the deductive-nomological model or the causal accounts. In this sense, what follows is thoroughly descriptive in intent and makes no pretence of providing a normative ideal for evaluating better or worse acts of explaining or instances of understanding.

Others before me have defended a pragmatic theory of explanation, most notably in recent times, Bas van Fraassen. In Chapter 8 I discuss his approach at certain length. On the one hand, I believe van Fraassen has gone astray in his account of explanation by failing to include – in a logically necessary way – the intentions of the explainer or the explainee. On the other hand, Peter Achinstein recognizes the importance of intentions, but I find that he misses an important point by demanding that an answer must be true to count as an explanation. In Chapter 10 I take a look at his approach and characterize the main difference between his view, which sees explanation as an *illocutionary* speech act, and my own, which sees it as a *perlocutionary* speech act. This latter view fits with my pragmatic-rhetorical approach to explanation.

The outline of the book is as follows. Chapter 1 presents a summary of how philosophers of science have regarded the topic of understanding, and its connection with explanation, in the light of earlier philosophical claims. Chapter 2 presents my view that understanding is a question of the organization of a belief system and contrasts it with what I call the "content view" and the "ability view." I also propose that understanding comes in two variants: embodied understanding, which we share with other animals, and reflective understanding, which can be associated with interpretation, explanation, and other forms of theoretical representation.

Chapter 3 focuses on interpretation in science. I argue that interpretation is concerned with understanding meaning and that it arises whenever we face a representational problem. Interpretation comes in two versions. In one we explain an already given meaning, and in the other we try to construct the meaning of something that has not yet been understood to our satisfaction. This leads us to Chapter 4, which is devoted to our concept of representation and to how theoretical constructions, like scientific models and theories, are used to express and produce scientific explanations. The main idea is that neither theories nor models are true or false by themselves. What may be true or false are the explanations that they enable us to produce. I argue that theories are nothing but linguistic rules which scientists use to describe the structures of idealized scientific models, which are abstract or concrete representations that may be good or bad, precise or imprecise. Based on these models scientists may produce hypotheses by which they explain concrete phenomena, and these are either true or false.

For a long time Hempel's 'covering law' or 'deductive-nomological' model of explanation was the received view of explanation. The main reason why it failed was because it regards *one type* of explanation, namely nomic explanation, as defining the form of *all* scientific explanations. As long as Hempel's model is confined to that subset of explanations expressed in terms of laws, and no pragmatic aspects of explanation are ignored, it may still give us a fair analysis of what an ideal *nomic* explanation might be. In Chapter 5 I deal with nomothetic explanations focusing on Hempel's covering law model and its requirement of deductive subsumption under a 'law' statement. To this extent I agree with Hempel's view that nomic explanations need not be causal explanations, but apart from this special (but very important) case of nomic explanations, I disagree with Hempel's logical way of seeing explanation as context-free, and I reject the view that this model covers all forms of scientific explanations.

As an alternative to Hempel's logical-formalist approach, in Chapter 6 I show how causal explanations are context-sensitive and that explaining something causally requires describing it from a certain standpoint. We never just explain something; we always do it from a particular perspective, focusing on some features and ignoring others. An explanation is always about something; it is never of something. Whether an explanation is good or bad depends on the interests and background knowledge of those evaluating the proposed explanation. If this is correct, it also means that the ontic view of causation, the view according to which causes explain their effects in virtue of a causal relation between them, is mistaken since causation takes place regardless of the explainer's interests and background knowledge. Rather, people offer causal explanations, and no fact explains other facts alone by itself.

Explanation is one of the most debated concepts in philosophy of science; yet, there is little consensus among specialists on how to characterize it. Three main approaches appear to be alive today: the *formal-logical* view, the *ontic* view, and the *pragmatic* view. Among these classes of philosophical theories there is little hope for agreement. In his review of the many rival models of explanation, Newton-Smith acknowledges that each provides insight into a different aspect of explanation, but he also claims that the lack of a deeper unifying theory is "an embarrassment for the philosophy of science."[3] There seems to be little common ground beyond the expectation that explanation is meant to provide understanding by providing some particular information about factual matters. The pragmatic view, however, has at least one advantage over the others, because this approach does not deny that the other conceptions of explanation aim at real and actual goals of scientific explanation, insofar as the explanatory context may make it appropriate and fruitful to pursue some of these goals. But pragmatists will deny both that the alternative models really define what scientific explanation is and that they can cope with all possible forms of scientific explanation. Rather their conclusion is that there are many different types of explanations, inside and outside science, depending on our particular cognitive interests. The hunt for the essence (so to speak) of 'scientific explanation' is a philosopher's error. Chapter 7 presents some of these other types of explanations with a focus on structural and functional explanations. It argues that even though causal understanding of the world underlies these forms of explanation, they are not reducible to causal explanation because they communicate information that fulfils different cognitive interests than purely causal information.

All theories of explanation must address these distinct features of explanation. As we shall see, philosophers who regard explanation as a logical argument also think that scientific explanations are distinct from everyday explanations. In contrast, philosophers who regard explanation as a communicative act think that everyday explanations and scientific explanations are not in principle different in kind. The difference between them lies in the emphasis that scientists put on rigour and generality. Other philosophers, like van Fraassen, hold that scientific explanations are responses to why-questions, and only to why-questions. In Chapter 9 I argue that responses to other sorts of questions such as how-questions and what-questions can also provide us with explanations as long as these responses convey the desired understanding. A question of any syntactic form can function as an explanation-seeking question in a given context if answering it requires that the explainer relates the topic of the question with some other information.

Of course we want our explanations to be true. But as mentioned above, an explanation need not be true to count as an explanation; even a false explanation may still be explanatory. To count as good (in contrast to true) an explanation must be informative, relevant, and convincing. A good explanation has to meet certain standards depending on the context defined by the topic of explanation, the background assumptions, and the communal and personal interests. Chapter 10 claims that *rhetoric* offers the best pragmatic approach to understand explanations considered as expedient responses to explanation-seeking questions. Here I introduce the notion of the explanatory situation, a notion that draws on Lloyd F. Bitzer's concept of the rhetorical situation.

In Chapter 11 I conclude by discussing pluralism in science. My pragmatic approach to explanation and understanding underscores pragmatic pluralism and perspectivism. However, because of the naturalization of human cognition I still believe that the positivistic 'unity of science' movement carried some valuable insight, though it held that this unity lies in the use of an observational language. As belonging to *Homo sapiens*, humans are formed by their biological evolution. Therefore, it seems natural to think that basic epistemic methodology is a result of this adaptive selection; it is ingrained in our neurons and manifested as 'cognitive schemas' which determine the acquisition of knowledge. This basic capacity reappears in practice while we are doing science, including the social and human sciences, and becomes formulated as methodological principles whenever we deliberate about scientific practice. Thus, like the positivists of old, I see methodology as the unifying principle in all sciences, but unlike them I ground the unity of that method not in the alleged common use of an observational vocabulary, but in the presupposition that we *Homo sapiens* all share common cognitive capacities shaped by natural selection.

Some of the ideas, which I will defend and develop here, have been presented in a couple of papers: "Explanation Explained" in *Synthese 120*, 1999, 61–75; "The Pragmatic-Rhetorical Theory of Explanation" in J. Persson & P. Ylikoski (eds) *Rethinking Explanation*. Series: *Boston Studies in the Philosophy of Science Vol. 252*. Dordrecht: Springer Verlag 2007, 43–68, and "Interpretation in the Natural Sciences" which appeared in M. Dorato, M. Redei, M. & Suarez (eds) EPSA07, *Proceedings of the First Conference of the European Society of Philosophy of Science*, Springer 2009, but most of the ideas and arguments have not been introduced before.

When I met my wife, Lisa Storm Villadsen, a rhetorician, it was at a time when I also became interested in the problems of explanation. Her lasting impact on me had not only consequences for my personal life but also for my views on explanation and interpretation. I realized that there was a strong rhetorical dimension connected to scientific explanation, which I recognized was of epistemic importance and had been relatively ignored by most philosophy of science. Here I wish to present her with my heartfelt thanks for the love she has given me in so many ways. As a result of my gratitude, I want to dedicate this new book to her.

Since then I have been working with explanation and interpretation in the context of both the humanistic sciences and the natural sciences, and I have had the luck to discuss problems concerning explanation and interpretation with many colleagues and students. In particular, I want to thank Alexander Bird, Finn Collin, Henk de Regt, Dennis Dieks, Olav Gjelsvik, Sara Green, Bengt Hansson, Michael M. Karlsson, Johannes Persson, Stathis Psillos, Matti Sintonen, Rebecca Schweder, Erik Weber, Petri Ylikoski, Eugen Zeleňák, and my students Thomas Basbøll, Jacob Birk Olsen, and Mads Sørensen. At one time or another we have discussed our common interest in understanding explanation. Ylikoski, Zeleňák, and Green have also given important comments on parts of the manuscript. I also wish to thank Mark Tschaepe for exchanging ideas of common interest concerning pragmatic theories of explanation. But I owe one person more thanks than anybody else. The generosity of my colleague and long-time friend Henry Folse has once again been immense as he took the time to polish my English and to make an overwhelming number of valuable comments. The benefit of his unselfishness is tremendous.

Notes

1. Zagzebski (2001), p. 239.
2. Writing about Roy Wood Sellars' evolutionary naturalism, Pouwel Slurink (1996) states the realist position behind Darwinism head on: "There has to be a reality outside the organism if the concept of adaptation is to make any sense, and the capacity for knowledge and even for science enables certain organisms to adapt themselves to their cosmic environment."
3. Newton-Smith (2000), p. 132.

1
Forms of Understanding

In every area of human discourse we meet with explanations. We find them in the church, at the marketplace, in hospitals, in locker-rooms, in news media, in public administration, in the parliament, in private companies, in laboratories, and at the universities. Successful explanations are attractive because they provide us with an understanding of what they are supposed to explain. Yet, the world appears quite differently seen from the church, or the man on the street, or a research laboratory. Because of these different contexts people addressing the same kinds of phenomena may propose different kinds of explanations. Thus the complex and multifarious notion of explanation becomes a source of heated debates and controversies.

Even when we restrict the conversation to *scientific* explanation, these debates have not produced, as Newton-Smith notes, "some deeper theory of explanation that explained what it was about each of these apparently diverse forms of explanation that makes them explanatory."[1] Today, although the consensus that explanations are explanatory because they yield understanding is growing, views about understanding remain very divergent.[2] Does all 'scientific understanding' have some distinct characteristic (or essence?) that distinguishes it from other forms of understanding? Is *scientific* understanding context independent in a way understanding in general is not? If scientific understanding is context independent, then every example of scientific explanation shares some common feature characterizing every scientific explanation in general independently of its aim, topic, or use. Furthermore, if understanding is often associated with knowledge, comprehension, intelligibility, and meaningfulness, do any of these associations help us to see what understanding, and in particular scientific understanding, involves? These are the questions this chapter addresses.

1 A subjective feature of explanation

Until recently it was quite common to argue that the notion of understanding is subjective and therefore could not help us in grasping the nature of explanation. Carl Hempel, for instance, assumed "such expressions as "realm of understanding" and "comprehensible" do not belong to the vocabulary of logic, for they refer to psychological or pragmatic aspects of explanation."[3] Despite his reluctance to deal with the pragmatic aspects of explanation, as we will return to in a later chapter, Hempel implicitly felt, so it seems, that the pursuit of science to produce explanations needed to be explained. Obviously in science as in everyday life, we pursue explanation because it provides us with understanding, but logic alone cannot give us an understanding of this phenomenon. Logic gives us a way to understand why the explanandum is expected when we are aware of the empirically determined initial conditions, i.e., it was logically necessary, given these empirical conditions, and naturalism presumes humans are hard wired to "expect" psychologically what is logically necessary. But Hempel's own opinion concerning understanding is that "understanding" simply refers to a psychological state, so the only explication of 'understanding' is a psychological one. Understanding in this psychological sense is caused by our expectation. He claimed: "the argument shows that, given the particular circumstances and laws in question, the occurrence of the phenomenon *was to be expected*; and it is in this sense that the explanation enables us to *understand why* the phenomenon occurred."[4] Indeed, such an expectation need not be subjective; it is objective or rational in the sense that being equipped with the relevant information about the particular circumstances and laws in question, including knowledge of deduction, most people, at least to the extent that they are rational, would not be surprised to know that the explained phenomenon did occur.

The fact that understanding comes with explanation is not a contingent feature of explanation but is the purpose of making explanations. Achieving understanding is the aim of explanation as of interpretation. Scientific explanation not only delivers one kind of understanding but also presupposes another kind. This presupposed understanding exists as background assumptions, beliefs, skills, and tacit knowledge. Moreover, since understanding involves persons, understanding is contextual. Its content may vary from one person to another. However, since we want understanding to have an epistemic role, and not to reduce it to a mere psychological state, we need to understand the cognitive *use* we make of understanding. Thus the proper notion of understanding is pragmatic in the sense that it depends on the function it serves for human agents. Although Hempel agreed that there are pragmatic elements in the way explanation functions in scientific practice, he held the pragmatics of an explanation had nothing to do with its quality as explanatory. Thus his covering-law model is abstracted from any pragmatic context. An explanation – nothing more than a logical argument – provides understanding of the phenomenon to be explained by showing that it was to be rationally expected. Any other characteristic of understanding is subjective and can play no epistemic role.

Like Hempel, J. D. Trout argues that we cannot separate understanding from the sense of understanding. As he claims: "The psychological sense of understanding is just a kind of confidence, abetted by hindsight, of intellectual satisfaction that a question has been adequately answered."[5] The sense of understanding cannot count as an epistemic virtue, because "confidence is, notoriously, not an indicator of truth."[6] Rather he believes that the best accounts of explanation are objectivist by eliminating those properties that are dependent of the psychology of the explainer. Concerning our sense of understanding Trout emphasizes that "the sense of understanding is not produced by a reliable relation between belief and truth."[7] We can have both true and false beliefs and both may give us a sense of understanding but it is only true beliefs that can give us genuine explanation.[8]

Against Trout, Henk de Regt has objected that since scientists like Galileo, Newton, and Bohr supported false theories, "[t]he relevant question is not whether these scientists were wrong but whether they were irrational."[9] This reflects the prevailing view that the epistemic question is not whether scientists possess false beliefs but whether they have acquired them in a rational or a non-rational way. I agree with this and with de Regt's remark that false theories can have explanatory force: "Today we can still explain phenomena with these false theories and rightfully experience a sense of understanding."[10] But I disagree with the weight de Regt throws on 'rational.' Did Kepler arrive at his laws "in a rational way"? If you believed in his cosmology, presumably it seemed 'rational,' but to us it hardly seems so. Unlike 'truth,' which most regard as non-relative, what is 'rational' depends very much on one's world view; an actor might regard a certain action as perfectly rational, while to a spectator with a different world view it seems devoid of reason.

Thus, Trout fails to recognize that although the sense of understanding is a psychological concomitant of epistemic understanding, this feeling of understanding is not the same as having genuine epistemic understanding. Sometimes one may enjoy having understanding without having a sense of understanding, as seems to be the case with our understanding of our practical skills. Also Trout fails to realize that an appeal to so-called objective factors is no guarantee of truth and that scientific education and training has formed the psychology of the individual explainer. Even if a scientist shares the same norms of understanding as other scientists of his community, he (and the community) may still have an understanding based on false beliefs about the facts. Of course, one can truly feel that one understands, although in fact one does not really understand, because that feeling of understanding may be engendered by false beliefs. Thus the Creationists certainly *feel* that they understand the living world and its diversity, but in fact they are wrong and this confidence that they understand is of course unjustified. As we shall see, whatever the 'objective' norms of understanding are, they are valid only with respect to a certain historical context of science or a particular context of problems.

There is no shared opinion among those who hold that explanation delivers 'objective' understanding. Some philosophers like Michael Friedman and Philip Kitcher think of understanding in terms of *theoretical unification*. Wesley Salmon holds that our understanding of the world consists of knowing the underlying *causal mechanisms*. However, James Cushing opts for *visualizability* as the marker of understanding, and others like Newton and Einstein have pointed to *locality* and *determinism*. Henk de Regt and Dennis Dieks have truly pointed out that these proposed 'objective' standards of intelligibility all fail to hold in every historical situation.[11] But they also fail because their proponents' use them to tell us what gives us objective understanding, but not why they give us that understanding. Answering this question requires that we know what understanding is. In the following three sections I shall analyse and criticize each of these proposals for being universal criteria of understanding.

2 The unification view

Michael Friedman's and Philip Kitcher's unification models of explanation have been developed in close response to the objections to Hempel's 'covering law' model. Friedman's notion of scientific understanding is 'global' by advocating that science encapsulates more and more phenomena into a single explanatory scheme, but Kitcher's notion is 'local' by allowing a plurality of explanatory schemes.

Friedman claims scientific explanation provides understanding by the unification of general phenomena:

> On the view of explanation that I am proposing, the kind of understanding provided by science is global rather than local. Scientific explanations do not confer intelligibility on individual phenomena by showing them to be somehow natural, necessary, familiar, or inevitable. However, our over-all understanding of the world is increased; our total picture of nature is simplified via a reduction in the number of independent phenomena that we have to accept as ultimate.[12]

Our scientific understanding of the world increases to the degree in which the total number of independently acceptable assumptions needed to explain the phenomena decreases.[13] A 'phenomenon' here is not an individual, but the consequence of a general regularity that can be expressed by law-like statements.[14] In fact, Friedman denies that explanation of one singular fact by reference to another fact is important in science. What is important is that empirical laws such as Kepler's laws of planetary motion and Galileo's law of the free fall can be derived from Newton's fundamental law of gravitation and certain additions. In order for Friedman's proposal to work, he needs a method for counting the acceptable assumptions in any given explanation. The rough intuition here is that each acceptable law-like sentence is an individually acceptable assumption, if there are sufficient grounds for accepting it, grounds which are not sufficient for accepting the others. Friedman's contention that the scientific enterprise seems to be the continuous effort to integrate such general phenomena into an overall explanatory schema is not so far from the covering laws view in the sense that Hempel and Oppenheim already argued that higher-level laws with a broader scope had the power of explaining lower-level laws with a narrower scope. The power comes from the deductive inference of the lower-level laws from the higher-level laws.[15] So, Friedman maintains, it is not expectation which gives us understanding, but the fact that a high-level law unites different low-level laws formerly considered mutually independent.

Friedman objects to the covering-law model on the grounds that contrary to Hempel's assumption scientific explanations rarely deal with singular events. Far more often science is concerned with explaining general regularities or patterns of events. (Astronomy and geology may be exceptions.) Hempel realized the covering-law model cannot easily cope with general regularities because the explanandum must be restricted to singular sentences. The covering-law cannot block the possibility that regularities, like Kepler's laws, are derivable not only from Newton's laws but also from a conjunction of Kepler's laws and, say, Boyle's law.

Friedman proposes that scientific understanding consists in having as few unexplained phenomenon or regularities as possible, and that explanation "increases our understanding of the world by reducing the total number of independent phenomena that we have to accept as ultimate or given."[16] Understanding cannot come with rational expectation since expectation is always at a particular time, but general regularities do not occur at any definite time. Scientists do not experience any expectation when these laws are unified under the Newton's laws of motion. Rather they get their understanding from the fact that Kepler's laws together with Galileo's law, Hook's law, and Huygens' law can all be united under one set of general laws of motion as specified in Newtonian mechanics.[17]

Thus, the goal of unification is cognitive economy. A successful explanation results in a decrease in the number of primitive or non-derivative postulates. But explaining so-called lower laws by higher laws confronts the inevitable logical point that the highest laws will have to go unexplained and simply taken as given, a brute contingency that flies in the face of 'scientific' explanation. Moreover, the unificationist criteria of understanding fail to do justice to singular explanations because in every single case we need at least one premise stating the initial conditions in order to be able to derive the fact to be explained.

Unfortunately, Friedman is rather vague on precisely what it is about unification which furnishes us with scientific understanding. It cannot be because unification embeds the general phenomenon in a causal pattern. The kind of understanding it provides has something to do with theoretical integration. Mill and Ramsay provide a hint: in characterizing laws of nature as statements, they point to a system of sentences expressing the fewest and most simple assumptions from which any regularity in the world might deductively be derived. David Lewis improves the unification view by maintaining that the system of laws must achieve the best balance between simplicity and strength.[18] These features can oppose one another; a theory may be strong in having a high amount of information content, at the expense of simplicity or vice versa. [19] The balance between strength and simplicity must be 'fair,' but what one considers 'fair' seems to depend on pragmatic considerations.

Obviously, Ramsay and Lewis regarded laws of nature as sentences and not as natural relations among phenomena. I agree that *some* laws are nothing but sentences about the explicit use of a language, namely those called principles or theoretical laws, but I disagree in holding that other laws are generalizations about causes among phenomena. So perhaps we can exploit their ideas to say something about theoretical unification.

But first I want to examine van Fraassen's disagreement with Lewis' analysis of laws.[20] He thinks that Lewis' account has problems with the identification of laws. He correctly points out that the notions of strength and simplicity are not straightforward and their degree is often reciprocally related. They also fail with respect to the historical development and the actual pursuit of science. Moreover, strength and simplicity are not translation invariant from one theoretical framework to the next. He concludes that Lewis' account of laws does not guarantee laws have explanatory powers. The only way to escape this objection would be to argue that laws have to be formulated in a correct language referring to natural classes. But van Fraassen rejects such anti-nominalism because actual science seems to move away from real distinctions in nature. We can easily imagine a new theory which is more informative than its alternatives but whose theoretical terms are not definable in terms of the old theory and therefore do not stand for natural classes.

Of course, the persuasiveness of van Fraassen's objection depends on one's understanding of information and explanatory power. Although I believe that van Fraassen is right in opposing essentialism, I am not convinced by his argument. First, if we assume one can create a completely new language which cannot be defined in terms of the old one, then one of the premises in van Fraassen's objection collapses, namely that strength and simplicity are not *translation* invariant. For if no translation is possible, it is absurd to complain that strength and simplicity are not invariant under translation. But if we assume a translation is possible, then those new terms which translate the old ones would stand for natural classes. Second, the practice of science requires that some fundamental concepts are carried over into the new theory because this is merely a general extension of our everyday understanding. Van Fraassen aims at something of importance in pointing out that theoretical terms need not refer to a real type of phenomena in the world. But the question of reference is a question about the difference between a situation in which these new terms can occur in the causal language of observation and experimentation, and one in which they occur only in the formulation of theoretical laws. Nevertheless, if theoretical laws don't refer to certain objective facts of nature, how can they carry information or have explanatory power?

Let us pursue this issue by distinguishing between two modes of unification. One concerns the *integration of principles* and the other the *integration of concepts*, both having a dimension of generality and a dimension of parsimony.[21] For instance, we may achieve a gain in generality by adding higher-order principles to our theoretical system without having a reduction in the number of lower-order principles. The gain in generality corresponds to a loss of parsimony. And the same holds for concepts too. Therefore, I propose that a theoretical system has the power of 'principle integration' if it reduces the number of low-level principles concerning different domains to fewer higher-order principles in which the various domains unify into one domain. Similarly, such a system has the power of 'conceptual integration' if it is able to systematize our knowledge in a manner allowing for fewer kinds of basic entities and properties than previous systems. Usually such systematization replaces a multitude of different concepts with a very narrow scope with fewer concepts of much broader scope. However, conceptual integration does not necessarily imply principle integration, although the implication seems to hold the other way around. For instance, the introduction of the notion of basic elements was meant to reduce all the various chemical forms of matter to compositions of these elements without evoking a reduction of principles. In fact its introduction required adding basic principles concerning atomic structure. But the capacity of deriving Kepler's laws, the law of a falling body, and the laws of tides from Newton's laws was possible only by introducing a concept of a universal gravitational force acting between heavy bodies.

Using this distinction we can see better how understanding rises from unification. Naturally we would expect conceptual integration to produce a change in the ontological commitments of the specific domain from a richer and more diverse ontology of the previously prevailing system to one which is more parsimonious and homogeneous. Once scientists believed the celestial laws were different from Earthly ones, but all agreed that our understanding of Nature greatly increased when Newton united both realms under a single set of laws. So understanding comes with the realization that a great number of seemingly different and unconnected phenomena can be accounted for in terms of general principles involving only a fewer number of basic entities. Thus, unification as integration of both principles and concepts advances theoretical understanding by providing explicit rules for how to use a uniform and more parsimonious descriptive language to cover many different phenomena.[22]

Although the minimization of formative concepts and constitutive principles is one part of the understanding upon which science relies, it is far from the sole aim of science. For instance, scientists may appeal to alternative models in seeking the best explanation of a certain phenomenon. The choice of any particular model depends on the exigency and the interests of the scientists. A variety of conflicting models might not be capable of being unified into one single scheme, and scientists use mutually exclusive models because they also prefer the most accurate and precise explanation rather than an explanation which is less accurate and less precise, but which can be covered by fewer unifying principles. The facts that (a) scientists do use alternative models, which according to their ontological (and logical) commitments cannot both be true, and (b) that they do so because such models allow greater accuracy and precision in empirical predictivity in different applications show us that while unification is certainly *an* epistemic value, other values related to empirical support can override the desire for unification.

Philip Kitcher offers a somewhat different approach. He focuses on the logical form of explanatory arguments rather than on the reduction of independent laws. Science does not explain each individual phenomenon by its own unique argument, as in Hempel's model, but similar phenomena are explained using the same explanatory pattern over and over again. An explanatory pattern is an argument structure in which the logical terms have been replaced by variables. As Kitcher says: "to grasp the concept of explanation is to see that if one accepts an argument as explanatory, one is thereby committed to accepting as explanatory other arguments which instantiate the same patterns."[23] For instance, Kepler's law explains the orbit of Mars around the Sun, but it also explains the orbit of all other planets. The pattern of the argument is the important feature because it is the logical form of the argument that gives its explanatory power. A theory unifies our beliefs by furnishing us with patterns of arguments, which can be used to derive a large number of acceptable sentences. The use of the same schematic argument, including filling instructions for the dummy variables, on different phenomena creates the explanatory unification. Kitcher, however, does not think that an explanation is identical with the argument itself, but it consists of an act which draws on arguments supplied by science.

Of course, Kitcher's philosophical values differ from Hempel's; he does not want to reduce scientific explanation to logical form, and he has no illusions about science as a body of propositions.[24] So his view should not be seen as a response to the problem with the covering-law model. It was problems with that positivist way of doing philosophy of science that precipitated a revolution making Hempel's project seem archaic against today's backdrop. Thus, Kitcher's theory of explanation is best seen not as replacing the old Hempelian theory, not as an alternate move in the same game, but rather as playing a different game altogether. Although this observation is true, I suggest that it makes sense to classify Kitcher's theory as an objectivist and non-contextual approach yearning for uniform criteria in terms of the schematic structure of an argument, which every scientific explanation must satisfy. But in contrast to Hempel, his view also incorporates an important pragmatic element of understanding because he does not exclude the possibility that the pattern of argument may vary depending on the class of explananda. Classifications are not always straightforward; for instance, Popper was not a logical positivist, but with respect to how he thinks of scientific explanation his way of doing philosophy was not so different from the style of logical positivism. While logical positivists searched for formal models of confirmation, Popper searched for formal models of corroboration and verisimilitude. In this perspective, it is reasonable to say that Popper and the logical positivists shared a formal-logical approach to philosophy of science. Likewise, Kitcher's unificationism can be thought of as defending and improving on the basic approach behind the traditional covering laws model. This seems also to be Kitcher's own view while he quotes Hempel as saying that scientific explanation aims at "an objective kind of insight that is achieved by a systematic unification."[25] He then adds: "This unofficial view, which regards explanation as unification, is, I think, more promising than the official. My aim...is to develop the view and to present its virtues."[26]

If we take a closer look at the higher levels of theory construction, we see that understanding does not necessarily come from unification under a common set of principles or a certain pattern of argument. There are different types of unification, bottom-up or top-down, and which one is preferred depends on one's cognitive interests. Long ago Einstein distinguished between *constructive theories* and *principle-theories*. Constructive theories "build up a picture of the more complex phenomena out of materials of a relatively simple formal scheme from which they start out," whereas principle-theories start out from "not hypothetically constructed but empirically discovered ones, general characteristics of natural processes, principles that give rise to mathematically formulated criteria which the separate processes or the theoretical representations of them have to satisfy."[27] The cognitive values, which characterize the constructive approach, are 'completeness,' 'adaptability,' and 'clearness,' while the values of the principle approach are 'logical perfection' and 'security of the foundations.' So theory-building may either take its departure from basic constituents and processes of nature, or it may start out by postulating certain principles which are arrived at by empirical means.

Unsurprisingly, Einstein saw the theory of relativity as a result of the latter view insofar as it had proven empirically impossible to find evidence in favour of theoretical asymmetries in electrodynamics or for the expected variation in the velocity of light with respect to the ether.[28] These negative experiments gave Einstein the insight to formulate his two basic principles of special relativity, namely that the laws of nature are invariant with respect to all inertial frames and that the velocity of light in a vacuum is the same in relation to all inertial frames. In contrast to his approach in special relativity, Einstein categorized the kinetic theory of gases as a constructive theory. Also the Bohr-Sommerfeld core model of atomic structure and Heisenberg's matrix mechanics might be said to belong to the former category. But the picture here is much more mixed.

Bohr began his program with two postulates, neither of which seem to be a result of empirical discoveries nor consistent with well-confirmed principles of electrodynamics. In fact they explicitly contradicted the claims of well-confirmed electrodynamic principles. Theoretical constructions on the basis of Bohr's model were set up in a determined attempt to understand an immense amount of spectroscopic data. Furthermore, the dynamical behaviour of the interactions of atoms and radiation were explained by Bohr's frequency law and – when this law was proven insufficient for understanding the emission spectra of elements other than the hydrogen – by Heisenberg's matrix mechanics. Both Bohr and Heisenberg have several times claimed that matrix mechanics, which involves only observable quantities, was established with the intention of obeying Bohr's correspondence rule, his methodological requirement on theory-building in atomic physics. In his Faraday Lectures from 1932, for instance, Bohr emphasizes: "A fundamental step towards the establishing of a proper *quantum mechanics* was taken in 1925 by Heisenberg who showed how to replace the ordinary kinematical concepts, in the spirit of the correspondence argument, by symbols referring to the elementary processes and the probability of their occurrence."[29] The correspondence principle holds that a transition between stationary states is allowed if, and only if, there is a corresponding harmonic component in the classical motion. Nonetheless, the replacement of the ordinary kinematical concepts gave the new quantum mechanics a much more constructive basis than the original core model.

Now, if we take Einstein's statement concerning the virtues of bottom-up or top-down explanation, one would imagine that it is a purely contingent matter which of them you follow, depending on the virtues that attract you most as a scientist. An interesting study has been made by Dennis Dieks showing that in spite of the fact that the special theory of relativity is considered as a top-down account of the famous length contractions and the time dilations, since these two effects are deducible from Einstein's two principles, these effects can also be understood in terms of a constructive, bottom-up account. In this approach one starts with atoms and forces following the same approach as H. A. Lorentz did when he explained the null result of the Michelson-Morley experiment. As Dieks says:

> This does not deny that the top-down derivations that have become standard in the literature also possess explanatory value and lead one to understand why these effects must exist according to relativity theory. The choice between these different explanatory strategies has a pragmatic character and depends on contextual factors. There is no clear-cut and general difference between the two types of explanation with regard to their power to generate understanding, because the notion of understanding is contextual in the same way explanation is.[30]

After Dieks has convincingly argued his case, he concludes that "there is no best way of explaining the relativistic effects." It is the scientist's interest that determines whether she prefers a top-down or a bottom-up schema of understanding as the best. Hence, explanation and understanding are relative to the questions we ask and the interest we have. I cannot agree more.

Sometimes scientists are interested in pursuing unification based on general principles and their exigencies will indeed reflect these interests, but sometimes they are interested in finding causal mechanisms based on an understanding of the constituents of particular processes, and at yet other times they look for a structural account depending on the explanatory situation in which they stand. Explanatory understanding in science and common life comes in many forms and can be reached in many ways. No form is more correct, or more objective than any other.

3 The causal-mechanical view

While the deductive-nomological approach continues to be defended (or assumed) by some philosophers of science, other philosophers have given up on logical-formal approaches to understanding. They may still agree with the formalists that the standards of understanding stipulate 'objective' virtues, but they reject the formal-logical approach because it appeals, they believe, to a metaphysically dubious realm of abstract ideas and structures. Instead, they propose that scientific understanding is something that appeals to causal mechanisms or other factual structures. This approach may be called the *ontological* view of understanding. Wesley Salmon called this the ontic form of explanation rather than the epistemic form, arguing that scientific explanation "consists in exhibiting the phenomena-to-be-explained as occupying their places in the patterns and regularities which structure the world."[31] In contrast to the positivists' logical approach, preserved by Hempel, the idea here is that facts (or events) explain other facts (or events), and if a fact or an event occurs in a causal process, this process explains it. In particular, causes explain their effects because a cause tells us why its effect happens. The explanatory strength of science comes from its insights into various causal processes. Information about such processes can be used to explain effects with reference to their causes. A scientific explanation is an objective account about causal connections between things (or events) in the real world. Human cognitive representation of the facts of the matter does not contribute to the meaning of explanation. An explanation is both true and relevant if, and only if, it discloses the real causal structure behind the given phenomena. Furthermore, an everyday account counts as an explanation only if it is reducible to scientific talk about causal processes.

Moreover, proponents of the ontological view also do not wish to count functional and intentional explanations among genuine scientific explanations. They see them as appealing to metaphysically questionable end states. They believe that such explanations are atavistic remains from a time when gods or goddesses were seen as the causal agents in natural processes, and thus belong to folk biology and psychology. Sciences that may use functional or intentional explanations do not meet the objective standards of proper science and are still to be regarded as immature. Thus, the scientific aim is to replace such explanations with genuine causal explanations.

Defenders of the ontic approach claim that the knowledge of causes or causal mechanisms provides genuine understanding of how and why things behave the way they do. So telling a causal story of what is going on makes these phenomena more intelligible and comprehensible than they otherwise would have been. Appeals to the appropriate causal laws and processes yield the desired understanding. The knowledge of causal processes provides grounds for holding certain otherwise unreliable beliefs. Justification of a belief requires having *good reasons* for that belief, and such reasons are what causal explanations give us. Thus, reference to causes plays an important role in justifying all scientific beliefs that require good reasons.[32]

Nevertheless, the ontological view has its own problems. First, our concept of cause is hardly transparent. The contenders for the 'correct' analysis of 'cause' are manifold. Since Hume's analysis of causation as a regularity of events, it has been realized how hard it is to provide a pure ontological analysis of causation without introducing some very heavy metaphysical notions such as natural necessities. Some philosophers are very open-minded and accept what may seem to others to be philosophical chimaeras, but those closer to traditional empiricism deny the existence of necessities other than as an abstraction from the uniformity of phenomena. Most accounts of causation turn out to involve conspicuous epistemic elements, a feature which does not fare well with an ontic approach to explanation. Furthermore, it seems to be a problem of any ontological account that causal explanations are highly contextual because 'the cause' is not a particular event which exists isolated from the field of causally relevant factors in which it happens. The 'cause,' although objectively appropriate to the case, is selected by us as the most prominent factor in the situation, a selection that depends on the situation.

Salmon, with inspiration from Phil Dowe, attempted to construct a non-contextual notion of causal processes in terms of conserved physical quantities like energy or momentum.[33] His hopes were that one could get rid of the use of counterfactuals in physics because the standards by which we judge one possible world to be more or less similar to another partially depend on subjective features such as the conversational purposes of asserting a particular counterfactual.[34] When physicists theoretically describe causal interaction between different systems, they often do not talk about "cause" and "effect" but, for instance, about "emission," "transmission," and "absorption" of energy. However, this does not make physics observer-free. One of the problems is that conserved physical quantities like energy and momentum are time invariant; hence it is impossible without any knowledge of the system to say whether it causally evolves forwards or backwards in time. We have to interact with the process to determine which are the initial states and which the final. But then we also have to know what would have happened if we had not manipulated with the process. Hence counterfactuals are sneaking in through the backdoor.[35]

Donald Davidson argues that causal relations are not identical to causal explanations because causal relations must hold no matter how the causal relata are described.[36] The relational expression "...causes..." relates two singular terms, referring to two events, and since these events are particulars, they can be described in various ways. Davidson claims that since properties have become simply our way of describing events the effects of an event cannot be caused by the properties of that event because properties are not objectively real. Events can be described in an indefinite number of ways, corresponding to an indefinite number of properties. The claim that c caused e in virtue of certain of c's properties would require an endless listing of necessary conditions for e, which always open loopholes, corresponding to the indefinite number of properties of c. This is certainly not good enough to pick out the causal relation between particular events. Instead by citing certain properties of c we are giving an explanation of why e happened. Thus Davidson concludes the explanation varies according to the description, but once a description has been selected, the choice of an explanation is objective.

Davidson's account opens the possibility for a pragmatized notion of explanation. Stating that c explains e because c caused e in virtue of certain of c's properties yields a case of causal explanation, but the selection of the particular properties we regarded as causal depends on the descriptions we have selected among indefinite many. But one need not accept Davidson's view to see that the ontic approach cannot cover all cases of explanatory understanding. Structural, functional, and intentional explanations seem to be irreducible to causal explanations, yet some sciences regard them as intelligible as causal accounts. Therefore causal understanding cannot be identical with explanatory understanding in general. No notion of causality is able to ground a satisfactory and complete account of understanding reality.

4 The visualization view

Scientists, in general, or even a particular group of scientists, are no different from any other social groups in that they also share particular norms of understanding, which not only involve, say, predictions but also call upon the whole practice of science or the entire practice of the particular discipline in question. Kuhn pointed this out long ago (indeed it is practically the basis for his diachronic model of science). To a certain extent such norms of intelligibility do change over time, but more importantly they may change from context to context. Lord Kelvin is well known for having said that the litmus test for understanding a phenomenon is that we can make a mechanical model of it, a view that was in high fashion (but also very successful) among nineteenth century physicists. Today no physicist would dare to make such a claim. Later purely mechanical representations were replaced by the requirement of visualization of a process in space and time. For instance, Bohr seems to have cherished some hope up to around the mid-1920s of creating a quantum mechanics which still would be in accord with the classical ideal of visualizability as *the* criterion of understanding by making it possible to trace quantum processes as a series of causal happenings in space and time. It was only in 1925, when Heisenberg convinced him of abandoning any visualizable picture that he began to advocate the view that visualizability could not be the criterion of understanding with respect to processes in which the quantization of action could not be ignored.[37] It is recognized that Einstein was repulsed by the lack of determinism in quantum mechanics, but what is sometimes less well recognized is that he and like-minded physicists such a David Bohm opposed quantum mechanics because they retained allegiance to the classical ideal of visualization as the criterion of 'understandability.'

'Visualizability' as a standard of understandability was originally a specific proposal *for physics*. To make it the basic criterion of understandability *in all science*, one must also adhere to the view that ultimately all science reduces to physics. Partially because biology and chemistry did not adhere to visualizability in their modes of explanation, yet were clearly progressing and enhancing our 'understanding' of the relevant phenomena, people became less enthusiastic about a universal reduction-to-physics. In any event, because visualizability was a specific model of understanding for physics erected on the then-reigning physical theory, the further development of physics could lead to its abandonment, and did so in the quantum revolution. Therefore the following analysis of the limitations of this visualizability approach requires a good bit of quantum physics, making this section distinct from the other two, where no special knowledge of science was presupposed. Thus, I want to warn the reader that this section will require technical knowledge not presupposed earlier or elsewhere in the book.

Visualization as understood in classical physics seems to be rooted in five principles, all of which hold classically but which are violated by quantum mechanics. They are the separability principle, the definite value principle, the causal principle, the determinism principle, and the continuity principle. So the dynamical behaviour of a system is *visualizable* if, and only if, it can be separated in space and time from other systems by having independent and determinate properties, and the trajectory of the system is continuous and determined by the system's definite state together with the action of external forces. Already Bohr's semi-classical model of the atom did not comply with some of these principles, but with Heisenberg's matrix mechanics the situation became even worse for any commitment to visualization.

Bohr and Heisenberg emphasized the methodological importance of the principle of correspondence in the construction of a quantum mechanics. We understand new phenomena if we can see them in analogy with how we see other related phenomena. A similar sort of analogical thinking stood behind the formulation of Bohm's dynamical equation of motion. In this case it was the Hamilton-Jacobi formulation of Newton's second law that constituted the paradigm of constructive understanding. In defence of Bohm's interpretation James Cushing defended causal visualization as an important norm of intelligibility. He believed a scientific theory must meet three distinct goals: empirical adequacy, formal explanation, and understanding. Orthodox quantum mechanics may give us formal explanation in the sense of the deductive-nomological model of explanation but such an explanation "does not in itself give us understanding of the phenomena subsumed under the laws in question."[38] We have explanatory understanding, says Cushing, only when these explanations stand up to certain regulative principles, ideals of the natural order beyond the purely epistemic. Nonetheless, these principles "may be era dependent."[39] One such regulative principle is the causal picture of local interaction: "a satisfying understanding may be possible for us only when we are able to tell a local, causal picture story of those processes. If this is so, then we are, by definition, doomed, according to the standard (Copenhagen) interpretation of quantum mechanics to be unable to understand quantum phenomena (and quantum mechanics)."[40] Cushing argues Bohm's empirically equivalent theory of the quantum potentials provides this form of understanding but Heisenberg's indeterministic theory dominates physics because it was invented thirty years before Bohm's deterministic theory.

Cushing's argument shows there is no best explanation of quantum phenomenon; physicists may use alternative theories, which are equally empirically adequate. The dominance of standard quantum mechanics may be because of contextual reasons such as scientists' training within a certain historical tradition. This education makes them pose their explanation-seeking question in terms of the orthodox theory. Nevertheless, no coherent extension of Bohm's theory into the relativistic domain exists, whereas standard quantum mechanics can be given a relativistic formulation. So there is an objective reason for why no one uses Bohmian mechanics other than the historicist interpretation Cushing gave it.

Finally, I think that we ought to be sceptical about the deterministic visualization of particle movements allegedly gained from Bohm's theory. To use the quantum potential to 'explain' the behaviour of any atomic particle, it must be understood realistically. The quantum potential appears to be a classical field but exists in a $3N$-dimensional space. So if the real position of a particle can be 'explained' with reference to the presence of a quantum potential, then the physical space in which the particle has its position must also be a $3N$-space. But such a space is not visualizable. Moreover, there are other problems with the quantum, so we have little understanding of what a quantum potential might be because it is not analogous to any other entity we know of.

But Cushing's point cannot be ignored. Scientists clearly *prefer* explanations that involve causal or deterministic descriptions. This bias in favour of determinism is due to the fact that causal explanations tie the explanans and the explanandum more closely than do probabilistic explanations because causal explanations effectively exclude contrast-classes and thereby exclude alternative explanations. But even more to the point, understanding comes only with coherence, which any explanation creates in our belief system. Causal and deterministic explanations make classical physics meaningful because they express how we are able to grasp conceptually robust connections in nature. So if by analogy we can use causal and deterministic explanations in quantum mechanics, then our beliefs about the behaviour of quantum objects can be made coherent with other similar beliefs about the behaviour of classical objects.

The history of atomic physics has seen a series of constructive representations starting with Thomson's model of the atom, Rutherford's planet model, Bohr's model of the hydrogen, the Bohr-Sommerfeld model, and the Bohr-Kramers-Slater model before Heisenberg's matrix mechanics and Schrödinger's wave mechanics. Bohm's much later theory of quantum potentials gives us such an interpretative understanding, but it is not the only attempt at explaining the indeterministic outcomes of quantum measurement. Both GRW, the model of GianCarlo Ghirardi, Alberto Rimini, and Tulio Weber, and the decoherence theory suggest different indeterministic mechanisms that can give us a sort of understanding about the experimental outcome. In GRW the object measured exists in an entangled state with the many particles of the apparatus, one of which is said to collapse spontaneously into a definite state and thereby trigger a collapse of the states of rest of the particles. The decoherence theory postulates that a system and its entire environment are in a superposition of many single states, but when the system interacts with the instrument in a thermodynamically irreversible way, their entangled states can no longer interfere with one another. This in turn stops the system from being in a state of superposition.

Nevertheless 'superposition' and 'entanglement' cannot be regarded as visualizable conceptions of the state of the atomic system. Rather these notions seem to give us an abstract, conceptual form of understanding of many similar systems. This makes us able to connect systematically the outcomes of different measurements of these systems.

The major theoretical problem in quantum mechanics with respect to visualization is connected to the physical interpretation of the wave function, viz. what does this mathematical symbol really mean? The wave function appears in the Schrödinger equation, which gives a deterministic description of how this function develops in time. Schrödinger originally believed that it could be associated with a charge distribution in the atom. However, Max Born quickly proposed what became the orthodox interpretation that the value of the wave function should be understood as a probability amplitude, where the square of the absolute value gives a probability distribution. Although Born's probability interpretation worked well in connecting the theory to its empirical foundation, it failed to provide the sort of visualizable representation that at least some physicists adhering to the classical ideal of understanding demanded. Thus one might say that a proper understanding of the physical representation of the wave function continues to haunt quantum mechanics today. This brings our discussion of visualization into the eye of the hurricane.

There are two opposite ways to understand the wave function: Either you can say that the time development of the Schrödinger function is nothing but a mathematical abstraction because it does not refer to any well-defined classical physical states between measurements. Or you could argue that it stands for a *concrete* state of affairs between measurements since it refers to some *real* complex entity that can be imagined as a wave packet formed by the superposition of many possible eigenstates in configuration space. A particular measurement discontinuously changes this objective state of affairs taking place in physical space. Bohr agreed with the first position, holding that the wave function is symbolic, whereas other physicists have assumed the second position, arguing in various ways for wave packet collapse, consistent histories, and many-worlds representation.

Bohr spoke about the state vector or the wave function in terms of a *symbolic* representation, which contrasts with literal language. A literal representation can be visualized in space and time, but the wave function does not represent visualizable states of the individual system that can be tracked in space and time as classical states can. Bohr claims this lack of pictorial representation is due to the fact that the mathematical formulation of non-commuting quantum states contains imaginary numbers. His view seems to be that imaginary numbers do not have real counterparts in physical space and time and therefore such magnitudes cannot refer to real, empirical properties in space and time. The ruling norm is that a literal understanding requires being able to picture a process, taking place in physical three-dimensional space. Thus, the state vector is symbolic, where 'symbolic' means that the state vector's representational function should not be taken literally but must be considered as an abstract *tool* for calculation of probabilities of observables:

> The entire formalism is to be considered as a tool for deriving predictions of definite or statistical character, as regards information obtainable under experimental conditions described in classical terms and specified by means of parameters entering into the algebraic or differential equations of which the matrices or the wave-functions, respectively, are solutions. These symbols themselves, as is indicated already by the use of imaginary numbers, are not susceptible to pictorial interpretation, and even derived real functions like densities and currents are only to be regarded as expressing the probabilities for the occurrence of individual events observable under well-defined experimental conditions.[41]

In many places Bohr talks about the mathematical formalism of quantum mechanics as the mathematical *symbolism*, and he talks about *symbolic operators*. Bohr also points to the fact that Minkowski's formulation of special relativity as a four-dimensional manifold contains a similar use of imaginary quantities as the non-commutative algebra of quantum mechanics. – This means, of course, that the four-dimensional representation of space and time does not give us a literal understanding of the empirical world.

The fact that wave functions in quantum mechanics are complex functions should make us understand from the very outset that we cannot interpret them as describing a real physical process in space and time because they do not represent any physical existence, as do the wave functions in classical mechanics. Bohr argues that any 'real' physical instrument cannot measure a complex number. The 'real' world consists of what can be represented by real numbers. Quantum wave functions are means for calculations which have a meaning in the frame of Schrödinger's equation where the fundamental connection between the properties of the wave function and those of the object being described is expressed as a probability density $P(x,t) = \Psi^*(x,t)\,\Psi(x,t)$ according to Born's probability postulate.

The opposite view, represented by Hugh Everett, Bryce Dewitt and many others, holds that the wave function represents a real physical state of an individual system before measurement and that the Schrödinger equation describes the dynamical development of this state in an abstract mathematical construction of a many dimensional configuration space. In contrast to Bohr's view, the meaning of the wave function is expressed in terms of Hilbert's complex vector space in which a ray or a vector represents a possible eigen-state of the quantum system. Thus, the unmeasured particle exists as a very complicated structure. Measurements require interactions with a quantum system in such a manner that these states drastically change from representing a superposition of many possible eigen-states in configuration space to definite states in many different space-times of as many different physical worlds. According to the proponents of this many world interpretation, the object and the measuring apparatus are multiplied during a measurement into just as many worlds as there are possible eigen-values of the system.

Everett and his followers seem to hold that the features of the mathematical formalism give us metaphysical understanding of the quantum world rather than as giving the physicist a tool for predicting the statistical distribution of the inputs from experiments and observation. How far can we claim to understand something about the world, as defenders of the many-worlds view believe, which is empirically inaccessible and stands in dramatic conflict with basic metaphysical tenets commonly accepted outside quantum mechanics, just because we understand the syntactic rules and the semantic structure of the mathematical formula? Many physicists are convinced that the many-worlds interpretation gives them a correct explanation of the measuring problem, but even if we grant this, the transformation from many worlds represented by Hilbert space to one world represented by real space is completely non-visualizable.[42]

Another non-visualizable aspect of quantum states is their entanglement, which implies that correlated particles do not act classically, as Einstein, Podolsky and Rosen had assumed, but behave so that the measurement outcome of one particle is dependent of the measurement outcome of the other. The measurement of particles in a superposition is outcome-dependent rather than parameter-dependent. The common interpretation of this phenomenon is that the two correlated particles do not have separated states (and thereby separated values of their state defining parameters) until measurements have been carried out. It is not the case, according to this interpretation, that correlated particles, whose quantitative properties are designated by non-commuting operators, are influencing the value of each other's properties via a non-local field. Such an interpretation is, indeed, very foreign to classical mechanics and impossible to visualize within standard quantum mechanics, but has been confirmed by experiments.

* * * * * *

Summarizing our discussion so far we notice that the epistemic aim of science is to enhance our understanding of the world in which we live, and the function of scientific practice is to accomplish such a goal. The sciences furnish us with explanations and therefore contribute to our understanding of nature and human experience. Most philosophers and scientists would agree with this assertion, but among them one finds as very little consensus on the nature of the sort of understanding provided by science. Some believe that explanatory understanding involves confidence and satisfaction, others see it as rational expectations of the phenomena to be explained, some maintain that understanding these phenomena requires that they fit into a general world-picture, and still others point to the knowledge of the inner mechanism of things in the world.[43]

Indeed, the characterization of understanding in terms of confidence and satisfaction may signify a psychological feeling of the individual scientists which may or may not accompany his or her explanation, but which presumably is caused by some cognitive features of his or her belief system. The same characterization may also refer to some pragmatic aspects of explanation with respect to how a particular explanation contributes to the common understanding of a scientific community according to its aim and interests. I take explanation and understanding, as well as the form they have, to be context-dependent because their achievement depends on what the scientist thinks is the most appropriate response to a certain type of epistemic challenge. Even though my view of the aim of explanations in science as context-dependent is much more humble than Salmon's goal of fitting all phenomena into a world-picture, the various proposals of scientific understanding considered in this chapter do not necessarily exclude one another. Depending on the object of study and interest of inquiry, each kind of understanding may be the one scientists are looking for. Only a few philosophers of science such as Charles S. Peirce, John Dewey, Michael Scriven, Nicolas Rescher, and partly van Fraassen have followed the same line of thought. In recent years, Henk de Regt and Dennis Dieks have argued independently that understanding in science depends on contextual and pragmatic factors.[44] This is also my view to which we shall return to throughout the book.

2
Understanding as Organized Beliefs

Explanation not only promotes understanding, but also presupposes understanding. Paradoxically, one must understand a lot to explain a little. To acquire understanding by means of explaining, one must already understand what it takes to give an explanation, what commitments the explanation requires, in what situation an explanation is requested, and the meaning of the words in the explanation. Furthermore, an explanation produces understanding only if one already understands what explains what, or if one offers a tentative hypothesis one must at least believe what can possibly explain what. In both cases presenting an explanation demands a fair grasp of the topic of explanation, as well as some understanding of oneself as a producer of explanation, and of the cognitive values of the persons or community to which one offers an explanation.

How did we get all this understanding in the first place? Scientists understand what it means to perform experiments, to interpret their results, and to offer explanations of these results, and can do this to their own and others' satisfaction. Scientists understand how to create and use particular theories, models, graphs, mathematical symbols, etc., and that a certain result is a suitable solution to a particular inquiry whereas another is not. Some of these cognitive competences have been learned by explicit instructions; some have been learned by copying the behaviour of peers, teachers, fellows, friends and associates; but most are the subconscious and involuntary result of participating in a certain social practice such as working in scientific research. When internalized as part of one's cognitive system, most of these skills and competences are not explicitly acknowledged or internally justified.

Just as understanding exists independently of explanation, it exists independently of any cognitive problem. As long as a person understands a certain issue there is no problem and no need for an explanation. So explanatory understanding is not a particular form of understanding but any understanding that is expressed by an explanation.

I begin by distinguishing between conceptual and semantic understanding, agreeing with those who distinguish conceptual (re)presentations from semantic representations. For instance, I understand the use of the word "cat" in the sentence "The cat is on the mat" as part of my grasp of English semantics. In this semantic sense of "understanding," a chemist understands how to use the chemical vocabulary to explain something, and a judge understands how to express his verdict in legal terms. On the one hand, understanding in this semantic sense can be partly characterized as a grasp of vocabularies and linguistic rules. On the other hand, our capacity for thinking and reasoning gives us conceptual understanding which might not be amenable to linguistic representation. In contrast, semantic understanding is always linguistically mediated even though the semantic representation need not be a part of the speaker's thoughts concerning the explanandum. The speaker's thoughts are his conceptual grasp, and indeed may be about semantic representation and expressed in terms of such a representation.

Understanding is also associated with *meaningfulness* or *intelligibility*. Making something meaningful or intelligible is the aim of both explanation and interpretation. In my opinion, understanding is a characteristic of people and consists of their capacity to make themselves acquainted with or to gain insight into any particular topic they wish. Thereby the association of explanans and explanandum makes the topic intelligible to whoever understands the explanation. But what becomes meaningful does not need to be meaning-bearing in order to be meaningful. Something can lack any semantic content and yet be grasped as meaningful. If B is explained as caused by A, B becomes meaningful in a conceptual sense but not in a semantic sense. So "meaningfulness" covers more than just the semantic content of a sign, a term, a figure, etc. Any state of affairs can be made meaningful just in case it is brought into the 'right' connection with whatever a person knows, believes, assumes, or feels, etc. Although understanding may have meaning as its object, it is not a necessity for something to have meaning in order to become meaningful.

So the object of scientific inquiry *partially* determines the sort of understanding we will find acceptable and illuminating. We seek to understand different kinds of phenomena, and whether it is meanings, feelings, purposes, physical processes, chemical structures, animated or unanimated things we actually study, the object of inquiry provides the content of our understanding. But our understanding is also partially determined by our particular interests in the object. The kind of understanding we seek depends on what aims we pursue, and scientific inquiry is focused accordingly. In this chapter I shall concentrate on our understanding of phenomena as cognitive schemata.

The thesis I defend is that understanding arises not from the content of our belief systems but from their organization. Fifty years ago Michael Scriven maintained that understanding could be identified with organized knowledge,[1] but in contrast to his view I hold that understanding is organized *beliefs*, not knowledge, and since the structure of the system is neither true nor false it does not count as knowledge. I claim that although the *content* of a belief system may be true or false, the *organization* of the system is neither. I claim that my view, which I call *the organization view*, has explanatory advantages over *the content view* and *the ability view*. It explains why people adhere to false belief systems and why people talk about, say, political, religious, philosophical, or metaphysical understanding when there are no empirical grounds for such claims. The organization view takes history of science seriously by, for instance, talking about the Aristotelian understanding of motion instead of the Aristotelian (pseudo)-'understanding' of motion and by allowing us to regard our present-day claims of scientific understanding as genuine even if they later turn out to be incoherent. Furthermore, it helps us see how explanation and interpretation are a means of organizing new beliefs with respect to one's background beliefs, assumptions, and knowledge.

1 The content view

The connection of understanding with the content of our belief system seems common among philosophers of science. For instance, Peter Achinstein holds that "explaining q has been defined as uttering something with the intention of rendering q understandable...Such understanding I take to be a form of knowledge."[2] In a similar way Peter Lipton holds that "understanding is not some sort of super-knowledge, but simply more knowledge: knowledge of causes."[3] I call this *the content view*: it takes understanding to be knowledge of the (explanatory) content of our beliefs. The specification of this (explanatory) content varies from author to author, involving beliefs about causal connections, covering laws, initial conditions, or unifying principles.

Thus, understanding consists of beliefs which are assumed to have objective truth conditions that provide them with propositional content. In other words, the content view presupposes that understanding has externally accessible criteria of success, so you can have understanding if, and only if, your alleged understanding has a propositional content, which is true according to how things stand in the world. As J. D. Trout argues, a subjective feeling of understanding is not an avenue to truth, but "only the truth or accuracy of an explanation makes the sense of understanding a valid cue to genuine understanding."[4] The content view holds that genuine understanding is objective: if one misjudges how things stand in the world, one does not understand. Misunderstanding is a case of misrepresenting the fact of the matter.

This view, therefore, makes understanding a species of knowledge. A person understands a set of laws only if she *knows* that these laws can be inferred from relatively few general principles; she understands why something happened only if she *knows* that this phenomenon was caused by a certain mechanism; or she understands the periodical system only if she *knows* that this system is structured according to certain selection rules, etc. Having the relevant understanding requires that the subject's beliefs are true and can be justified by reasoning and/or observation.

In all theories of knowledge, a belief must be true to have the status of (propositional) knowledge. So if understanding is a species of knowledge, then it must by necessity be true. A person understands a phenomenon if, and only if, that person holds relevant beliefs about the phenomenon, which are in fact true beliefs. If the relevant beliefs are factually false, then it is a case of 'misunderstanding' the phenomenon, even though the subject, who holds the beliefs, may have a subjective feeling of having understood the phenomenon. Earlier claims of understanding that turned out to be false in the course of history of science are in fact cases of misunderstanding. For some this is the basic ground for making truth a necessary condition for understanding. 'Knowledge' as it is used both in everyday discourse and in philosophy is an alethic (truth seeking) concept; i.e., no one would want to say they 'know' what they believe to be false. So we have to reconsider what understanding can be in order to allow for understanding to be 'false'. Does it suffice to say that we *understand p* in case we are justified in our belief that *p* is true but that *p* may nevertheless be false?

Since truth, even in science, is often impossible to reach, the appeal to truth seems to be misleading as a norm of genuine understanding; the feeling of understanding should be a signal of genuine understanding even though the beliefs on which that feeling is based are fallible and sometimes false. I shall presently try to show that the feeling of understanding is not cued by truth but by some structural features of our beliefs that have given rise to a sense of confidence and satisfaction. It is certain properties of our belief system other than truth that cause our subjective feeling of understanding and so should be nominated as the necessary and sufficient condition for genuine human understanding.

Philosophers of science typically identify understanding with knowledge, so it may come as a surprise that practically every epistemologist, who has seriously contemplated the issue, concludes that understanding is not a species of knowledge.[5] However, there are different motives for this unanimous claim: some hold that understanding is immune to Gettier problems, while knowledge is not, some that understanding is transparent, while knowledge is not, and others that understanding is possible in the absence of truth, while knowledge is not. Also it seems that understanding may come in different degrees, whereas knowledge cannot. All these contrasting features, if true, would jeopardize the assertion that understanding is a species of knowledge.

I hold that neither does knowledge imply understanding, nor does understanding imply knowledge. They are closely, but not logically, connected. They are closely connected because both concepts are associated with beliefs.[6] Each of us knows many things that we don't understand because we cannot *explain* why they are as they are. We may know singular facts such as who we are, but in most cases it does not make sense to say that we understand who we are. Scientists also know, for instance, that the velocity of light in empty space is constant but cannot explain that constant. Facts such as these are brute contingencies, which go without explanation. So we can possess knowledge without having factual understanding. Yet, knowledge requires having some form of conceptual understanding other than factual understanding of what we know. So what about the other way around: does understanding entail knowledge?

Philosophers of science can find good examples of subjects having both conceptual and factual understanding without having knowledge. First, understanding not only relates to factual matters. Understanding has much to do with making sense of something. A conceptual grasp of how to represent factual matters is also a form of understanding. Having knowledge presupposes this conceptual form of understanding. Second, it may sometimes be impossible to distinguish conceptual from factual understanding. Various philosophical theories of space-time substantivalism and space-time relationism involve both conceptual and factual understanding of the nature of space-time, but since these theories are empirically underdetermined, they cannot both be claimed to express knowledge of this nature.

It is uncontroversial that we can understand what we believe to be false, as in cases of historical understanding of discarded theories in science or as in instances of 'common knowledge' of former times such as the 'absolute' direction of up and down. Imagine a scientist facing a choice between rival explanations; she can decide rationally, only if she has some 'understanding' of both, prior to any commitment concerning which is true. This kind of understanding concerns the meaning of the various theories. Is this conceptual understanding so different from explanatory understanding? I don't think so, because understanding natural phenomena presupposes understanding concepts and the use of language. If an Aristotelian understands and believes his explanation while a critic understands but disbelieves the Aristotelian explanation, they both understand (in order for the dispute to be rational) the same thing, one accompanying his understanding with epistemic commitment and the other not.

First, we need an account of 'understanding' covering understanding with respect to both explanation and conceptual meaning because an explanation can have meaning as its topic. When Stephen Grimm claims that our understanding of natural phenomena is "arguably the paradigm case of understanding," he seems to be historically incorrect. Traditionally understanding was strongly connected to meaning as something mental, a view which originated in Kant's thinking about the faculty of understanding as that which makes the act of judgement based on some *a priori* principles.[7] Second, both conceptual and factual understanding come in degrees. Before Fermat's theorem was proved, most people neither understood nor knew it, but mathematicians who had studied the subject for years certainly could be said to have 'understood' it better, though they did not yet know it. Now that it has been proved, the mathematicians understand it better, no doubt, but their state of mind prior to that point was certainly not the same as a total lack of understanding, which was the case with most people. Similarly although the theory of relativity overturned classical mechanics, we would not deny that before Einstein physicists 'understood' the movements of the planets. Today physicists understand these movements better than ever since Newton. Third, any notion of understanding that reflects actual cognitive practice must recognize that both conceptual and explanatory understanding depend on the epistemic context. Finally, in contrast to knowledge, not all understanding has internally accessible criteria of success.

Linda Zagzebski argues that understanding is transparent, whereas knowledge is not.[8] She thinks that understanding consists in grasping how pieces of information relate to one another, seeing how they can be connected so they hang together. Her reason is that understanding has internalist conditions for success, whereas knowledge claims, even viewed internally, somehow involve an extrinsic evaluation. Understanding not only has internally accessible criteria but also these are transparent: it is impossible to understand without understanding that one understands. In contrast to the usual internalist assumption, Zagzebski departs from most internalists in holding that it may be possible to have knowledge without knowing that one knows. Although I agree that knowledge is distinct from understanding, I am critical of her reason for this distinction. Her stance towards understanding is normative, mine is from a naturalistic standpoint. My approach does not demand understanding be transparent. I think it is a mistake to make transparency the feature, which differentiates understanding from knowledge. It is reasonable to believe that certain animals possess some kind of understanding, but it hardly seems reasonable to believe that they are in an epistemic state of understanding that they understand. Animals understand, but the criteria of success are their behaviour of reaching a certain goal as a consequence of such an understanding. I also believe that as far as animals possess concepts, which I think their understanding of factual matters sometimes shows they do, they do not understand that they use concepts.

I hold that the reflexive states of knowing that one knows (K-K thesis), or understanding that one understands (U-U thesis), are possible only for self-conscious beings, which excludes most animals. Both cases would therefore suggest that the K-K (or U-U) criterion is not a necessary condition for having knowledge or understanding. In fact, as we shall see, the U-U condition is satisfied for self-conscious beings only if these beings are occupied with what I shall call reflective understanding.

However, Stephen Grimm argues that to count as understanding, in contrast to conceptual understanding, factual understanding must result from beliefs that are true. Grimm holds that factual understanding is true or false because the beliefs on which it is based are true or false. If so, genuine understanding must be a species of knowledge. Take the following example: you enter a hotel room and want to watch TV. You try to switch it on by the remote control, but nothing happens. You look around in an attempt to solve the problem, and you realize that the TV is unplugged. The moment you see this, you seem to understand why the TV does not work. As part of your background beliefs and assumptions you know that a TV needs electricity. So when you connect this information with the piece of information about the unplugged TV, you feel certain why it doesn't work properly, namely in virtue of being disconnected from the electric circuit. In this situation it seems that you have a good understanding of the problem. Now, assume that you then plug in the TV, but that it still fails to work. Again, you become puzzled because your understanding of the problem up to then was not equivalent to knowledge. Perhaps it is dead batteries in the remote control, and so you try the on and off button on the apparatus. Nothing happens. You then call the receptionist who tells you that the TV in your room broke down this morning, and that the maid must have disconnected it so the repairman could replace it, but he never showed up. Again you claim to understand why you cannot watch TV. The problem is: did you understand at the beginning? Grimm denies that you had any understanding because your 'understanding' was based on false beliefs. In this situation "you don't understand. You've mischaracterized how things stand in the world."[9] Consequently, factual understanding cannot be transparent.

But I want to point out that neither is conceptual understanding transparent: you can be wrong in claiming to understand the meaning of a description because you may falsely believe (feel) that you understand its semantic content since you falsely believe that some internal criteria of understanding are met. The same holds with respect to factual understanding although this is not due to the factual dimension of understanding. After the receptionist told you the TV was broken, it would be strange to claim you understood why it wasn't working from the beginning; now you recognize what you then thought was a misunderstanding. But imagine that you received a phone call before you were able to plug in the TV and thereby to ascertain that the apparatus didn't work. In this case we might say that you had a sense of factual understanding but you nevertheless didn't really understand (the truth of the matter?). Indeed, you didn't understand the truth of the matter; but since no one is omniscient, a distinction between feelings of understanding and genuine understanding may make sense *only retrospectively*. However, until then your understanding is not just a feeling; you are justified by internal criteria in claiming that you understood the problem, and your feeling is based on the fulfilment of these criteria.[10] Understanding, much more than knowledge, need not be stable over time, since, as I shall argue, it need not be true or false. You can correctly claim to understand the explanandum as long as your background knowledge, beliefs and assertions fitted nicely with all evidence available at a certain time. Indeed, your understanding may fluctuate while you get more and more information about a situation. Further new evidence may add or subtract to your understanding over time, so understanding comes in degrees.

With hindsight you can recognize that your former understanding was incorrect, that in retrospect it was a misunderstanding. Because 'truth' is a non-epistemic, metaphysical concept, and because it is built into our concept of knowledge, it is impossible to have revisable true beliefs.[11] But understanding is different. If we can connect certain puzzling facts with our background beliefs in a successful manner, we are right to claim that we understand what is going on until we are forced to change opinion.[12] Since we also want to have understanding, which is correct, we may be able to elevate our beliefs to the level of knowledge under the right circumstances where we are justified in ascribing truth to them.

Intuitively we may have a strong inclination to say that you didn't understand since you didn't know why the TV was out of order in this *particular* case; you first understood only when you got the causes right. This is because commonly scientific knowledge is by definition true. While it is certainly common to say knowledge must be 'true' by definition, it is not common, or even a well-formed phrase, to say understanding must be 'true.'[13] The modifiers 'true/false' are normally applied to 'belief' or its synonyms, things that can be said to be believed. However, since 'understanding,' is a second-order epistemic capacity, 'true' or 'false' do not seem to be the appropriate predicates but rather 'correct' or 'incorrect', or perhaps 'precise' or 'imprecise', in the sense that one could answer a question with a true statement but it still would be an incorrect answer to the question posed. In that case one has 'misunderstood' the question. The difference between 'belief' and 'understanding' therefore seems to be that the truth or falsity of a belief, or of a statement expressing this belief, can be associated with its meaning or semantic content, whereas one's understanding can be said to be correct or incorrect not because of its semantic content but thanks to the relation of its structural form to reality. Whereas knowledge is 'true' by definition, understanding is not 'correct' by definition.

In the TV-case we can say afterwards that originally you *misunderstood* the problem, where we mean that your understanding failed to be correct; what you understood included false beliefs; you didn't imagine things as they were. You were successful insofar as you gained some degree of understanding, but you were unsuccessful to the extent that your understanding didn't correspond to knowledge by matching all of the facts of the matter. This, I think, shows that factual understanding as such does not come with the objectivity of the content of our belief. Understanding arises merely from how well the content of a particular belief goes with the subject's belief system. So the same organization of beliefs may count as understanding at one time, but may be seen as misunderstanding at another time. For one person a number of beliefs signal understanding, for another it indicates a misunderstanding.

Now, the crux of the matter is that adherents of the content view, like Grimm, agree with Hempel's strategy of explicating a non-pragmatic concept of explanation while advocating an objective, non-contextual notion of understanding. The attempt to see understanding as unification or as insight into a causal mechanism is an effort to produce an account which gives us (1) a link between explanation and understanding; (2) an objective and non-variable notion of understanding; and (3) a notion of understanding according to which whether or not a proposed explanation gives understanding is open to rational examination.[14] Obviously, the opponent of the content theory of understanding must deny that (2) holds strictly because she maintains that thinking, and therefore understanding, is context-dependent. But the general claim that understanding is context-dependent does not exclude the possibility that some types of understanding cover many contexts equally and consequently are more 'objective' than others. Similarly, no adherent of a pragmatic theory can accept (3) as stated because in order to rationally examine an explanation, you must already understand what the suggested explanation intends to explain and how this intention is carried out.

It is not merely incidental that explanation may provide understanding; rather, I hold that providing understanding is the inherent function of explanation. The pragmatic view that I defend rejects the traditional assumption that there is one and only one proper way of understanding a given problem, or scientific problems in general, and that the correctness of every proposed explanation can and must be measured against the same ideal notion of understanding. Even Hempel saw a connection between explanation and understanding; but understanding in his sense of rational expectation formed no part of his notion of explanation. One obvious way to integrate understanding and explanation is to take Hempel at his word and search for a pragmatic theory of explanation and understanding. Michael Scriven pointed the way by emphasizing that an explanation starts with some request for information and is "a symbolic vehicle of conveying understanding... [which] is acquired whenever the capacity for solving a certain appropriate range of problems is achieved."[15] Doubtlessly, some kinds of understanding can be associated with skills and problem-solving capacities, but we need to ask what gives rise to what: do skills and problem-solving capacities build on understanding, or vice versa? This is the issue to which I will now turn.

2 Skills and intelligibility

Although explanation and understanding do go together in my view, understanding can be separated from explanation. Certainly, if we can explain something, then we understand it, but do we also understand something, *only if* we can explain it? I think not. Explanation is an answer meant to convey some sort of understanding. It expresses the explainer's grasp of the explainee's epistemic problem, and hopefully helps to solve it by giving the explainee an understanding of the problem similar to the explainer's understanding. But it seems that one can have understanding even though one cannot explain what one understands. We may have skills and abilities demonstrating that we have understanding without being able to explain the subject of our understanding. Explanation defines only one level of very high and reflective cognitive interaction with the world.

In a couple of recent papers, Henk de Regt has developed a pragmatic view of scientific understanding centred on intelligibility.[16] When Trout says, "I understand why planes fly if I know Bernoulli's principle,"[17] he suggests two possibilities of having understanding. Either one has knowledge of the principle together with relevant information about dimensions and weight. This is objective understanding independent of the subject. Or one might experience a subjective sense of understanding. But de Regt disagrees that understanding reduces to having knowledge: a knowledgeable person must be able to use the theory of aviation, which requires possessing specific skills.

He distinguishes between (1) feeling of understanding, (2) understanding a theory which is equivalent to 'being able to use the theory,' and (3) understanding a phenomenon which he takes to mean 'having an appropriate explanation of the phenomena.'[18] The first psychological form accompanies understanding in the second and third form. De Regt also thinks that scientific explanations bring the explanandum into a broader theoretical framework. So the epistemological goal of explanation presupposes the scientists' ability to use a theory. The choice of an explanation depends on both the scientists' skills and the strength of theory: "[I]t depends on whether *the right combination* of scientists' skills and theoretical virtues is realized."[19] The theories are selected based on different criteria such as visualizability, simplicity, preciseness, unification and fruitfulness. A scientist's preference for a theory depends on his background knowledge and the intelligibility of a theory. Says de Regt: "I define the *intelligibility* of a theory as the positive value that scientists attribute to the cluster of theoretical virtues that help them in their use of the theory."[20] Intelligibility is a value projected onto a theory enabling the scientist to anticipate the qualitative consequences of using the theory. Thus, de Regt concludes knowing certain theoretical principles or natural laws and the relevant background conditions are not sufficient for having an explanation. As he says, "The extra ingredient needed to construct the explanation is a *skill*: the ability to construct deductive arguments from the available knowledge."[21]

I call this *the ability view* of understanding to distinguish it from the "content view." It is a very instrumental view of understanding, regarding it as nothing but our abilities to perform some actions to reach certain cognitive aims most effectively. Solving certain cognitive tasks successfully demonstrates that one understands appropriately. In de Regt's case this is the ability to construct explanations in terms of deductive arguments. Unfortunately, this view applies only to physics, which can explain phenomena in a deductive nomological form, and it does so only partially.

At first glance it might seem reasonable to characterize understanding in virtue of certain cognitive skills. It admits acquired embodied capacities as forms of understanding. Practical skills involve some form of understanding, which the skilled person may not be able to articulate verbally. A language user may not be cognitively aware of how she must form her sentences grammatically, but in actually producing the sentences she demonstrates that she understands the grammatical rules.

Although skills are connected to understanding, I disagree that understanding reduces to skills. First, understanding can be correct or incorrect, but skills cannot.[22] If we go back to Michael Polanyi we see that having a skill is to have "tacit knowledge."[23] But again a skill cannot be ascribed a predicate like 'correct' or 'incorrect' (nor, for that matter, 'true' or 'false'). Second, providing explanation may be a skill in the sense that when you explain something you manifest both your understanding of the cognitive problem of the explainee and your knowledge of the matters of fact on which the explanation depends. But when the explanation is successful and changes the explainee's state of puzzlement into a state of understanding, it seems incorrect to say that the explainee has acquired a particular skill other than the ability to restate the explanation. The explainee may still not see the practical consequence of her newly acquired understanding. So, it seems that you can have understanding without having the corresponding skills.

Imagine, for instance, that I get some information from which I understand clandestine motives behind George W. Bush's wishes to go to war against Iraq, but I am not able to solve anything: I may not even be able to solve any epistemic problem because his motives never had come to my mind (before I was told) and therefore were never an exigency for me. Similarly, we may be able to solve some problem without having any deep understanding of it. I could solve a problem by sheer accident. For example, my TV starts to work after I have slapped it, but I do not understand why it began to work. Only if we think of solving a problem entirely in epistemic terms does it make sense to say that understanding is equivalent to a capacity of problem solving. I could understand how to solve a problem but not be able to solve it, for that very understanding might indicate the need for resources vastly beyond my powers or perhaps not even technically possible.

Skills are merely functional, giving you the capacity to do some particular thing. A skill makes you able to realize some specific goal, but understanding need not be functional in the sense that it has a practical purpose or actually leads to a goal. You can understand a verbal order even though you are unable to carry it out. You may understand a joke or a paradox without acquiring any skill other than the ability to repeat it. You may understand the instructions by reading a manual but may still not know how to carry them out in any practical way. Also one may argue that machines have skills without understanding. A machine capable of translating an English text into German displays certain skills without having any understanding. As far as I see, this corresponds to de Regt's second sense of understanding. But the machine intelligence is grounded in the manipulation of symbols according to explicit rules where it is the form of the symbol – or the proposition in which it takes part – and not understanding the meaning of a text that is the basis of its capability. This excludes understanding since even factual understanding requires a conceptual grasp of the matter. Understanding cannot reduce to a cognitive skill because the ability to construct an explanation, or to make deductive inferences, presupposes an understanding of what an explanation is, or what deduction is.

Finally, a person can learn a procedure of calculation well but not understand the appropriate context in which it applies. Is this deficit also a lack of skills or merely a lack of understanding? Apart from having a skill, you must understand the limits of its application and be able to identify a situation in which it is applicable. Errors are often a result of the use of previously acquired skills in new situations because having a skill does not necessarily provide an understanding of the circumstances in which that skill can be successfully employed. Thus, in contrast to de Regt, I think that understanding can give rise to skills, and that cognitive skills are based on understanding but these skills are not equivalent to understanding. The concept of understanding is not reducible to the concept of skill.

3 The organization view

Michael Scriven once claimed: "Understanding is, roughly, organized knowledge, i.e. knowledge of relations between various facts and/or laws. These relations are of many kinds – deductive, inductive, analogical, etc."[24] So, according to him, understanding consists in *beliefs* about relations. Apart from the fact that he construed understanding as a particular kind of knowledge, which I would reject, I agree that understanding is associated with our ability to relate things together to our rational satisfaction. We know the date of our birthday, but we understand the calendrical system.[25] Scriven holds that we may in principle have knowledge without understanding, but have no understanding without knowledge. In opposition I hold that understanding is organized *beliefs* rather than knowledge, and that you may have organized beliefs that are not true and therefore do not count as knowledge. However, I do agree that knowledge and understanding are interdependent: the more knowledge one has, which is related to the topic, the greater one's understanding is likely to be. Reciprocally, the more one understands about a topic, the greater one's knowledge is likely to be. So to avoid confusion, it should be emphasized that while our understanding (and misunderstanding) follows from the actual organization of our beliefs, and therefore does not have any truth value, understanding gives rise to beliefs about these connections, and those beliefs are either true or false.

Other philosophers agree with Scriven but base their reflections on a more empirical approach. Both David Perkins and Raymond Nickerson discuss experiments with trained college students attempting to establish their ability to solve simple problems in physics or mathematics in which incorrect answers were evidence of lack of understanding. These experiments led Perkins to describe understanding as knowledge of relations. He concludes, "understanding involves knowing how different things relate to one another in terms of such relations as symbol-experience, cause-effect, form-function, part-whole, symbol-interpretation, example-generality, and so on. Broadly speaking, understanding something entails appreciating how it is 'placed' in a web relationship that gives it meaning."[26] But if understanding requires knowing a web of relations to be adequate, some other features must characterize this knowledge. Perkins holds one feature is 'coherence', knowing how well it hangs together and how "it coheres with the external world." But knowing a coherent web of relations is not enough for having understanding; understanding must be 'generative' and 'open-ended'. A person must be able to use her understanding for practical purposes and demonstrate that she in fact possesses the alleged knowledge by doing the 'right' things; understanding is open-ended in the sense that her knowledge of the web of relations can always be enlarged or made more accurate, and from the various enterprises in which she participates the standards of coherence change – how well the web fits together or how closely it corresponds to the world.

The features that Perkins identifies are important for a theory of understanding, but regarding understanding as a form of knowledge, namely knowledge of relations, is a regression to the content view applied to relations. Raymond Nickerson also associates understanding with relations. For him understanding is an active process that connects new information with old knowledge in a cohesive way: "Understanding is an active process. It requires the connection of facts, the relating of newly required information to what is already known, the weaving of bits of knowledge into an integrated and cohesive whole. In short it requires not only having knowledge but also doing something with it."[27] Apparently, Nickerson, in contrast to Perkins, separates knowledge and understanding. However, he wrongly thinks of understanding as relating facts and, moreover, wrong in identifying understanding with adding new information to one's body of knowledge. For instance, if you explain something to somebody, you often express your longtime understanding of the subject. His characterization holds only in connection with the acquisition of *new* understanding. But Nickerson correctly observes that understanding is context-dependent. Understanding a game of chess differs fundamentally from understanding the concept 'first lady.' Likewise he correctly observes that understanding can vary in degrees and completeness. Consequently, he notes "one's understanding of something should probably not be thought of as right or wrong but rather more or less right or more or less complete."[28] This seems to imply that understanding is not knowledge since 'to know' is an all-or-nothing affair in which one's belief is true or false.

Recently, Zagzebski observed that understanding emerged from the Platonic notion of *episteme* as "a state gained by learning an art or skill, a *techne*." Understanding is knowing how to do something well; and it "is not directed towards a single object, but involves seeing the relations of parts to other parts and perhaps even the relation of [a] part to a whole."[29] Zagzebski rejects the knowledge view of understanding as do I. Modern epistemology associates knowledge with propositional knowledge. In contrast, Zagzebski argues that understanding is seeing how parts of a body of knowledge fit together, where this 'fitting together' does not have a propositional form. Hence, although our mental representation of our understanding may include accepting a series of propositions, the object of understanding is not a discrete proposition. "Understanding deepens our cognitive grasp of that which is already known. So a person can know the individual propositions that make up some body of knowledge without understanding them. Understanding involves seeing how the parts of that body of knowledge fit together, where the fitting together is not itself propositional in form."[30] Zagzebski emphasizes that sometimes understanding can be expressed only by such representations as maps, graphs, diagrams, or three-dimensional models. In general, "understanding is the state of comprehension of non-propositional structures of reality."[31] According to this characterization, "we can understand such things as an automobile engine, a piece of music, a work of art, the character of a human person, the layout of a city, a causal nexus, a teleological structure, or reality itself."[32] The upshot is that understanding is not carried by propositions or states of belief but is "a property of persons."

I agree that Zagzebski's analysis is insightful, in recognizing that understanding does not consist of propositional beliefs but is an embodied cognitive capacity. If 'understanding' is considered to be a cognitive capacity of organizing a set of beliefs, it is reasonable to say that this faculty cannot be a belief itself. The process of interrelating beliefs does not require a particular 'organization-belief' to connect discrete beliefs to gain understanding. Taking this insight seriously is the only way to avoid Aristotle's the third man argument; namely the puzzle that relating this organization-belief to other beliefs seems to require a further organization-belief and so on *ad infinitum*. But this organization view does not deny that understanding gives rise to beliefs. Hence understanding itself can and should be belief-generating. For instance, I may understand how I get from *A* to *B* and I may express this understanding in terms of certain propositions. These are two different mental acts. Since there is no one correct way of expressing a mental state of 'understanding,' having a belief as a result of my understanding does not imply that this understanding had a propositional content before being represented in terms of some beliefs.

However, Zagzebski's insight has its shortcomings too. As discussed above, she assumes that since understanding is 'transparent,' only persons can have understanding because we understand if, and only if, we are aware that we understand. I take her to mean that if one is in a state of understanding, one must be consciously aware that one is in this mental state. In contrast, Zagzebski believes that one can know without knowing that one knows. But I cannot see why understanding should be distinct from knowledge at this point. We can consider understanding to be organization of beliefs, without necessarily being aware of the organizing relationships. If we can have beliefs, which we do not know we have, it seems also that these beliefs can be related without our being aware of these relations. In depriving almost every animal from possessing understanding, Zagzebski's view not only is very anthropocentric, it also does not fare well with empirical observations and ethological experiments.

I hold that animals may have understanding without being conscious of their own comprehension. Here it is important to keep various senses of "consciousness": (1) an animal or a human being is conscious if it is awake; (2) some creatures are conscious by being conscious of things and state of affairs; and finally (3) mental states may be either conscious or unconscious. It is generally agreed that the notions of consciousness in (2) and (3) denote two different things, since (2) characterizes a first-order property of a subject, whereas (3) refers to a property of a mental state which is a second-order property of the subject having the mental state. The main question is: are these different notions of consciousness related, and if so, whether and how one of them can be used to explain the other. Recently there have been strong attempts to explain (3) in terms of (2).

To develop these attempts successfully there must be general agreement over what makes a mental state a 'conscious' state, since we cannot say that (2) explains (3) unless we know what is being explained. Two features have been in focus. Thomas Nagel points to the phenomenal character of conscious states: there is something it is like for the subject to have a particular experience. The other feature is what has been called self-presentation: a mental state is conscious because the subject is aware of being in that state. This feature is essential to higher-order representational theories of consciousness and basic to the phenomenological tradition's conception of pre-reflective self-consciousness. However, first-order representational theories of consciousness will generally tend to deny the existence of self-presentation. Higher-order representational theories hold that for a mental state to become conscious its presence must be represented by another state, whereas first-order representational theories hold that the content of a mental state has various functional roles, which are enough for it to be conscious.

Understanding and consciousness seem to follow each other. We understand something in the world because we can become consciously aware of it through the interaction of our body and brain. Understanding emerges from this interaction between brain, body, and world. We can see that if one claims that understanding is cognitively transparent, as Zagzebski does, one can mean one of two things: either one is aware of one's state of understanding because this state has the quality of self-presentation, or one is aware of one's state of understanding because one is able to represent it by reflection. I claim that understanding does not require a creature to be in a state of self-presentation nor that it be capable of introspective awareness of its own cognitive states. As we shall see, there are lower levels of understanding in which the organism is not capable of representing its conscious states of understanding; yet, these are cases of understanding because of the functional aim of the organization.

Nevertheless, only when we become aware of knowing something and understanding something do we begin to seek normative criteria for knowledge and understanding. Until then the successful criteria for knowing and understanding rest solely on whether these cognitive states lead to practical accomplishments and are beneficial for our survival. The turn from external to internal criteria for success appears the moment we distinguish between knowing and believing that we have knowledge, and between understanding and being aware that we understand.

4 Embodied and embedded understanding

Gilbert Ryle famously distinguished between "knowing that" and "knowing how": one requires learning a *fact* and the other learning to *act*.[33] We can know that Albert Einstein is the creator of the General Theory of Relativity or that England is a kingdom, but we also can know how to answer a question or to drive a car through heavy traffic. Already in his *Human Nature and Conduct*, John Dewey introduced this distinction, identifying "knowing how" with habitual and instinctive knowledge and "knowing that" with predicative thinking involving reflection and conscious appreciation.[34] Ryle attacked the intellectualist doctrine denying any basic differences between practical knowledge and propositional knowledge. Intellectualists hold that even practical activities are accompanied by some internal act of considering propositions. As an anti-intellectualist, Ryle argued that people have practical knowledge such as skills, abilities, powers or aptitude, none of which refers to any particular belief. Practical knowledge is manifested in our actions quite independently of whether any internal process accompanies them.

Philosophers have debated whether 'knowing-how' is logically distinct from 'knowing-that', i.e. distinct from propositional knowledge. There are good reasons to separate them because they are established and grounded in different ways. Propositional knowledge, or 'knowing-that', is partly infected with normative virtues; it is communicable by language and must be true and justifiable. One gains it through perception or by learning from others. Practical knowledge, or 'knowing-how', is acquired differently, from learning by doing, and we cannot attribute truth-value to such a cognitive state. If I say, "Tom knows how to speak English properly," I'm not necessarily attributing to Tom any particular *belief* about English pronunciation, which is true and justifiable. Nor do I see any reason to suppose that Tom, as a foreign speaker, could have learned to speak English properly by just reading an English dictionary and an English grammar. Tom learned English correctly by practice.

Two conclusions follow: first, understanding does not have a propositional form; second, practical knowledge may be non-propositional as well. These two conclusions may seem to imply that understanding is a form of practical knowledge or perhaps that understanding is equivalent to abilities. But drawing such an inference would be too hasty because there are cases of understanding that cannot be identified with practical knowledge. These cases are either based on interpretation of how things fit together or based on explanation given by somebody else. Thus understanding is *sometimes* revealed as 'practical knowledge,' sometimes as superimposed structures of reflective thinking. This is the difference between non-representational and representational understanding.

An adequate picture of science as a social and cognitive activity requires relating it to the non-representational part of scientific activity. Taking part in scientific practice requires scientists to rely first and foremost on an embedded form of understanding learned by doing rather than having something explained; scientific explanation and interpretation provide scientists with a reflective representational form of understanding. Therefore, interpretation and explanation build on an established practice of understanding and then eventually contribute to this practice. Scientific practice emerges from a shared understanding *embedded* in human agents whose actions make up this practice. It is learned by doing, imitating others, or being taught by peers. This practical non-representational understanding is a prerequisite for producing explanations and interpretations.

If understanding is a form of practical knowledge, can this kind of immanent knowledge be made public in terms of knowing-that phrases? In many cases it seems very difficult. Imagine a surgeon performing an operation. How much of her skill is due to her state of 'knowing-that' and how much is the result of 'knowing-how'? It may be difficult to decide, and the answer may even change from surgeon to surgeon depending on the degree to which their actions are routine. Today we can construct robots able to perform many human-like actions, including surgery. In order to do so, computer scientists and robot engineers must translate their descriptions of the surgeon's state of 'knowing-how' into their descriptions of states of 'knowing-that.' But these descriptions express the computer scientists' and robot engineers' theoretical comprehension of the matter, not the practical understanding of the surgeon. So what counts as non-representational and representational understanding depends on who you are and your level of training and expertise. No sharp border line distinguishes these two kinds of understanding, but for each individual, culture, and society, there is always a vague line between the kind of understanding embedded in a practice and the kind of understanding that comes with imagination and reflection. Only an outsider may be able to explicate 'knowing-how' as 'knowing-that' by representing what is immanent and non-representational for the performer.

Searle defended a principle of expressability, holding that "whatever can be meant can be said."[35] Paraphrasing his dictum, we might say, "whatever can be understood can be said." But Michael Polanyi countered by saying "that one can know more than one can tell."[36] Polanyi called this more "tacit knowledge" or "personal knowledge." He emphasized the significance of non-formalizable features in how scientific understanding is manifested in skills, intuition, and judgment. This personal dimension of science cannot be rendered into specific formal rules of action or thinking. It is neither propositional nor internally justifiable. Tacit knowledge lies in our ability to use it. Scientists are trained in many different ways which make them capable of seeing possible errors in a data set, drawing consequences from the adjustment of the equipment, grasping a new order in what appeared to be a random pattern of observations, recognizing conceptual inadequacies, and exhibiting competence in describing physical problems in terms mathematics. Polanyi holds that this sort of knowledge is empirically demonstrated by psychological experiments in which "we apprehend the relation between two events, both of which we know, but only one of which we can tell."[37] Polanyi's description of tacit knowledge looks strikingly like a characterization of understanding: "Since tacit knowing establishes a meaningful relation between two terms, we may identify it with *understanding* of the comprehensive entity which these two terms jointly constitute."[38] Polanyi does not deny that a performer's tacit knowledge can be made explicit by other scientists. Rather, if a psychologist can establish that tacit knowledge is embodied cognition, it should also be possible to explain in detail how it comes about, but this is because of the difference between representational and non-representational understanding.

Thus, I will specify that *embodied cognition* contains whatever knowledge or understanding an *individual* person has learned through participating in an information acquiring activity. It stems from the interaction between that particular person and the world and makes its presence known by the ability of that person to carry out a particular action, whose operation is not activated through an act of interpretation.[39] Embodied cognitions should be distinguished from embedded cognition. On the one hand, *embodied* cognition is personal in the sense that it involves experience and imagination of our environment and possible actions given such an environment, including our particular linguistic skills and social competences. On the other hand, *embedded* cognition is *institutional* and constitutes a social practice. We learn a social practice by learning to act according to the understanding embedded in this social practice. For instance, we learn a language by imitating other persons' speech, and thus acquire a 'tacit knowledge' of the linguistic rules, embodied in our linguistic acts, which we may not have learned explicitly. When a group of people shares the same embodied cognition, they are agents in the same emerged social practice. Hence, embodied and embedded cognition are closely related: the embodied cognition of a group may give rise to a common understanding embedded in a social practice, and this practice may be learned by single persons by being imprinted on their action as embodied comprehension.

But before anybody can participate in a social practice, one needs a non-propositional holistic framework of embodied understanding. One needs skills and sensory capacities by which one can distinguish and identify items in the surrounding environment in order to produce observational sentences or theoretical propositions by which one can report what one experiences as happening in the world. As *living creatures*, we already have a perceptual standpoint in the world; the sensory world is already understood as something bodily comprehended before it is linguistically represented. Our body is our first way to engage the world, as embodiment is a necessary condition of human existence. And everyday objects in the world, given our bodily constitution, arise as *gestalted phenomena*. It is by their *gestalt* that we can analyse objects and properties, and if we are successful in our analysis, we can make possibly true statements about these objects and properties. The *gestalt* must be prior to our analysis of the objects as we can only analyse phenomena that are cognitively decomposable. Therefore, our first understanding of actions must have its point of origin in our embodied experience of the world.

This naturalized notion of embodied cognition takes understanding to be the given outlook by which we conduct our philosophical investigation of the world in empirical science. All other methodological strategies have to take some standards as already given and elevate them beyond our experience of the natural world as being epistemologically prior. However, it is not an a priori claim to take a naturalized phenomenology at face value. If a priori claims have any meaning, and I think they do (but only inside a specific perspective), they have to signify something that is not apparent in embodied phenomena and is not dependent on it in any way.

So my suggestion is that this embodied understanding is an important ingredient of scientific practice, lying beneath our reflective judgment. It is a non-propositional form of understanding enabling individual scientists to participate successfully in this practice. Embedded cognition adds to the scientists' background knowledge, theoretical assumptions, and professional beliefs. Much of what we call "background knowledge" constitutes scientific practice by defining the intentional behaviours of human agents, whose actions form this practice, and whose existence is a prerequisite for constructing scientific explanations.

Another terminological point is that I take "cognition" and "comprehension" as distinct. 'Cognition' includes both knowledge and understanding, i.e., a whole system of beliefs where the particular beliefs and the beliefs about their interrelations are true. In contrast, 'comprehension' refers to understanding alone, i.e. to the organization of some beliefs without any regard to the objective content of those beliefs.

So by participating in science the individual scientist depends on a first order comprehension consisting of *embodied understanding*, which is *non-representational*, and does not come with scientific explanation. Embodied understanding is not consciously acquired beliefs, but the organization of uninterpreted bodily stored information. Scientific explanation provides us with a second-order reflective understanding based partly on our embodied understanding but at the same time *representational*. The distinction I want to make is between *embodied* understanding, which is unintentionally acquired and grounded by its utility, and *reflective* understanding, which is intentionally acquired and empirically or theoretically grounded. Thus, most of our embodied understanding of the world is embedded in various practices and proves its existence in virtue of our faculties of acting, perceiving, identifying, interpreting, judging, describing, explaining, and predicting. It is both spontaneous and non-interpretative, not arising from inferences or interpretations. Mastering a language is itself a form of embodied cognition; however, unlike most embodied understanding, language can be used to express reflective understanding. Some of our understanding of the world stems from our ability to construct mental representations of the world in the form of theories and models, which then give us a new form of theoretical comprehension. This abstract form of understanding is the result of a purely reflective, intentional capacity of thought. It is due to an interpretive act of reasoning and our ability to form representations. In science we confront both kinds of understanding. In actual research they are often intermingled, and what is at one time a reflective form of understanding may become over time embedded in practice like any embodied understanding.

There are two sources of non-representational understanding: one springs from genetic encoding and one comes from learning by experience. Parts of embodied and reflective comprehension are learned, but the part usually called "instinct" is inherited. Cranes (*Crux crux*), for instance, have a cognitive capacity to migrate from Africa and Spain to Sweden encoded in their genes, not learned. But a part of their migrating capacity is also learned. Some years ago a flock of cranes migrating over the expanded fields of northern Germany was observed every spring to change its course directly above a solitary tree in the middle of a field. One year the tree was cut down during the winter. So the following spring when the flock arrived at the place where the tree had once stood, they began to circle over the spot where they had 'expected' the tree to be. Apparently, they had by experience internalized this specific German landscape, and the juveniles of the preceding summers had learned to change course over the missing tree from the older adults of the flock. It is only this learned capacity that I call "embodied cognition," whereas the inherent capacity may be named "adapted cognition."

Using planetariums ornithologists have established empirically that nocturnal migrators orient their directions with respect to the stars. This seems to be a clear example of adapted cognition. The nocturnal migrators have a star map genetically encoded in their brain together with an encoded sense of the appropriate direction of migration with respect to an annual and a diurnal clock. One may object that migrating birds possess no understanding but have only an embodied capacity. But since I allow understanding to be non-propositional, and since the criterion of success in reaching understanding is the ability to solve a cognitive problem, it seems that there is no good reason to deny migrating birds a form of understanding inherited from their parents. Certainly, birds are not aware of their inherited capacities, and certainly they do not understand that they understand. But nothing demands cognitive transparency must accompany every state of understanding. Instead I suggest that cognitive transparency characterizes only higher levels comprehension.

One might object that ascribing any form of understanding to animals on the grounds that thoughts are bounded to concepts and concepts are bounded to language. In this discussion Gerald Edelman distinguishes between "primary" and "higher-order" consciousness.[40] He defines primary consciousness as the ability to construct a mental scene in the present that requires neither language nor a true sense of self. Primary consciousness exists in animals with brain structures similar to us, but these animals do not have linguistic capacities or self-consciousness; they are not conscious of being conscious. Higher-order consciousness, which he claims we find only in humans, is defined by linguistic capability and possessing a concept of time beyond the present; i.e., a grasp of future and past tenses. However, recent research shows that some species of monkeys communicate not only through very elaborate vocabularies of sounds and gestures but also deploy grammar in order to create sentence-like messages wherein a difference in structure conveys a difference in meaning.[41] They are also capable of temporal and spatial displacement, which means that they can use language to refer to something, which is not present either at a given time or at a given location.

However, there are also many experiments demonstrating that even without a language animals have thoughts, and therefore possess concepts and forms of understanding.[42] Let us define a concept functionally as an organism's ability to recognize individual tokens as being of certain types. Experiments with prairie dogs show that adults have different warning calls for a coyote, a badger, a snake, or a hawk, but juveniles lack these different warning calls; they must learn to distinguish between them before learning to associate a particular alarm signal with each species.[43] This behaviour is explicable only if we assume that prairie dogs eventually acquire a concept of each species and therefore acquire a correct belief whenever they experience an example of one.[44] Thus it seems that prairie dogs understand the connection between seeing a predator and possible dangers and can warn other members of the colony of what kind of predator to look for. Prairie dogs have learned from each other's mistakes and are now capable of imagining what can happen during the actual encounter and can conceptualize the possible outcome to avoid it.

Similarly, several different researchers have shown that different species of magpies, crows, and ravens can carry out logical reasoning in using tools, in figuring out what other crows will do, and in solving complex tasks.[45] These experiments also demonstrate corvids' cognitive flexibility because crows can establish general rules based on acquired knowledge and react according to new information; they provide evidence of rich imagination in cases where crows can comprehend situations and solve tasks in thoughts before acting, and finally they exhibit the corvids' sense of time in crows' use of earlier experiences to predict events taking place in the future (up to 5 minutes). All these ethological experiments bear witness to monkeys, jays, magpies, and crows' high intelligence, which is possible only if they possess concepts, knowledge, as well as understanding. These examples show us that while humans definitely show the most advanced development of the neural structure of the brain, we have no reason to think that there is a clear-cut demarcation between primary and higher-order consciousness. Thus, though there is no doubt that animals possess understanding *in some sense*, it is equally obvious that human understanding is *in some sense* 'higher' insofar as it leads to scientific knowledge. So the interesting question ought not to be "Do animals understand?" but rather "How is human understanding (a precondition for science) different from the kind of understanding to which animals are limited?"

5 Levels of understanding

Our observations thus far teach this lesson: Understanding is a part of adapted, embodied, and reflective cognition. On each level understanding appears in different forms. The first variant of embodied understanding occurs on the level of experience. I call this *experiential understanding*. We share it with higher animals. Experiential understanding is revealed in the ability to see things as causally connected, or spatially and temporally related. It also appears in recognizing that certain things are separated from other things in having length, depth, and height, in changing or being identical over time, in belonging to a certain type or in being part of a whole. Experiential understanding is a cognitive skill, integrated with our capacity to form epistemic perception, and appears in all situations where we immediately recognize what we experience.

Distinct from experiential understanding, but related to it, is *imaginative understanding*, the capacity to organize visual elements into mental pictures. We use it finding our way through a city and when we read or create maps, graphs, illustrations, and symbols. But this capacity also exists in corvids, monkeys, apes, and cetaceans as shown by their ability to plan actions and solve tasks where the goal is out of sight.

Another variant of embodied understanding is *linguistic understanding*, which allows us to put what we see (or imagine) into particular words. Indeed, human language is far superior to any communication system discovered in animals, but empirical evidence seems to show that a rudimentary and referential language exists among some higher animals.[46] Experiential, imaginative, and linguistic understanding are acquired as 'tacit' skills while we focus on other things.

Instrumental understanding is present whenever we perform an action with the intention of reaching a goal and have the ability to do it correctly. Indeed this form is closely integrated with both experiential and imaginative understanding. This is another cognitive capacity we share with higher animals.[47] We understand by showing *how* to do things, and at least humans also can do this with words. Indeed, this form of understanding is often learned not by being told but by copying what more capable people do. It appears whenever we play tennis, ride a bike, operate a tool or a machine, whenever we handle numbers or an equation in physics, or when we exhibit by our actions how far we have to stand from another person in a certain culture when speaking to each other. A physicist may manage to use an experimental apparatus without any detailed knowledge of why it functions as it does, and is able to use Schrödinger's equation with rather little understanding of *why* it works for prediction. Both experiential and instrumental expertise come in degrees. These cognitive and practical skills cannot be isolated from a person's background knowledge, assumptions, and beliefs.

Experiential, imaginative, instrumental, and linguistic understandings are all forms of 'pre-reflective' embodied understanding, but the following four variants are forms of reflective understanding, presupposing reflective or higher-order consciousness. *Explanatory understanding* is the capacity to ask and answer explanation-seeking questions concerning what causes what, why something is as it is, why something occurred, how something works, or how this is possible, etc. Moreover, it is the understanding acquired from answers to explanation-seeking questions. Explanatory understanding is often obtained in science through models representing causal factors or mechanisms behind observed phenomena. It is also the type of understanding provided by nomic explanations subsuming a particular phenomenon under a covering law or by an overarching unification of a set of laws. These two forms of explanation differ in that one is concerned with particular causal processes, the other with general formal principles. The kind of methodological approach we find most fruitful depends on our epistemic interests and cognitive norms. Thus explanatory understanding leads to the theoretical level.

Scientific knowledge requires truth. However, *theoretical understanding* concerns only the domain of phenomena about which scientists *claim* to have some 'understanding,' a domain certainly much larger than the domain of phenomena of which we *know* their explanation. We have quite detailed understanding of, say, the Big Bang, but I do not think even the most brazen cosmologists would want to say we know the explanation of the origin of the cosmos. We understand the Big Bang in the sense that cosmologists can apply the principles of general relativity to models of the expanding universe and then consider the earliest states in the light of the Standard Model for elementary particles. This gives us a conceptual grasp of what happened, but nobody really knows whether it is true or not, or indeed why these theories are the way they are.

So we demonstrate the 'theoretical level' of understanding whenever we describe a phenomenon by using a theoretical representation, by interpreting a model in terms of a theory, or by making a judgment about the appropriate fit between a model and observation. But if you realize that you don't possess accurate theoretical understanding of a phenomenon, you have either to redefine the phenomenon or to find a better theoretical representation. In either case you look for an interpretation of the observations that may yield the looked-for understanding. Interpretation is concerned with making sense of what we do not understand. Like explanation, the job of interpretation is to deliver a kind of *interpretive understanding*. We acquire this by seeing a problem or phenomenon in analogy with familiar problems or phenomena, or by identifying and classifying things in a different way.

Finally, we have *metaphysical understanding* in those cases where we say what scientific theories can tell us about the structure of the world at the most fundamental level. This form of understanding is also norm and interest driven. For instance, philosophers and scientists may prefer deterministic world-views over indeterministic world-views, an Aristotelian-like world-view over a Platonist world-view, etc.

Of course, these distinctions we have drawn between levels of understanding are drawn as a philosophical abstraction. Thinking and acting draw on cognitive states, which are reflective or representational, as well as states that are embodied or non-representational. In the real world various kinds of understanding are intermingled in supportive cognitive networks and can only be analytically separated. Solving an advanced scientific problem is no exception. It involves both embodied and reflective cognition.

6 Norms of understanding?

All embodied or non-representational understanding is externally grounded in a physical or a social practice, but reflective or representational understanding is internally grounded in certain epistemic standards. For instance, explanatory understanding is a cognitive state in which a belief concerning a state of affairs is connected with similar beliefs according to some epistemic *standards*. An explainer can convey his understanding to the explainee, only if both parties share the same *explanatory* standards and the proposed explanation adheres to these two explanatory standards: The first one is the specific form of the relationship that the explainer believes holds with respect to the topic in question and therefore determines the type of explanation he selects, which may indeed vary from topic to topic, from one domain to another. The second feature is the specific content of the beliefs about the topic involved in the explanation. The resulting understanding can be regarded as relevant or irrelevant, appropriate or inappropriate, depending on the particular beliefs induced by the form and the content of the explanation. For example, an account in terms of an intentional relationship will not generate an acceptable understanding according to the norms of intelligibility in the physics community, whereas an account with the form of a causal or a nomic relationship normally will do so. But the right form alone does not suffice. The content of the explanation must also be acceptable. An explanation of a problem in electrodynamics in terms of a purely mechanical description will not be regarded as offering an appropriate understanding with respect to the physics community. Both form and content of an explanation are determined by the standards the appropriate scientific community considers adequate; thus, it follows that explanatory understanding is dependent on a context of problems, theories, acceptable methods, metaphysical world-view, and personal and communally shared interests.

Belief systems may be organized on a local or global scales, depending on what we think is relevant for an appropriate understanding of the topic with which we are concerned. On the global scale we gain understanding of the phenomena in the world in terms of fitting them into a comprehensive, coherent, and general worldview, the content of which may change through history from one person or culture to the next. Mechanical descriptions were provided acceptable physical understanding during the seventeenth century, but they were not successful as purely naturalistic explanations. God still had an interacting role in preserving and stabilizing the solar system. Today physicists accept explanations violating the standards of seventeenth century mechanism, but they are purely naturalistic. Science accepts only natural causes as intelligible. Hence, scientists increase their insight by connecting as many beliefs as possible by accepting some naturalistic viewpoint over any other. Naturalistic understanding on a global scale is in some respect similar to the conception of explanation as unification. Though apparently similar, the two are really different because, unlike unificationists, contemporary naturalists do not insist that science is able to establish a single over-arching understanding of all *natural* phenomena. Science produces many belief systems which are only loosely connected or connectable. Even physics is not, as some physicists believe, ultimately able to reduce all of its different theories to one single theory of everything.

On the local scale we want to understand how things work and what they are made of. Understanding is achieved on a local scale when we, say, investigate the mechanism of an old-fashioned mechanical watch, seeing how each part functions in relation to all the other parts. The ability to provide a causal description of how the springs and gears work constitutes our local understanding of this phenomenon. Salmon holds that global and local understanding are the two conceptions of explanatory understanding we have.[48] But explanatory understanding need not be limited to grasping internal deductive relations, as in understanding by unification, or to comprehending external relations as in the causal-mechanical account of understanding. Many other relations such as inductive, analogical, mereological, and structural relations may help to organize and bring coherence into our system of beliefs. Different kinds of explanations connect phenomena in distinct ways, and these different notions of how the phenomena are related, which the various explanations use, provide different kinds of understanding.

We shall return to the role of interpretation in understanding in the subsequent chapters. For now we can conclude tentatively that understanding depends on whether our beliefs hang together in the light of our background assumptions. In dealing with the world we form certain beliefs about natural, social, and cultural facts. Some will be true, some false, as determined by how the facts really are in the world, or how we agree they are. Some may not be regarded as true or false in the sense of corresponding to reality; nevertheless, they may cohere with what we otherwise believe. Regardless of whether our beliefs are true or false, we claim to understand things, concepts, or facts if we can fit our beliefs about them into the wider framework of our other beliefs. By explanation we show how our beliefs about the explanandum cohere in acceptable ways within a system of beliefs, but these beliefs can be individually false even if they are internally consistent as a group. For instance, a belief in action at a distance was false within Newton's mechanics even though it fitted in quite nicely.

Furthermore, our beliefs are usually organized such that some are seen as more basic than others, so that minor beliefs are subordinate. These basic beliefs form a worldview. But scientists do not all necessarily share the same worldview; therefore, their understanding differs. Such a suggestion immediately raises a problem: All a creationist needs to do is to show how his beliefs concerning the creation hang together with what he otherwise believes or claims to know even though the biblical story is probably false. Does this not entail relativism? I don't think so. From a scientific perspective the creationist's understanding is not as good as the scientific account, but this is not because the Bible's explanation is not coherent and thereby meaningful for the creationist. Scientists reject the explanation given by *Genesis* because it does not accord with scientific norms of intelligibility such as employing only natural causes. Nor can such an explanation meet the methodological prescriptions we value in science such as testability, predictability, and empirical adequacy. These standards and prescriptions are intersubjectively recognizable, and have proved to be epistemologically superior in scientific practice by providing a more accurate, detailed, and coherent worldview than the Biblical one.

No single norm of objective understanding predominates in all sciences; rather the criteria of understanding vary according to the epistemic context in which scientific thinking takes place. Thus, our norms of empirical understanding differ from those of conceptual understanding, and those of physics almost certainly deviate from those of literary criticism. Even within one science like physics, some scientists might demand virtues like mathematical rigor, unification, simplicity, and consistency, whereas others will prefer causal mechanism, prediction, and experiment.

For instance, Einstein mentioned completeness, adaptability, and clearness on the one hand, and logical perfection and security of the foundations on the other. But the existence of many different standards does not establish that it is impossible to sketch a formal characterization of understanding applying to all the different forms in which scientific understanding is manifested. Indeed, since understanding is a quality of people with different cognitive interests, different epistemic standards of understanding are to be expected.

From what we have seen so far we can characterize *reflective* understanding formally as follows:

(U) A human agent A understands a state of affair, P, in a certain context, C, in the terms of a theory, T, if, and only if, A's belief regarding P connects (in the epistemically correct way for A; i.e., in accordance to A's epistemic norms, N, of understanding) with A's cognitive system, including A's background knowledge, beliefs, and assumptions (A's worldview).

In A's worldview there are many beliefs and assumptions which do not count as pieces of empirical knowledge but which accord with A's system of beliefs, including, for instance, conceptual beliefs and assumptions. Thus A may truly claim to understand P, and therefore to *assume* that a belief regarding P is true, but not to *know* whether her belief is true.

The organizing principles connecting A's beliefs about P with some of A's other beliefs, or perhaps structuring the entire set of A's beliefs, include many kinds of cognitive relations such as deductive and inductive inferences, explanatory schemata, conceptual representations, structural similarities, and analogical comparisons. The cognitive principles used to organize our beliefs depend on the particular kinds of beliefs and our particular kind of interest in the subject of our beliefs. Nevertheless, no matter what our interests might be, the organizing principles need to meet certain internal conditions of rationality for our understanding to be successful.

Let me draw attention to two further points. First, the role of theory in the above definition allows the individual scientist to conceptualize a certain domain of phenomena and enables her to express herself such that her understanding is consistent with that of other scientists who either use the same theory or, if not, at least can discuss it rationally (*pace* Kuhn). Second, if the purpose of explanation is to produce understanding of P, and understanding consists in relating a certain belief about P to other beliefs, forming part of A's cognitive system, we have to establish standards which can make such an organization rational. There are at least four internalist conditions for having *reflective* understanding: A's conception of P must be (1) consistent, (2) coherent, (3) relevant, and (4) sound with respect to A's background assumptions.

If A's conception of P is inconsistent with his other beliefs, we would not say that he understands P. He might not 'see' this inconsistency. Blinded to the inconsistency, he *feels* he understands, but this does not imply that he genuinely does understand. Many of our background beliefs are stored in different cognitive departments between which there is little or no explanatory or inferential connection: say, between physical, political, and religious beliefs. We often act as though these different belief systems have only little relevance to each other. We tend to compartmentalize them. So the mere presence of an inconsistency somewhere in the whole system as grounds for saying that A doesn't understand P is too strict, because if it were accepted, then really no one would understand anything. We need to hedge it in a bit, such as allowing no inconsistency with his other beliefs *qua* scientist; as for trans-scientific beliefs, an inconsistency with them could be tolerated and we would still say as a scientist A understands P. Even so, probably there exists a certain amount of inconsistency within each and every cognitive department. Sometimes we want to get rid of inconsistencies in the battle for obtaining understanding, whereas at other times we may regard them as the inevitable result of different cognitive perspectives on the same phenomenon. The upshot is epistemic pluralism if not the same in metaphysics.

My formal definition of reflective understanding underscores the fact that a person's understanding relates to a context, since people's understanding varies with the subject of their understanding. One might think that understanding so defined consists of only relations internal to one's system of beliefs. But contexts involve external relations which can still be handled within the framework of beliefs. Take Coulomb's law. Here a physicist understands this law because he knows the context in which it applies, namely to *represent* an electrostatic system consisting of two point-like charges, and in which context it doesn't apply (in case the charges are not point-like). Everything takes place harmoniously inside the cognitive system.

In general, understanding emerges from the epistemic structure that our beliefs form with other beliefs. But explanatory understanding makes the explanandum intelligible by relating beliefs according to some norms and standard ways of relating things. Thus the various genres of explanation reflect different cognitive *schemata* for relating beliefs to make them *coherent with* other of our beliefs.[49] At the same time such schemata pose standards of intelligibility. If I am told that each time I take a sip of tea, my heart beats, it does not give me any understanding of anything because the temporal relation between my heartbeat and taste of tea does not relate the events in a manner such that the occurrence of these happenings becomes meaningful. To make their co-occurrence intelligible would require appealing to a causal or functional relationship between them.

7 Cognitive schemata

Nobody, I imagine, would deny that our ability to think is inherited. But there is no reason to deny that some forms of our thinking are inherited. But if cognitive schemata of understanding do exist, what are they?[50] As cognitive-neurological patterns they must result from biological evolution, during which our ancestors related things in uniform ways of practical benefit in their interaction with their environment. Little by little our hominid ancestors experienced a sense of understanding in cases where somehow they were able to connect things successfully, imagining, predicting, and acting. Imagination allows one to connect one's experience with what is not present; prediction allows one to relate current experience with what one is going to experience in the future; and action allows one to direct behaviour to realize certain imagined goals. Over millions of years human understanding developed by this process of rewarding cognitive success. Our ancient forebears adapted mentally as well as physically to their environment, giving them a huge advantage in the competition of surviving and reproducing. But hominid progress was also possible because much of their environment exhibited regularity and stability on a temporal basis and therefore in what was foreseeable or imaginable.[51]

Hence, I propose that human understanding arose through our adaptation of a capacity to create cognitive connections between what we did experience and what we did not experience. Those connections, which were constantly cognitively rewarded by the environment, were entrenched as cognitive matrixes in the brain and thereby established the cognitive schemata by which we organize our beliefs. Schemata are pre-organizing modes of cognition allowing us to receive insight and introduce intelligibility into our thinking. The different types include, for instance, the *schema of causation* by which two things are understood such that one of them appears due to the other.[52] Another example is the *schema of function* by which things are understood in relation to their actual aims. A third example is the *schema of intention* by which things are understood in relation to their intended aims. A fourth one is the *schema of analogy* by which things are understood in virtue of their similarity to other things. A fifth is *the schema of structure* by which things are understood as taking part in a whole. A sixth is the *schema of imagination* whereby things are understood as representations of something else. Finally, a seventh example is the *schema of inference* by which we understand that one belief can be derived from another. I do not claim that this catalogue is exhaustive, but it includes some of the basic cognitive schemata. What is common to all schemata is that they generate understanding by establishing, based on visual information, a connection in thoughts between different things.

The operation of these cognitive schemata can be illustrated by a simple example. I observe that my daughter's knee is bleeding, although I do not understand why. I do not understand because I cannot relate my belief about her bleeding knee to any other belief in my belief-system. When she selects a particular schema of understanding by telling me that she fell on the pavement in the schoolyard, I understand why she has skinned her knee. The causal story she gives uses a cognitive schema of understanding which is also one of my schemata of understanding. Now I can fill in the blanks because I know that tissue bleeds when the skin scrapes on a rough and hard surface. Her answer is explanatory because it provides information closing the gap in my knowledge by adding a new belief and connecting it, according to an acceptable schema, with the whole constellation of other beliefs I already have. By closing the epistemic gap in terms of a causal explanation, what happened with her knee becomes understandable to me.

A naturalistic account of our reflective understanding is grounded in our adapted and embodied understanding, and the use of cognitive schemata determines this grounding and gives reflective understanding a biological foundation. These schemata are the connecting link between concrete and abstract thinking; abstract thinking alone would never yield understanding. The reason why we don't feel that we 'understand' quantum jumps or quantum entanglement is that these ways of connecting phenomena are not grounded in innate schemas. However, the use of cognitive schemas as the foundation for reflective understanding is not intentional. Rather this happens automatically as a result of our evolutionary heritage.

It makes sense to say that we can have explanatory understanding without knowledge. It is much better to say that understanding is organized *beliefs* and, therefore, that adapted understanding shapes an explanation by giving it a certain form, an explanation imparts explanatory understanding by connecting a certain belief to a system of beliefs. For example, Aristotelians gave an explanation of the movement of an arrow which we know today was wrong because it rested on a misconception. They understood the flight of an arrow in terms of a causal schema within the framework of their beliefs, but from the framework of our contemporary beliefs they were wrong with respect to the particular items they put into the schema. The truth-value of our beliefs is not so important for having explanatory understanding. Seen in this light, we can talk about explanatory understanding in mathematics, philosophy, and theology without presupposing that these areas of thinking have intuitive access to improvable or unverifiable truths.[53]

Evolution hard-wired these cognitive schemata in our brain, so it is impossible to say whether schemata are true or false because truth-values apply only to the beliefs the schemata lead us to embrace. Reflective understanding is representational, implying that this part of our understanding can be represented by a propositional structure. Reflective understanding is representational not because it represents anything, but because the understanding it imparts is expressed by theories, models, pictures, graphs, maps, symbols, languages, or mathematics. These different types of representation can be used to express our understanding and as such they may be true or false. Understanding need not be *sound* in the logical sense that all beliefs on which it is based are true. Many of our empirical beliefs stem from inductive inferences, and many of the premises of our deductive inferences need not be true because they are themselves inductively derived or they built on postulates or pure speculations. If our representation of understanding is based on (or results from) sound reasoning (or inference), it means that all *relevant* beliefs are in fact true and that all inferences are valid. We may therefore, rightly or wrongly, feel entitled to hold that the stock of beliefs is true because connecting a particular belief to this stock by an explanation gives the stock epistemic support. So we want explanatory understanding in an actual explanation to be true, simply because our actions, intentions, and strategies will be successful, only if they rest on a robust representation of empirical facts. At least we want our understanding to be correct for pragmatic reasons. I don't think of this cognitive norm as a condition for having understanding, since the relevant responses to explanation-seeking or interpretation-seeking questions may be false but still provide understanding even though it is an 'incorrect' understanding.

In sum: As part of their training scientists have a concrete and embedded understanding, which does not rest on interpretation, but which interpretation presupposes and from which it departs. Concrete understanding results from our existence in the world (as Heidegger claimed) and from our scientific socialization into the practice of a certain discipline. This kind of understanding constitutes the scientist's competence as a particular scientist. It does not appear from a distinction between representations and the represented. If you consider only embodied understanding, it is not possible to uphold such a distinction.

But scientists also have reflective understanding: they form new theories or new models in order to explain new phenomena. Before they can do this, they must set up theoretical representations through interpretation of their experiences of the world. Only after the scientist's experiences have been properly interpreted can we separate the abstract representation from what is represented because the representation goes beyond our immediate understanding of perceptual and experimental experience. This reflective form of understanding is neither embodied nor based on tacit knowledge. It is explicitly expressed. Often scientists come to this kind of understanding by an appeal to analogical thinking; they gain reflective understanding with either a fruitful conceptualization of a certain unfamiliar phenomenon – in order to bring it within the scope of a conceptual framework – or a successful explanation of it as the causal effect of other familiar phenomena to which it is nomically connected. Reflective understanding consists in consciously grasping how pieces of information relate to one another, seeing how they can be connected so they hang together.

3
On Interpretation

A scientist's training requires learning a certain practice with its own specific language, epistemic goals, methodological standards, and a metaphysical outlook. As a result scientists acquire a huge repertoire of skills and domains of competence needed to act as a member of the scientific community. For the most part scientists participating in this practice know what to do, how to do it, and how to explain to outsiders why they are doing it. Doing all these things demonstrates an immediate understanding of the phenomena being studied. But if a particular phenomenon, an experimental anomaly, or a theoretical problem is not understood by anyone in the community, they must try to make sense of the phenomenon, the anomaly or the problem by *interpreting* it.

There are at least two notions of interpretation in science, which trade on an ambiguity similar to its use outside of science. One use is in the sense of mental *construction* by which we try to grasp a certain phenomenon, or classify and identify certain objects, or create a new theory or conceptual framework through which we understand a certain area of human experience. But in another sense interpretation is a specific kind of *explanation* concerned with the meaning of data, or the content or metaphysical presuppositions of a theory. Although I shall distinguish 'interpretation' as *construction* from 'interpretation' as *explanation*, both are equally important for understanding what goes on in science. Here I will explore this distinction and discuss how a pragmatic and contextual theory of scientific understanding can account for both uses.

1 Two notions of interpretation

From medieval times we have inherited the distinction between *subtilitas intelligendi* and *subtilitas explicandi*. Both are equally important for understanding the practice of science, but today our use of 'interpretation' seems to cover both senses. Therefore I shall distinguish between two notions of interpretation: one of which I call *construction of meaning* and the other *explanation of meaning*. In general, however, any interpretation is a response to a representational question requiring a hypothesis connecting the representer and the represented. Usually our representational understanding begins as a constructive inquiry, and only later it may be used in an explanatory response.

We ask for an explanation and engage in interpretation for a similar reason: both hope to gain understanding. Interpretation is commonly associated with understanding meaning. The objects we interpret are either intentional objects or have intentional properties. Therefore, interpretation is seen as a process that leads us to understand persons, actions, or consequences of these actions, such as linguistic expressions, texts, painting, sculpture, music, film, dance, plays, and social institutions. What we understand in these cases is the meaning expressed by these objects, and an interpretation shows the way to this meaning through a hypothesis. In my opinion this view of interpretation as a response to questions about intentional meaning is too narrow and simplistic.

First of all, not all objects of interpretation need be of human origin. In scientific inquiry any natural phenomenon can also become cognitively meaningful. For instance, causally produced effects become meaningful if they are designed as data or are considered as evidence of their causes. In addition to being natural phenomena, they also have gained a cognitive status by being conceived as 'data' or 'evidence.'[1] We thereby impart a certain meaning to them in the sense that we take these phenomena to be informative about something they are designed to represent. So data and evidence are interpreted whenever the scientist doesn't understand what they inform us about or are intended to represent.

No phenomenon out of a given context of inquiry points to anything beyond itself. It does not reveal how it must be understood conceptually. When the object of an interpretation is considered to represent something other than itself – such as data, signs, symbols, and symptoms – the interpretive question is what are these phenomena evidence of, what do they stand for, refer to, or what caused them. But at other times when the objects of interpretation are types of phenomena lacking an appropriate classification or conceptual representation, the interpretative question concerns how they can be made the subject of representation. Finally, there are times when the interpretative question is about whether a particular phenomenon belongs to this or that category.

So when the interpreter addresses questions like "What does X stand for?" "What does X represent?" "What is X evidence of?" etc., the 'interpretations' become explanations of cognitive meaning, and the interpreter intends to answer them by a response in the form of a hypothesis. Therefore I shall suggest that like other types of explanations interpretations of this kind can be considered as the interpreter's response to a question expressing a lack of understanding.[2] Both question and answer presuppose that the phenomenon X can be understood as intended by somebody to represent something or can be seen as evidence of something which X represents. Interpretations of this sort may also be called *determinative* interpretation since the interpretation determines by explanation the cognitive meaning, which a culture, a group of people, or an individual person in a culture associates with the phenomenon under consideration. Explanation of meaning is indeed possible only if the interpreted phenomenon *already* carries a meaning independently of the interpreter's act of interpretation because it has previously been interpreted to have such a cognitive meaning.

However, in other acts of interpretation the interpreter gives meaning to the phenomenon in question. Here the focus is on a phenomenon Y, which the interpreter finds unintelligible as it appears and therefore does not understand. In these situations we ask, "How can Y be understood or represented?" I call responses to such questions *investigative* interpretations. In these cases the act of interpretation makes the phenomenon Y intelligible by assigning a certain representational understanding to it. This happens by proposing a new form of classification, conceptualization, or schematization, and then by bringing the resulting beliefs into a coherent connection with the interpreter's belief system. In this case the understanding of Y depends on the interpreter's own invention.

Therefore I am proposing a much broader perspective that sees 'interpretation' as offering both explanation and conceptual comprehension in both the humanities and the natural sciences. A determinative interpretation proposes a deliberately formulated hypothesis concerning what it is that a representation really represents. Here the interpretation is a form of explanation, which explains *the representational role* of some phenomenon. An explanation of meaning in connection with the consideration of effects, data, evidence, signs, symbols, images, graphs, models, and theories is called for when a subject regards a phenomenon as representing something else, but he is unsure what it represents. In its non-explanatory, investigative form 'interpretation' is a proposal for classification, conceptual representation, or theory formation. We respond intentionally with an interpretation only when we do not possess enough information to understand what a puzzling phenomenon represents or how it should be represented.

Pragmatists recognize that determinative interpretations are context-dependent responses to questions about the meaning of a phenomenon. Because interpretations are a kind of explanation, a pragmatic and rhetorical theory of explanation will apply to interpretations as well.[3] According to this unified theory of explanation and interpretation, the form of an interpretation is produced deliberately to answer a question about the meaning of a phenomenon. The interpretive process results in certain statements concerning a representational issue, whereas the process itself is the communicative action leading to these statements. But how this result turns out depends on the context and therefore, among other things, on the aim and interest of the interpreter.

I suggest that the type of interpretation is determined partly by the interpreter and partly by the object of interpretation. Interpretation requires an intentional agent, but the object plays an important role in the interpreter's selection of the relevant type of interpretation. The interpreter constrains her interpretation in accordance with her grasp of the object by choosing the type of interpretation accordingly. But the interpreter's beliefs about the situation, and her goals and interests, also partially determine the form of interpretation. Thus, the person's background assumptions and beliefs about the object influence the hypothesis she generates. This applies not only to the form of the hypothesis, but also to its content. This content is as much context-dependent as is the form. But, again, the object also imposes some constraints on any possible understanding of the content.

Therefore, to summarize the two notions of interpretation introduced above, on the one hand the *determinative* interpretation signifies a situation in which the interpreter explains the meaning of what he regards as a representational phenomenon, usually evidence, an image, a graph, a sign, a symbol, or something similar, by a hypothesis concerning what this phenomenon represents. The *investigative* interpretation, on the other hand, signifies a situation where the interpreter constructs a hypothesis endowing the phenomenon with the meaning he wants it to represent, in case he does not already grasp the meaning of this phenomenon. However, I shall also hold that the result of an investigative interpretation may change status and become part of a determinative interpretation whenever the members of a scientific community reach a common agreement that a certain conceptual construction of an investigative interpretation forms their general understanding of the phenomenon in question.

Strangely, philosophers of science, occupied with explanation, have shown little interest in characterizing interpretation in spite of the fact that they themselves frequently speak of interpretation. The lack of interest among analytic philosophers is partly due to their intuitive assumption that these two concepts belong to opposite sides of Reichenbach's famous distinction between the context of discovery and the context of justification. Thus, interpretation has to do with the context of discovery, whereas explanation belongs to the context of justification. This led philosophers like Popper and Hempel to develop the deductive-nomological model of explanation. They simply ignored 'interpretation' as being too much of a psychological notion closely tied to 'meaning' and 'understanding;' they seem to have tacitly accepted the hermeneutic division between 'explanation' and 'understanding' as important for characterizing the difference between natural sciences and humanities (*Geisteswissenschaften*).[4] 'Explanation' could be analysed logically; 'interpretation' involved a subjective synthesis of understanding. In contrast to analytical philosophers, hermeneutic philosophers have dealt with 'understanding' and 'interpretation,' but paid little attention to 'explanation.' For them interpretation was concerned with meaning and intentionality. But both interpretation and explanation are parts of the same scientific practice, and we cannot understand this practice as a whole if we cannot give a unified account of both aspects of this practice.

2 Interpretations in the natural sciences

Certainly, we talk about interpretation in the natural sciences. Think of interpreting visual phenomena, experiments, measuring effects, data, formalisms, and theories. Interpretation takes place in cases where we want to understand what is going on in astronomy, physics, chemistry, and biology, too. If this is true, then the objects of interpretation need not be intentional objects. I assume that scientists use the word "interpretation" to refer to some procedure of getting representational understanding, regardless of the nature of the object under investigation. What turns a question about a phenomenon into one about interpretation is not the kind of object but the kind of epistemic context in which this question appears. A meaning-seeking question arises in an epistemic context in which the interpreter faces a representational problem.[5]

In the natural sciences we find scientists interpreting phenomena on both empirical and at theoretical levels. We also see scientists entertaining different kinds of interpretation. On each level we meet at least two kinds of interpretations reflecting the ambiguity in how the term is used outside the sciences. What is called interpretation in one sense is a kind of construction in which we try to make sense of a certain phenomenon, where we struggle with identifying or classifying certain objects, where we create a new theory or conceptual framework in attempting to understand an area of reality. But in another sense what is called "interpretation" is a kind of explanation, for example when scientists explain the meaning of data, the content of scientific theories, or the metaphysical understanding of a theory.

Let me give some examples to illustrate this account of interpretation. A classic example of an investigative interpretation is the definition of a planet. In ancient Greek the word "planet" meant a wandering star. Eventually astronomers became aware that these wandering stars were not stars at all, but objects orbiting the Sun. Around 1900 no exact definition of "planet" existed, but there existed a common agreement that the word applied to a small number of objects in the Solar System, since smaller objects like moons, comets, and asteroids had slowly been given their separate classifications. Immediately after the discovery of *Pluto* by Clyde Tombaugh in 1930, this object was classified as a planet. But where Uranus and Neptune have nearly circular orbits, large masses, and proximity to the ecliptic plane, Pluto defies these characteristics, and its orbit even crosses Neptune's. These facts were first discovered later but had no immediate implication on Pluto's status. The astronomy community began to discuss reclassifying Pluto when it was discovered that objects in the Kuiper belt were similar to Pluto in composition, size, and orbital characteristics. In 2005 Mike Brown discovered the trans-Neptunian *Eris*, an object larger than Pluto, forcing the International Astronomic Union to review the definition of "planet." A decision was reached only by compromise between three rival proposals stipulating a new definition of "planet" combining several characteristics: a celestial body that: (a) orbits the Sun, (b) has sufficient mass for its self-gravity to give it a nearly round shape, and (c) has cleared the neighbourhood around its orbit.[6] This definition of a planet was supplemented with definitions of a "dwarf planet" and of a "satellite," changing Pluto's status from a planet to a dwarf planet.

This is an example of constructing a new meaning to meet the challenge facing the astronomical community: how to represent celestial bodies conceptually so that "planet" was not arbitrarily but continued to have a scientific meaning. This concept is not a social construct in the sense that it is based on an arbitrary social decision.[7] The concept of a planet is defined in terms of objective, observable features like orbiting around the Sun, nearly round shape, and clearing the neighbourhood. These properties are selected from many others because they are typical of a certain category of objects and relevant for explaining other features of these objects. Although some astronomers remain critical of the definition, their disagreement is based on theoretical, not social, objections.

The second example is Vesto Slipher's discovery of the Doppler shift in galaxic velocities. In the beginning of 1912 he began observing the lines in spectra from galaxies (then called "spiral nebulae") beyond the Milky Way. First he found the spectral lines from the Andromeda Galaxy were blue shifted – indicating that it was approaching the Milky Way at −300 km/s. Slipher concluded: "The magnitude of this velocity, which is the greatest hitherto observed, raises the question whether the velocity-like displacement might not be due to some other cause, but I believe that we have at the present no other *interpretation* for it."[8] To which of the two kinds of interpretation is Slipher referring? Since he already understands the blue shift as a Doppler Effect, his use of the term "interpretation" must refer to a *determinative* interpretation. He explains the blue-shift as an 'effect' of velocity, or perhaps as 'empirical evidence' for holding that Andromeda is approaching the Milky Way.

Four years later Slipher announced that the blue shift of Andromeda was exceptional, most other spiral nebulae showed a significant red shift. For a long time these 'nebulae,' interpreted as the early stage in the evolution of stars, were thought to belong to the Milky Way.[9,10] However, Slipher interpreted his red shift data as evidence that the spiral nebulae recede with an average velocity of 700 km/s, many times higher than that of any observed star. Slipher's data did not imply a hypothesis about the cosmological red shift and the expansion of the universe because at that time he had no means to measure the distances of the various nebulae, but did allow him to question the common interpretation of nebulae: "For us to have such motion and the stars not show it means that our whole stellar system moves and carries us with it. It has for a long time been suggested that the spiral nebulae are stellar systems seen at great distances. This is the so-called 'island universe' theory, which regards our stellar system and the Milky Way as a great spiral nebula which we see from within. This theory, it seems to me, gains favor in the present observations."[11] Thus, Slipher used his data to cast doubt on one interpretation, representing spiral nebulae as inside our own galaxy, while also supporting another interpretation, representing them as galaxies external to the Milky Way. Both interpretations are examples of the investigative notion of interpretations.

The third example of interpretation also involves both determinative and investigative aspects at the same time, and it shows that interpretative responses need answer not only what-questions but also how- and why-questions. In the year 1572 Tycho Brahe saw in Cassiopeia a bright point of light, which he knew from his background knowledge had not been there before. His data revealed the object exhibited no parallax from which he inferred it didn't belong to the spheres of the planets. The question "What is it then?" became equivalent to a question like "What does the unexpected appearance of a hitherto unobserved star mean?" Tycho interpreted it to mean it was a new star, a 'nova.' We now know his interpretation was mistaken; it was a star dying of age, a supernova.

Observational data of such very bright objects were first interpreted by Wilhelm Baade and Fritz Zwicky to imply a distinction between common novae and supernovae based on their brightness. The changing luminosity of observed supernovae and Einstein's mass-energy relation allowed them to calculate the amount of mass being dispersed into space. They concluded "that the phenomenon of a super-nova represents the transition of an ordinary star into a body of considerably smaller mass."[12] But initially they did not account for the nature of this object; later in a consecutive paper just a year after James Chadwick discovered the neutron, they proposed it might be a neutron star: "With all reserve we advance the view that a super-nova represents the transition of an ordinary star into a neutron star, consisting mainly of neutrons."[13] Thus, it could be said that their initial supernova hypothesis was a determinative interpretation, whereas the later neutron star hypothesis counts as an investigative interpretation. Their theoretical deduction that such a star might "possess a very small radius and an extremely high density" suggested the hypothesis that the leftover of a supernova explosion was possibly a neutron star because neutrons can be packed much more closely than electrons and nuclei.

The first proper explanation-seeking question might be expressed as "How does an ordinary star become a supernova and transform into a neutron star?" This question requests a factual explanation, not an explanation of meaning. Assuming that light from a supernova provides evidence about this process, a possible response is to ask: "How does the light of a supernova vary over time?" In 1934 Baade and Zwicky did not have the data to answer this question, but by 1941, Rudolph Minkowski, using an improved observation technique, was able to interpret the data as suggesting a division into type I and type II according to their distinct spectra and light-curves. About the same time the first explanatory answer was finally offered by constructing a theoretical model based on various astrophysical hypotheses proposed during the preceding years, which showed that a massive star becomes unstable whenever the thermal pressure outward is not strong enough to counteract the inward gravitational force. This implies that further matter is added to the core, and this massive star crushes under the pressure. Then a supernova occurs through an explosion created by the gravitational collapse of infalling matter, which liberates potential energy, heating and expelling the star's outer layers.

Although many puzzles remained, now astronomers had reached a level of understanding allowing more minute explanations to be introduced. After the standard model of stellar nucleosynthesis had been developed, by 1960 Fred Hoyle and William Fowler could construct a quantitative model of supernovae.[14] According to this model, the explosion of type I supernovae is caused by a process which differs from the explosive process in type II supernovae. By 1970 a reasonable model of both supernovae was available, although by 1990 the explanatory advantage of making a finer distinction among supernovae (type Ia, Ib, Ic, IIP, and IIL) led to a revised and even better understanding.

Thus, when we ask for an interpretation, we pose a representational problem. We request an interpretation whenever we lack the information necessary to solve a representational problem, but we believe someone can provide a suggestive clue. An appropriate response is generated by the interpreter based upon a certain understanding of the cognitive problem raised by the interpretation-seeking question. The supernova example shows how a relevant response is constrained by our background assumptions. Tycho assumed that if a bright object in the sky didn't show any visual parallax, it belonged to the stellar sphere, an assumption we still accept. Similarly, Baade and Zwicky presupposed that Einstein's mass-energy formula is correct and that the electrically neutral particles Chadwick discovered had no electrostatic forces between them and so could be densely packed. The relevant hypothesis in Tycho's understanding was that the sudden appearance of a star presented a birth, whereas in Baade and Zwicky's understanding, the same phenomenon pointed to a star dying of age in transition from the stage of an ordinary star to the stage of a neutron star.

3 Roads to interpretation

Is there a difference between conjectures and interpretations, and if so, what is it? Are conjectures more tentative than interpretations? Or are conjectures more arbitrary than interpretations? Both conjectures and interpretations have in common that they are intentional acts of reasoning. Apart from that, conjectures seem to be mere uncertain guesswork.

It is well known that Popper maintained that there is no logic of discovery and that the construction of scientific hypotheses is a matter of bold guesses, intuition, and creative conjectures. This view can be criticized because it regards hypothesis formation as arbitrary and not rooted in scientific reasoning. Nonetheless, conjectures seem less systematically constructed than interpretation. One may say that 'conjectures' can be proposed more or less without much theoretical background – it's the sort of thing done by the ancient Milesians – whereas 'interpretation' virtually demands it. An interpretation is a hypothesis about a representational issue intended to solve a problem of understanding. It may be an explanatory hypothesis about what evidence or intentional objects stand for, or it can be a constructive hypothesis about how a certain phenomenon should be represented by an intentional object. In contrast a scientific conjecture is not necessarily occupied with representational issues but may be concerned with factual ones.

If an interpretation is an explanation of meaning, there is no reason to hold that the hypothesis involved in the interpretation is not based on common systematic procedures such as simple induction, demonstrative induction, and abduction. For example, Baade and Zwicky realized a remarkable difference in the observed intensity of the light coming from those stellar objects whose light flares up to a maximum and then fades away. As part of their background knowledge, Baade and Zwicky knew that if these objects were in a nebula, they must be outside our own Milky Way, and the red shift of a nebula enables one to calculate its distance. They knew common novae are rather frequent in our galaxy, ten to twenty flashes per year and a similar frequency appears in the bigger Andromeda galaxy.[15] Their absolute brightness at maximum is 20,000 times that of the Sun, making them absolutely the brightest stars in our galaxy. Moreover, due to their brightness they can also be observed in nearby galaxies.

Baade and Zwicky interpreted their data as revealing a new group of objects that did not fit their knowledge of common novae. To find an interpretation of these data that explained their meaning in relation to the data from common novae in our own galaxy Baade and Zwicky asked: Do these data show a systematic pattern and mutual interrelationship distinguishing them from the data of common novae such that they can be interpreted as evidence of a new kind of object? They discovered that this data is found not only in the nearest galaxies but also apparently from all systems of stars within the observational range. The evidence on the photographic plates showed these events were rather scarce compared to common novae and that at their maximum brightness these objects emit nearly as much light as the entire host galaxy, and most of the visible light emitted during a period of maximum brightness, which lasted for 25 days. The two astronomers calculated that this amount of light is what the sun would radiate with its present brightness over ten million years. They concluded that a stellar object of this sort disperses most of its energy into space within a very short interval, and using Einstein's mass-energy relationship to determine how much of the object's mass was thrown out from the object, they calculated that the leftover was a very small object compared to the original star. Finally, they pointed out that Tycho Brahe's "Stella nova" was apparently such an object, since no visible traces of it can be found today.[16] Thus, Baade and Zwicky interpreted their data as evidence of a new group of stellar objects that are like novae but many times stronger. They interpreted the data as representing "supernovae": *"The phenomenon of a super-nova represents the transition of an ordinary star into a body of considerably smaller mass."*[17]

This example clearly shows that the interpretation of evidence always occurs within the context of a particular scientific practice including shared understanding. Baade and Zwicky realized the photographic data were similar to those of common nova in some respects, but dramatically different in other aspects, and these differences were sufficiently robust to warrant hypothesizing a completely new class of novae, which we now call "supernovae." Thus they had to construct a new concept, which gave meaning to all the relevant data. But not all explanations of meaning require constructing new concepts. This happens, for instance, whenever a teacher explains the meaning of data, traces, graphs, models, concepts, and mathematical formulas to his students. But if data or evidence cannot be understood in terms of already well-accepted concepts, then scientists must indeed introduce a new concept, but to be useful a new concept has to be accurate and consistent with background beliefs.

How do scientists come up with exact and precise concepts? Hempel suggested two ways: by nominal definition or by a real definition.[18] In nominal definitions, one stipulates how the concept is to be understood. In the real definitions, one tries to establish a natural class by listing the necessary and sufficient properties shared by instances falling under the concept. However, Hempel noted that real definitions can be problematic because of the implausibility of assuming the existence of natural classes in many areas, and because of the high probability of ending up with incomplete or vague definitions. However, Baade and Zwicky's introduction of the new concept of "supernova" does not correspond to either of Hempel's suggestions. Their arguments for the interpretation of the data show that their reasoning did not seek necessary and sufficient conditions for being a supernova. Rather they were searching for an entity whose structure and behaviour would explain the then unexplained patterns of spectroscopic data of galaxies, regardless of the ontological status of this entity. Their reason for introducing this kind was its suitability for induction and explanation.

Turning to the opposite use of interpretation, scientists construct concepts or classifications to endow our observations of natural phenomena with scientific meaning. As in the cases where we explain the meaning of intentional objects like data, evidence, or theoretical signs, we find that our beliefs, assumptions, and background knowledge guide our comprehension. Historically one of the greatest challenges for scientific interpretation has been establishing an appropriate taxonomy within a certain field of investigation. Aristotle set up the key for any system of classification by dividing things into groups, subgroups, and other relevant distinctions based on what was considered to be their common properties and what marked their divergences. Since his time, discussions in the sciences have swayed to and fro over what should be the defining features of a particular taxonomy. Scientists initially focused on immediately observable features, but later they turned to more theoretical ones. In biology, for instance, Aristotle categorized animals according to their shared physical characteristics; he looked for communality of functions, parts having the same kind of attributes, and the correlations of these parts. He put all quadrupeds that had fur, and blood, were terrestrial, and gave birth to live young into one group, excluding seals, whales, and human beings from the class. In the late seventeenth century the English scientist John Ray revised Aristotle by introducing characteristics like the function of reproduction, blood circulation, respiration, and locomotion as definitional, leading to defining "mammals" as the group of animals whose females are viviparous and have "mammae" by which they can nourish their newborns. This categorization includes seals, whales and human beings.

Although Ray's taxonomical system was said to be "at once the shortest and most comprehensive," a hundred years later classifying the platypus became a real challenge.[19] An animal with a bill like a duck, web between its claws, covered with fur, oviparous, nearly invisible mammary glands, and having a common urino-genital passage joined to a cloaca, it seemed to fall between birds, reptiles, and mammals. The naturalist community took years to reach any consensus about the classification, partly due to the lack of enough anatomical evidence and reliable field studies, but also because so much about the platypus defied hitherto well-established taxonomic criteria. Today the platypus is classified as a mammal, but this giving up Ray's proposal that all mammals are viviparous. The new defining characteristics allowing the platypus to be a mammal required only that the female have mammae that produce milk to feed its offspring. Moreover, this taxonomic conception was supported by the fact that like other mammals the platypus has seven cervical vertebrae, but unlike them it bears cervical ribs.

The platypus ended up as being interpreted as a mammal because it shares two distinctive features with other mammals; features which were then taken to be the defining criteria of this class. The criteria work both inclusively, by bringing the platypus into the class of mammals, and exclusively in eliminating the platypus from membership in any other subclass of the phylum *cordata*. But that interpretation required biologists to relinquish the traditional criterion that mammals are necessarily viviparous. Although the interpretation of the platypus was not arbitrary, it seems that one could just as well have argued, as Lamarck did, that the platypus and the echidna formed their own class Prototheria.[20] Here we are not facing a natural class defined by its essence. Even though Lamarck was wrong in assuming that the platypus didn't have mammary glands and so couldn't be a mammal, there is no necessity in taking the feature of having mammae rather than being oviparous as definitional. Taxonomists might have created two classes, both having mammae, but today *Prototheria* is considered a subclass of *Mammalia*.

Interpretation always takes place against a background of a conceptual framework that constrains possible interpretive outcomes. Sometimes the interpretation seems reasonable in isolation, but does not make much sense in a broader perspective because it violates some basic scientific assumptions. To overcome such a problem, an interpretation must involve not only interpreting the evidence but also constructing a new representation before a phenomenon becomes intelligible to the scientist.

Bohr's 1913 model of the atom provides an illustrative example.[21] The background included the following: In 1896 Sir J. J. Thomson discovered the negatively charged electron, and Henri Becquerel discovered spontaneous radioactivity; in 1899 and 1900 Ernst Rutherford and Paul Villard had noticed that this radioactivity consists of three different kinds of particle, which Rutherford named "alpha," "beta," and "gamma" particles. In 1907 Rutherford and Thomas Royds established that alpha particles were Helium ions. During their 1907 experiment, Rutherford and Royds used the ability of alpha particles to penetrate a very thin glass wall of an evacuated tube to capture a large number of the hypothesized helium ions inside the tube. To show their presence Rutherford and Royds caused an electric spark inside the tube, providing a shower of electrons, which would be taken up by the ions and form neutral atoms of a gas, the spectra of which showed it to be helium. Thus the alpha particles were indeed the hypothesized helium ions. It was also a well-established belief that atoms were electrically neutral, so the negative charge of the electrons had to be balanced out by positive electrical particles.

The interpretive question then was how to represent the structure of the atom in accord with the available data. In 1904, Thomson had suggested a model where electrons were embedded in a sphere of positive electric charge like raisins in a plum pudding. In 1909, Rutherford decided to test this model by bombarding a very thin gold foil with alpha particles. His assistants Hans Geiger and Ernest Marsden observed that most of the alpha particles penetrated the foil, but sometimes the particles were deflected at extreme angles, from which Rutherford inferred that Thomson's model could not be right, but he initially offered no further interpretation of the evidence. Finally in 1911 Rutherford produced his nuclear model of the atom to explain the cause of this observational data.

Rutherford's interpretation of the evidence made a lot of sense from an experimentalist's point of view.[22] If one could accept the surprising hypothesis that most of the volume of the atom was empty space, then Rutherford's nuclear model could be made consistent with both the observational data and the basic classical mechanical description of the alpha particle scattering processes. He reasoned that some of the alpha particles are deflected from the atoms because atoms have a hard massive kernel of a considerable size compared to the alpha particles, such that they can interact with the kernel. Also this core needs to have the same electric charge as the alpha particles so that when the alpha particles hit the core they are reflected due to the repulsive force between like electric charges. Moreover, since the nucleus cannot be electrically neutral, but atoms must be, the electrons cannot be part of the nucleus, and finally, because most of alpha particles fly through thin materials, the space between the atomic nucleons has to be more or less empty. Putting these steps of reasoning together, Rutherford could explain the meaning of Geiger and Marsden's data by constructing a new representation of the atom. Rutherford's model provided a way of interpreting the data making it intelligible instead of mysterious, as it had seemed. However, from a theoretical point of view, Rutherford's model is incoherent. Classical electrodynamics implies that such a model could not be a stable structure, since electrons orbiting a positive charged nucleus would be attracted by the much heavier nucleus and would quickly be swallowed by it while radiating energy. To overcome this instability, in 1913 Niels Bohr added some non-classical features to Rutherford's model.[23] Bohr accepted Rutherford's basic model because of its ability to explain the meaning of Geiger and Marsden's data. Drawing on Planck's discovery of the quantum of action in 1900 hypothesizing that radiative energy is quantized in lumps, Bohr suggested that electrons can orbit the atomic nucleus only with an angular momentum which is an integral multiple of Planck's constant. While an electron is in one of its orbits, Bohr postulated, it does not change energy and therefore does not radiate energy. However, an electron can jump from one orbit of a higher energy to one of lower energy while it emits radiation with a frequency equal to the difference between the energy of the two orbits divided by Planck's constant.

Bohr's drastic reinterpretation of Rutherford's model made Rutherford's empirical results intelligible in spite of introducing revolutionary features. But more importantly Bohr's model of the hydrogen atom gave physicists an interpretation that enabled understanding its various series of spectral lines. Likewise, Bohr's model permitted deriving Rydberg's formula, describing a constant relationship between the wavelength and the wave number. Hence this new representation made a lot of spectroscopic data intelligible.

From what we have argued so far, we see that interpretations are not wild guesses but based on what we already know and understand. Scientists' background assumptions and beliefs guide their interpretation of evidence and their construction of representational meaning. Indeed, no algorithm can help them in gaining this understanding, so we cannot say that the background assumptions *determine* these interpretations. This indicates that there is a certain amount of "free play" in getting from the background to the interpretation. Some of that is no doubt influenced by non-scientific, social factors, but even so it seems there is still a residue, as it were, which is determined by neither social nor background factors – that's the arena of creativity – and there is a sense in which that is indeed arbitrary. Had Einstein or Bohr not come along at just that point in time, no doubt the history of physics would not be identical to the story we now tell. And that's the "arbitrary" factor. Scientists must attempt to grasp things in the form of analogies to old theories and in the form of making changes in preconceived criteria of meaning. But does this mean that there were no methodological clues, which could help them in forming such new representations? We shall turn to this problem in the next section.

4 Interpretation and the discovery of scientific hypothesis

Empirical evidence is not the only thing that guides us in constructing meaning. From empirical evidence we need some method to help us to formulate a constructive hypothesis that indicates how we may grasp unidentified phenomena. Interpretation is not simple guesswork where our chances of being off target are just as great as or greater than being on target. Investigative interpretations offer new constructive hypotheses in response to representational problems and therefore somehow they are connected to the methods of discovery. The methods of discovery include all kinds of ampliative reasoning such as induction, abduction, and inference to the best explanation, methods which operate only within a system of background assumptions and beliefs. Simple induction, for instance, may be used to develop a scientific hypothesis such as finding evidence that a number of individuals share the same features so that these features can serve as criteria for classifying these individuals together. But some other forms of ampliative reasoning may generate semantic hypotheses concerning how incomprehensible phenomena could be understood, thus directly helping to establish new classifications or representations.

The distinction between constructing hypotheses concerning conceptual matters and generating hypotheses concerning factual matters may seem clear. But when conceptual issues focus immediately on representations of factual matters, the distinction blurs because we understand such issues only in terms of concepts that are themselves understood in terms of their contribution to our factual understanding. We can raise questions about purely conceptual issues or purely factual issues, but when it comes to a representational problem, the solution refers to the interconnection between the conceptual and the factual in the representation. Nevertheless, I shall continue to distinguish between *scientific* hypotheses and *semantic* hypotheses. Scientific hypotheses expresses factual matters and do not involve interpretation as such, whereas semantic hypotheses say something about the concepts in terms of which scientific hypotheses are expressed. Using this distinction I shall argue that all interpretive work includes a significant element of constructing semantic hypotheses to grasp unknown phenomena. But if constructing semantic hypotheses has a modicum of rationality, which science obviously implies, then the reasoning must be constrained by some methods.

Among ampliative inferences, the method of abduction guides us in performing investigative interpretations. Abduction is the method that helps to extend the conceptual framework of science. My use of "abduction" here is different from C. S. Peirce's original definition. He believed abduction couldn't be construed as literally accounting for the sequence of steps leading to the discovery of a new hypothesis.[24] In this respect Peirce does not diverge much from the traditional view of Popper, Hempel, and others insofar as the generation of hypotheses involves a distinctly psychological component. Rather "abduction" signifies a procedure of enumeration of hypotheses from which only one at a time is drawn, examined, and hopefully entertained. This is what I would call "inference to the best explanation."

Any discovery process requires modifying some background knowledge, assumptions, or beliefs. Simple induction works within the framework of a single well-established conceptual scheme, whereas abduction introduces new concepts by which we attempt to understand what couldn't be understood on the old concepts. Thus, simple induction is an inference from some data to a hypothesis, where the hypothesis language is *similar* to the data language. In contrast, abduction is then an inference from some empirical data to a hypothesis where the hypothesis language is *different* from the data language. So I hold that abduction is a method of discovery distinct from simple induction, in virtue of its proposing a new conceptual understanding of a set of phenomena. In general, such an improvement of our understanding requires consideration of ontological principles, methodological prescriptions, epistemic goals, and semantic definitions. Most cases of abduction operate within a certain paradigm and do not violate its ontological principles, or its methodological prescriptions, or any cognitive goals in a revisionary sense. The scope of most abductive reasoning is much more limited. Most cases of abduction are local instead of global: they depend on changing our semantic conceptions.

Thus, the ontological principles of a scientific discipline are usually left untouched by abduction. Almost all scientific abductions happen within the existing metaphysical framework. By the same token investigative interpretations generally do not question accepted ontological principles, because these principles already establish our general understanding of a scientific discipline. So abduction to a semantic or a scientific hypothesis requires a background of ontological principles which are presupposed in the general practice of individuation and identification embedded in our pre-scientific and/or scientific language.

An illustration of the ontological principles is the principles of individuation in classical physics. These are: (i) the principle of congruence in space and time: something existing in space and time is identical with itself; hence, two physical objects of the same sort occupying the same place at the same time are actually one object; (ii) the principle of separation in space and time: the assumption (a) that since spatially separated objects have real separated states, each defined by definite properties, such objects exist separately in space and time such that they are localizable and countable, and (b) that causal processes connecting changes of these states happen in space and time; (iii) the principle of causality: every change of a property has a cause; (iv) the principle of determinism: the values of every later state of a system are uniquely fixed by those of any earlier state; (v) the principle of continuity: all processes which are distinguished by a difference in the initial and the final state have to go through every intermediary state; (vi) the principle of conservation: certain basic properties of an isolated system can be transformed into various modes but are never gained, lost or destroyed; (vii) the principle of causal closure: the cause and the effect belong to the same realm of reality. We abstract these principles from our ordinary experiences of interacting with nature and our practical ways of being informed about the surroundings. These classical tenets of individuation form the conditions for naming, reference, and predication, and therefore for all objective description of nature in physics. Eventually they became the principles upon which classical physics was constructed.

Therefore one would expect that constructing new cognitive representations by abduction violates as few as possible of these ontological principles, and in most cases it actually does. Abduction attempts to be conservative as long as it does not interfere with experience. But there are exceptions. Bohr's semi-classical model violated some of the above principles. The transition of an electron from one orbit to another orbit within an atom happens spontaneously and discontinuously in space and time, and thereby such a "quantum jump" violates at least (ii), (iii), and (v). When Bohr's model was replaced by Heisenberg's quantum mechanics, some of the other principles were put into question. It would be better to see the principles of individuation as still valid presuppositions for the identification of a certain domain of objects that are subject to classical physics. We understand that at least some reflect basic conditions for all human knowledge, including the comprehension of an ontological separation of the known object from the knowing subject, and the grounding of this separation in practical actions and observation. The principles themselves may change as our knowledge enters new domains, but some concepts of the revised framework remain basically the same.

Indeed, the principles of individuation would not have been violated by quantum mechanics if our knowledge of quantum objects had used quite different concepts from classical mechanics. But such a situation would be quite incomprehensible because – as Bohr maintained – all understanding of nature must be interpreted in terms of the fundamental categories already formulated and elaborated in pre-scientific experience as part of our general practice of separation and orientation in space and time. Those categories are, for instance, location, time, object, and cause. Following Bohr we can say that concepts in classical physics like 'time,' 'position,' 'momentum,' and 'energy' are more or less explicit formulations of such basic notions.

Since quantum mechanics violates the classical principles of individuation, we must introduce new ways to single out atomic objects, and therefore new constraints on attributing to such objects the dynamical and kinematical properties in terms of which classical mechanics describes macroscopic objects. In fact, we have at least two options: one can say that the violation of the principles implies either orthodox quantum mechanics is incoherent or incomplete, or some new operational principles of individuation according to which an atomic objects can be singled out as individuals only in relation to actual performed experiments.[25] Whether one prefers the first or the second of these two options depends very much on one's cognitive goals. If one believes that truth may depend on something other than the evidence, like the nature of an underlying metaphysical reality, one will opt for the first. But if one settles for more lenient goals like empirical adequacy, simplicity, and/or coherence, to mention a few typical goals advocated, one will find the second option more attractive.

Physics, like all sciences, adopts principles of classification normally constraining any possible understanding of sorts, types, and kinds, but sometimes, when confronted with anomalous and inexplicable phenomena, some scientists will come to doubt these classificatory principles. The specific principles, which might be revised, depend on the object of study. In the case of the platypus there was a lot of conflicting evidence forcing naturalists to revise their principles of classification.

But this case also shows that there is not only one 'correct' way to classify the platypus, merely one that is most conducive to our goals. On a theoretical level, modern evolutionary biology has changed its principles of classification, and nowadays biologists speak in terms of lineages from a diachronic perspective rather than species from a synchronic perspective. However, scientists are interested in knowing not only what things are but also what things do. For example, we might want to know the causal role of certain types of objects. In general, classification is of less importance in contemporary science; more important is finding out how objects contribute to a dynamical understanding of nature.

The adoption of principles of classification is not identical with the historical effort to determine the essential characteristics of things, but most scientists would recognize that our choice of classificatory principles changes with our cognitive interests and the aim of investigation. For instance, what is called the species problem shows a uniform and universal taxonomy of species does not exist. Some definitions apply only to sexually reproductive organisms; others apply only to specimens with a cluster of similar phenotypes, and again others apply only to organisms sharing certain genetic similarities, etc.[26] If one works as a field zoologist and investigates the behaviour of lions, one may look upon lions synchronically and categorize them according to Linnaeus' classical manner of classifying species in which an individual is a lion if it sufficiently conforms to an ideal set of properties defining what it means to be a lion. But if one works as an evolutionary biologist, one would rather adopt a diachronic perspective and talk about lineages where a group of organisms either shares a pool of similar genes or has a common ancestor. Thus, we may talk about typological species, evolutionary species, phylogenetic species, ecological species, biological reproductive species, etc. It does not make sense to highjack one of them and claim that the ransom is paid for the true one only. Depending on their research topic and theoretical approach, biologists choose whatever system of classification is most convenient.

5 Interpretation and the construction of scientific concepts

An interpretation that constructs meaning aims to generate a semantic hypothesis representing an unintelligible phenomenon. Apart from being constrained by ontological principles, cognitive aims, and methodological prescriptions, abduction also operates along a semantic dimension to reach its aim. Abductive reasoning consists of interpreting data as evidence for something unobservable. Simple induction reasons from the observed to the unobserved, whereas abduction takes the step from the observed to the unobservable. In inductive inference the theoretical language and the observational language are the same, but in abduction the theoretical language is different from the observational language.

Thus I will say in inductive reasoning the theoretical language is *semantically invariant* with respect to the observational language if, and only if, the terms contained in the hypothesis have the same intension and extension as the terms of the observational language. On this level of reasoning we do not have to interpret the evidence or the world around us, but the case of abductive reasoning is significantly different. Here some conceptual innovations must be employed, and this forces introducing one of several forms of semantic hypotheses.

First, there is the semantic hypothesis based on a definition of *semantic variance*: the language of a scientific hypothesis H varies semantically with respect to the language in which the evidence E is expressed if, and only if, some of the terms contained in the language of H do not have the same intension but the same extension as those terms contained in the language of E. Each time a new theoretical term is introduced into a hypothesis, its extension is covered by some observational terms but its intension differs from theirs. A simple illustration of this definition stems from Hempel's discussion of the Theoretician's dilemma.[27] Wood floats on water, and iron sinks in it. Our hypothesis H is that a certain object floats on a liquid if its specific gravity is less than that of the liquid, and it sinks in the liquid if its specific gravity is greater. Here "wood," "water," "iron," "sinking," and "floating" are terms of the language of E, whereas the terms in the language of H are "specific gravity," "weight," and "volume." Eventually, the language of H can become part of the language of E with respect to a further discovery. In fact, the example of "specific gravity" should perhaps be analysed as a two-step discovery procedure. First, weight and volume are ascribed to every kind of physical stuff and object; these terms are then included in the language of E, and based on them the term "specific gravity" is defined in the language of H.

It should also be noted that simple classification of objects is developed according to a definition of semantic variance. Based on abstraction and induction from observation of certain objects and a language, E, used to report the observation of these objects, we generate a classificatory hypothesis, H. In biology, for instance, the observation of various organisms gives rise to the idea that some are more similar to others and that we can think of variations of traits among those individuals, which are in certain respects most similar to each other, as contingent properties. Thus, we abstract from these contingent properties, described by the language of E_1, to get the first set of classificatory hypotheses H_1, which groups individual organisms into 'species', a group of organisms whose members all share some common properties. Then this new semantic level of H_1 including the term "species" forms the language of E_2 for the next set of classificatory hypotheses, H_2, where 'species' are put into 'genera'. Again H_2 is reached via abstraction and induction. The methods of abstraction and induction are once more used to reach a higher level. A new evidence language, E_3, is supplemented with the language of H_2 to form a new set of classificatory hypotheses in the theoretical language of H_3 that put the various 'genera' into 'families.' And so the work of classification continues to create new semantic levels of 'orders,' 'classes,' 'phyla,' 'kingdoms,' 'domains,' ending with 'Life'. The higher semantic levels become increasingly abstract so that even today there is no consensus among taxonomists on how many kingdoms or domains into which life can be divided.

Second, there is the semantic hypothesis based on a definition of *semantic inclusion*: the language of a scientific hypothesis H is semantically enlarged with respect to the language of evidence E if, and only if, the terms contained in the language H have the same intension as those contained in the language E, but where the extension of the language of E is a proper part of the extension of the language of H. Examples from physics illustrate semantic inclusion. Initially "electromagnetism" was confined to some electrical phenomena, but after Maxwell developed his theory of electromagnetism in 1865 it came to include light and all forms of electromagnetic radiation. Another example would be "the quantum of action," which Planck employed in 1900 only to explain black body radiation, but around 1913 it was extended to spectroscopy, the structure of the atoms and specific heat, and later from 1928 onwards to molecules (chemical bonds), atomic nuclei, stellar energy, and so on. A third example would be the notion 'electron' which has two different origins in electrodynamics, one in the theory by Lorentz and others, and the other in the interpretation of cathode ray experiments reported by Thomson. Very soon physicists discovered electrons in radioactivity, the Zeeman Effect, metallic conductivity, etc. A final example from chemistry: originally organic and inorganic bonds were considered to be different, but around the 1830s it became clear that carbon-bonds were no different from other bonds and that the theoretical notion of *vis vita* apparently had no referent in reality.

82 The Nature of Scientific Thinking

Third, there is the semantic hypothesis based on the definition of *semantic disjointedness*: the language of a scientific hypothesis H is semantically disjoint with respect to the language of evidence E if, and only if, some of the terms contained in the language of H have a different extension but the same intension as those terms contained in the language of E. An example of semantic disjointedness can be found in the change of the application of the term "natural movement" from Aristotelian to Newtonian physics. Natural motion was the kind of movement that required no physical explanation. For Aristotle such a force-free motion was vertical movement because 'up' and 'down' were the 'absolute' directions and he assumed the basic elements would tend to move towards their natural places. According to Newton, natural motion was uniform motion in a straight line due to the absence of forces. Another example has been reported by Hacking.[28] In 1936 two groups of experimentalists discovered in cosmic rays a new kind of energetic particle with a mass between the electron and proton. The particle was dubbed "mesotron," soon shortened to "meson." A year earlier Hideki Yukawa had developed a theory of the strong nuclear force explaining the interaction between protons and neutrons in which a particle having a mass in the same range as the mesotron acted as the messenger particle. Bohr and others identified this new particle with the mesotron although there were problems in matching the empirical evidence with Yukawa's theory. First, after World War II, other particles were discovered which fitted much better with Yukawa's predictions. These were called π-mesons, and the particle discovered in 1936 then became known as a μ-meson. Eventually, physicists realized that π-mesons and μ-mesons were two very different kinds of particles. Therefore it became quite confusing to group both together as "mesons." Ultimately the name "meson" was restricted to only the post-war particles, and the pre-war particle, the one that was originally named "mesotron" (meson), was renamed "muon."

Fourth, there is the semantic hypothesis based on the definition of *semantic expansion*: the language of a scientific hypothesis H is a semantic expansion with respect to the evidence language E if, and only if, (1) the language of H includes the relevant terms of the language of E; and (2) the language of H furthermore includes terms such that: (a) the intension of these terms is not included in the language of E, and (b) the extension of these terms is not included in the language of E. Briefly we may refer back to the discussion of Yukawa's discovery of the mesons. Yukawa defined it theoretically according to the basic assumptions of the quantum field theory, when no empirical evidence indicated the existence of what we now call mesons. Indeed, later empirical criteria of its identity were established in the evidence language of experiments thereby connecting the language of H and the language of E. Therefore, we can say that semantic expansion takes place whenever a scientist expands the language of a theory (hypothesis) by elaborating its mathematical structure (syntax) without any clues from observations and experiments.

Finally, there is the semantic hypothesis based on the definition of *semantic generalization*: the language of a scientific hypothesis H_2 is semantically generalized from language of the scientific hypothesis H_1 if, and only if, (a) the language of H_1 is a semantic variant of the language of E_1; (b) the language of H_2 results from a semantic variance with respect to the evidence language E_2 and E_1; and (c) H_1 and H_2 have the same truth value with respect to a common empirical domain D described by both E_1 and E_2, but may possess different truth values outside D. Examples of semantic generalization are the conceptual move from classical mechanics to either relativity theory or quantum mechanics. Instead of arguing, as Kuhn did, that the languages of these different theories are semantically incommensurable, i.e. untranslatable from one to another, because they are expressed in very different concepts, I will say that the languages of relativity theory and quantum mechanics are "generalized" from the language of classical mechanics in order to cope with new data expressed by the language of E_2 and still able to express meaningfully our understanding of phenomena in terms of the language of E_1. Semantic generalization is the procedure, I think, which Einstein, Bohr, and Heisenberg followed subconsciously while creating the theory of relativity and the theory of quantum mechanics respectively. Semantic generalization is a conservative way to interpret phenomena in a new light so that we can still meaningfully compare the old and the new hypotheses against available evidence. It makes sure that some of the basic concepts by which a science already expresses its observations continue to be used in reporting the bulk of evidence within this field. Bohr's explicit commitments to the use of classical concepts together with the methodological prescription, which he called the correspondence principle, became guidelines for Heisenberg's construction of matrix mechanics.[29]

Abductive reasoning helps us to interpret phenomena which we neither understand nor know how to represent. The goal is to formulate semantic hypotheses. Abduction goes beyond what we can possibly observe and therefore requires the invention of a different and more abstract language than the one used to report our experience. The specific kind of semantic hypothesis needed depends on the kind of representational problem the scientist is facing and the context in which the problem appears. Finding a constructive solution to a representational problem can be quite methodologically sound even though it sometimes involves revising our background assumptions and beliefs. When a semantic hypothesis eventually proves to be successful in virtue of providing us with the ability to classify and understand previously incomprehensible phenomena, we may then continue, based on the new scheme of representation, to formulate scientific hypotheses about purely factual matters.

4
Representations

The science is that human enterprise in which scientists cooperate in performing experiments and collecting observations, in reasoning and theorizing about the world, and in discussing and making public their results. This social phenomenon may be divided it into smaller cognitive units called *scientific practices*. These are defined by groups of agents within science who use and discuss similar representations of the world, experimental methods, mathematical canons and standard notation, standards for recognizing correct and incorrect problem solutions, standard idealizations, and heuristics of thought. Many of these are never explicitly stated but are present only as tacit understanding underwriting scientists' concrete practices. As I use the term, "scientific practices," roughly covers Kuhn's "paradigms" and Laudan's "research traditions," but I include all the embedded understanding carried over from one practice to another.[1]

Beside its embedded, non-representational part, scientific practice consists of constructing the representations scientists use to express their thoughts about the world. Representations are composed of the various constructions scientists use when acting, describing, and explaining the world, i.e. in scientific reasoning and theorizing. These constructions include natural language, mathematical formulae, physical models, graphs, charts, maps, drawings, and computer simulations. None of them represent, describe, or explain anything in isolation, but they are objects or tools scientific practices use to represent the world.

The act of representation is an intentional activity where an agent proposes some symbolic construct to represent a phenomenon for a certain purpose. Like other intentional activities, representation is context-dependent because its content or meaning is determined by the use we intend to make of it. Adopting the terminology in which the agent is the "representer," the object used to represent is the "*representans*," and the phenomenon represented the "*representandum*," the pragmatic view of representation holds that whether a representans can be said to represent the *representandum* depends crucially on the purposes of the representer. In science the major purpose is to provide explanation and understanding.

To say an object is "representational" does not mean that it is itself a *representation*, i.e., that it plays an independent representational role, but only that it is an object or tool for representing the world in scientific practice. However, *some* representational components *are* intended as representations. Two salient categories of scientific representations are *hypotheses* and *models*. "Hypotheses" are linguistic entities postulating something about the world (usually sentences, such as "The velocity of the mass m at time t is v," or "The particle accelerates because it is affected by gravitation.") Hypotheses are used to make predictions, descriptions, and explanations. "Models" are symbolic devices representing some specific part of the world or kind of phenomena. The many forms of models include physical (the Crick-Watson DNA-model), graphical (force-diagrams, maps), abstract (the Einstein-solid), and formal/mathematical (Friedman's cosmological models).[2]

Models and hypotheses differ in that hypotheses are truth-apt (have truth-conditions and are capable of being true or false), but models are not. The representational function of models is closer to that of pictures or maps: A picture of some scene represents the depicted scene, and can do so more or less precisely. But it makes little sense to say that a picture is "true": There are no plausible truth-conditions for this depiction. What can be done, though, is to articulate various statements about the depicted scene based on the perception of a picture, and those statements are truth-apt.

Analogously, the hypotheses we infer from models are the representational activities of a scientific practice most plausibly understood as truth-apt. Models are tested through hypotheses: hypotheses (in the form of observational predictions) are compared to the available data. Insofar as science gives true (or empirically adequate) descriptions of the world, these are also most plausibly construed as hypotheses. Models do not aim to represent the world perfectly, but rely crucially on idealizations and even *mis*representations. Nonetheless, models play a crucial role in scientific reasoning and theorizing: they mediate scientific reasoning because hypotheses are rarely deduced directly from other hypotheses (e.g. predictions are not derived directly from "laws of nature," understood as fundamental postulates about the world).

1 What is a representation?

In the beginning of *Scientific Representation* van Fraassen writes: "A representation is made with a purpose or a goal in mind, governed by criteria of adequacy pertaining to that goal, which guide its means, medium, and selectivity."[3] I fully agree, but the same thing could just as well as have been said about explanation and interpretation. The explainer also put forward an explanation with the purpose of addressing the explainee's question in order to improve his or her understanding of the matter, and this purpose determines the adequacy of the explanation.[4] So reference to the agent does not help distinguish representation from explanation. I propose that the purpose of representation is to produce a cognitive tool for explaining what we don't understand. Thus, a representation acts as a platform to enhance our understanding of the representandum by attributing certain meaningful features to the representation.

Although van Fraassen does not advocate a positive theory of representation, he presents a *Hauptsatz* that says what a representation is not: "There is no representation except in the sense that some things are used, made, or taken, to represent some things as thus or so."[5] I agree with van Fraassen that representation is conceived as an intentional activity. A representation must be intentionally used to represent; nothing in the representation itself shows that it is a representation. A representans may be either man-made or natural, but in neither case is it required that the representans is somehow similar to the representandum. Take a blurred photograph: It might show little resemblance to its object; it could still be taken as a representation even though it is completely out of focus. We may take it as a representation for the same reason that footprints on the beach or tracks of elementary particles are representations. They are traces, semi-permanent effects in a medium, causally connected to the object that made them. Traces are used whenever they are considered as evidence of what leaves behind the traces. Hence as *evidence* traces functions as representations of their maker, fitting with van Fraassen's suggestion that measurements are designed as representations.[6]

By definition a representation differs from what it represents, and therefore the representans needs not to show resemblance to the representandum, and may rest on natural causes or on conventional stipulations, but in all cases its 'representationality' is still a product of human intention. As van Fraassen notices, "at least if taken entirely literally, it has no room for the notion of mental images or mental representations, whether taken to be the brain states or something more ephemeral – for no such things, if they exist at all, are *used* or *put to use*, or *taken* in one way or another."[7] There are some consequences of this pragmatic view, which van Fraassen doesn't really consider, because his description of a representation allows us to bridge the gap between seeing concepts as *representation* of reality, as Descartes and Locke did, and concepts as rules, which specify how something is to be done, as did Kant and Hegel.[8]

Words, pictures, maps, graphs, signs, and mathematical formulas are good examples of kinds of representans. We can clearly distinguish them from what they stand for, and we may directly or indirectly compare their form of representation with that of the representandum. If this were not so, we would be unable to talk about a representation as more or less adequate for particular purposes. Hence, we could not pinpoint misrepresentations. But do concepts represent anything? Whether they exist prior to any representation or result from a representation depends on our definition of "concept." The short answer is that only if concepts can be intentionally put into some use do they have the power to represent anything.

One may take "concept" to mean "the semantic content associated with well-established words or sentences," or one may consider it to mean "a mental vehicle for reasoning and thinking about types in the world." In the first sense it does not add any further entity to the list of possible representans because any representation establishes a semantic content which we can identify in order to enable understanding what the representans mean. Having a concept consists of an understanding according to which a certain representans, say a word, relates to the representandum, i.e., what the word stand for. To possess a concept like 'dog' gives us the ability to use the term "dog" in the right circumstances because it demonstrates our understanding that "dog" refers to dogs. However, concepts in the second sense result from our endeavours to understand the world and do not rest on any linguistic skills. In this case also a concept consists of an understanding according to which the thought (the representans) gets its content by being mentally related to the representandum. A concept in this sense is what a person possesses whenever he understands a linguistic expression, in some cases formed prior to any representation.

Concepts in the sense of mental representations are not without problems. Sometimes it seems right to say that we create a concept to represent something separated from it. Take, for example, concepts like that of 'phlogiston' or 'the collapse of the wave function.' Eventually scientists discovered that there was nothing that corresponded to the concept 'phlogiston', but originally it was supposed to represent an element or substance, which was released during combustion. In quantum mechanics a concept like the 'collapse of the wave function' may be considered to represent a real event in which the superposition of quantum states change during a measurement to a non-superpositional classical state. Heisenberg, who coined the concept, used it to represent a process in abstract configuration space, GRW uses it for representing a cascade process taking place in real space, while both Bohr and the proponents of the many world interpretation, for different reasons, deny that the concept represents anything at all. So what the 'the collapse of the wave function' represents – if anything at all – is a discontinuous change in the state of the system. Such a concept is intended, and functions as a mental representation of, a complicated, theoretically defined, physical process.

Our everyday concepts like 'red', 'water', 'rock', 'man', 'women', 'horse', or 'tree' are so ingrained in our way of experiencing things in the world that we cannot separate our visual experience from how we grasp what we see. Here concepts are not a veil between the world and us; they do not exist in virtue of an interpretation of the world. It seems beyond the reach of our intention to bestow a representational role on these concepts. When we see a horse, we cannot avoid seeing it as a horse. Under normal circumstances in our world, this does not require a hypothesis about the type of object we are seeing. There is little room for a separation of the concept 'horse' and our seeing it as a horse. The concept of a horse is nothing but seeing a horse as a horse; nothing 'extra' is added to our perception. In ordinary cases we do not use the concept 'horse' intentionally to interpret what we see because a horse usually presents itself to us as a horse. Thus we draw a distinction between 'presentation' and 'representation'. Presentation belongs to the embodied, non-interpretational parts of cognition; it characterizes the state where we automatically experience familiar things in our surroundings. In contrast, representation belongs to the reflective, interpretational parts of our cognition, resulting from an act of will to adopt certain entities to represent other entities. By intentionally using one group of entities to represent the other group, we construct concepts as rules for applying the first group to the second group.[9]

Scientific representation as an integral part of scientific practice can only be isolated from that practice by philosophical abstraction. So I use "scientific practice" more broadly than Kuhn's "paradigm," Lakatos' "research program," or Laudan's "research tradition." These well-known characterizations of scientific practice are defined in terms of cognitive representations and explicit knowledge, rather than in terms of skills and tacit knowledge. A scientific practice includes not only various theoretical representations of the world, mathematical canons, standard notions, standards for recognizing correct and incorrect solutions, standard idealizations, and pervasive heuristics of thought, but also experimental techniques, methods, skills, habits, rules, implicit assumptions, tacit beliefs, and embodied understanding. Much of this cognition is embedded and therefore not explicitly expressed even though it guides scientists in action. Embedded cognition is the *census communis* of actual practice of science, establishing the concrete daily practical activities of scientists. Scientists learn these practices as part of their socialization as a physicist, chemist, geologist, biologist, etc.

Embodied cognition is the important *individual* element of embedded cognition, not derived from or a result of a representation. It is neither inferential nor based on interpretation, yet we demonstrate its existence whenever we perceive something familiar and immediately present. For instance, when I see my PC in front of me, it is part of my visual, auditory, and tangible awareness. I cannot deny to myself or talk myself into a belief that it is not a PC I am writing on right now. Part of my experiential capacity enables me to identify and use a PC. As Wilfrid Sellars emphasized in *Empiricism and Philosophy of Mind*, whenever we perceive any object we immediately acquire experiential beliefs, not inferred from other beliefs.[10] But he also emphasized that any belief presupposes other beliefs. For example, I acquire directly the belief that this is a PC in front of me, but my ability to form such a belief presupposes that I have many other beliefs that can be connected to this belief, such as that there is a text on the screen and that I can do certain specific things with it.

Moreover, my experience of a PC does not require an act of interpretation, or – what I think in this case is the same – a construction of a representation. Withdrawing my belief that I experience a PC demands giving up other beliefs connected to this particular belief. Indeed, identifying what I see as a PC demands having the concept of a PC. Without such a concept I might still see something, of course, in the form of simple seeing advocated by Fred Dretske and van Fraassen.[11] But as soon as I have acquired the concept, I cannot avoid seeing it as a computer. It is beyond my cognitive power to represent what I see as a book or as a cat or a table. Therefore, the concept of a PC does not act as a representans. If it did, what would be the representandum? Thus, I neither face a representational problem when I see a PC nor do I have to interpret what I see when I see my personal computer. Indeed, acquiring a belief concerning the computer is formed by unifying a semantic template with visual inputs. It takes place automatically in the brain via a cognitive processing of the sensory information. However, this cognitive processing is not the same as interpretation. Interpretation does not happen automatically; it is willed and works on reason. The interpretative action begins only when we want to give meaning to something we don't understand.

It is impossible to distinguish sharply within a particular scientific practice the part consisting of embodied understanding and the part based on representation and interpretation. Historically our beliefs change their epistemic status from being a result of interpretation to being learned through socialization. Beliefs that originally resulted from theoretical interpretation, like the meaning of certain lines on a computer screen, eventually become part of the scientists' background knowledge, allowing the lines to be immediately experienced as tracks of specific subatomic particles in a collision experiment. In this situation, Sellars argued, scientists do not derive what they experience from more basic beliefs. He held correctly that a physicist sees these lines as tracks of well-defined particles; the physicist does not infer their existence from an 'interpretation' of the visual lines and how these lines are produced.[12] According to Sellars, when a visual experience is regarded as an 'observation,' i.e., as non-inferential knowledge, the observer is able, under the proper circumstances, to report his experience, and he knows what he is reporting. Thus a physicist, looking at a certain pattern on a computer screen, sees these lines as subatomic tracks because he has learned by training that this is how tracks of electrons, positrons, mesons, etc. look. Recognizing a certain pattern on a computer screen has become part of the physicist's immediate non-interpretive understanding.

Indeed, there was a time when the most common tracks were not immediately seen by physicists as particular types of tracks but were inferred from explicit knowledge about the function of the apparatus and an interpretation of the particular phenomena in question. The immediate embodied cognition found in the physicist's ability to see a meson track contrasts with the abstract sort of cognition involving 'interpretation.' Thus a cognition originally requiring interpretation may eventually be transformed by learning into embodied cognition in which understanding is not based on a reflective separation between representans and representandum.

2 Representing and understanding

Representations and understanding go hand in hand. Understand a problem is just being able to describe it in the appropriate technical vocabulary. And solving a problem is the same as understanding what represents a solution to the problem. But as we have just seen, representation implies understanding, but the reverse does not hold. This leads us to ask how does an abstract representation provide new understanding? The answer is twofold according to whether the focus is on representans or representandum.

The first point is that representation presupposes understanding. Representation is an intended action requiring that the representer already understands what it means to represent something and understands something about the means of representation. We can use signs or objects to represent other signs or objects because various representans work as external systems expressing our internal understanding of other internal or external things. A representans is such only in relation to a representer or a community of representers. To avoid representational problems, a representation usually builds on well-established social conventions. Admittedly, some people create their own memotechnic representation to help in calculating or remembering, and others may create new kinds of signs to represent something, but this is still only possible if they know what to represent. So a representans does not directly represent the representandum; it does so only indirectly via the representational role the community (or an individual) gives it and the actual use the individual representer makes of this role.

Second, since a representans need not be similar to the represented, how can a representation offer understanding of the represented? It can do so only because a representation results from an intentional act of representing. As intentional agents, it is scientists, not representations, who do the representing; it is scientists who use a representation by establishing a relationship between the representans and the representandum, and it becomes a "representation" only in the context of the scientists' use of it as such. This is a contextual relationship, not a natural or a structural relationship.

In physical science, Talal Debs and Michael Redhead have defended the view that representation contains both a social and a formal dimension.[13] They claim that one should not try to find a foundation in either of these dimensions, but that only the combination of both together gives the representation its importance. This suggestion accords with Ronald Giere's, when he emphasizes that it is the scientist who represents the phenomenon, not the model the scientist may use.[14] The social dimension is as important to understanding the nature of representation as the formal, but Debs and Redhead's account also identifies three types of relations: the representational relation, the informational relation, and isomorphic maps.[15] Somewhat confusingly, the three types of relations are not always located in the same domain of scientific representation – only the representational dimension and the isomorphic maps are situated within the domain of formal representation, while all types of relations are included in the domain of informational relations. This is explained in the following quotation:

> There are two mutually interacting ways of analyzing scientific representation. One involves structures, models, and the mathematical and logical relations between them. The other involves a path of transmission, or how these structures have come to have some representational interpretation; in other words the way in which they come to be taken by someone as representing a feature of the world through rhetoric and the mediation of networks of social agents.[16]

So for Debs and Redhead the formal dimension of representation revolves around structures and models. Importantly, because their interest is in the mathematical and logical relations between such models and structures, they refer mainly to idealized structures and models, not those actually used by scientists. This is important to note because their focus on abstract models and structures misleads, creating problems in understanding what they want to say with their theory. I shall return to this later.

The social dimension is concerned with how different types of relations are used in representation, how the scientists use and interpret them, and how scientific practice is socially mediated. Since neither the social nor the formal dimension alone permits a sufficient explanation of scientific representation, Debs and Redhead propose a 'combined account' of representation, making the two dimensions inseparably interacting with each other, yet at the same time constraining and facilitating the action of each other. "Scientific representation does not decompose neatly into two non-interacting domains. In reality, they interact with one another and each places constraints upon the other."[17]

What does it mean that the domains interact and constrain each other? One example could be, as the authors say, a large part of what a scientist does is intended to convince others that his or her model of choice represents some real world feature.[18] Of course, this activity is partly rhetorical. A scientist's approval of the model does not alter the model as such but may determine whether and how it will be used.

Understanding how the representational relationship functions requires knowing the particular elements being related. Here Debs and Redhead argue: "The representational relation...is a conceptual relationship that holds between two structures such that one is considered to be a model of the other; thus this relation links a model with an original or a token with a type."[19] Hence a representation is a structure that is conceptually related to another structure, but "conceptual relationship" is not defined. I take this term to refer to the logical and mathematical relations mentioned earlier. Apparently, the whole area of intuition, understanding, interpretation, etc. regarding structures and models is relegated to the social domain. However, presumably because of their focus on the logical side of representation, Debs and Redhead fail to flesh out this domain.

A relation between elements is called a "structure," which Debs and Redhead define as "an entity, (or object) comprised of other sub-entities that bear specific relationships to one another."[20] The scope of this definition is very broad, including physical objects, mathematical entities, works of art, and narrative accounts. However, all such types of structure can be relevant for understanding representations in physics.

All of the many different structures can be divided into just two kinds: a) *concrete structures* consist of "any particular set of elements, plus the relations between them."[21] (For example, a number of physical objects in a particular spatial relation can be defined as a concrete structure. To give another example, a number of points in a geometrical framework can be said to comprise a concrete structure.) And b) *abstract structures* are "simply the structure (elements and their relations) that is shared (and hence defined) by a collection of concrete structures."[22] An abstract structure can be understood as the selected common features of several concrete structures. Since it does not exist independently of the set of concrete structures, it becomes similar to the structure from which it is made. But the particular similarities, I want to insist, are selected according to our actual purposes and what is relevant for a particular use: Human beings delineate their experiences of real-world phenomena into categorical entities corresponding to the common conceptual framework of the community, or if that is put into question, into new interpretative hypotheses.

Of course, we have to have good reasons to include all the physical elements in a spatial relation under the same heading as part of the same particular structure. An act of interpretation and codification by the observer determines which elements are included in the structure. Two modes of representational relation between structures exist: (1) a token of a type relation – a token stands for a category of structures or a type of structures. This relation may exist between both concrete and abstract structures; (2) a model of an original relation – this model carries or expresses the most salient features of an original structure. This relation exists only between concrete structures, since it starts with a particular original structure. These two modes of representational relations may overlap. A visual representation of some physical entity, such as a molecule, may simultaneously stand as a token of one particular type of molecule while also being a model of a particular molecule involved in a specific chemical process.

Debs and Redhead do not spend much time on the concrete structure mode of representation, choosing instead to focus on the second mode of relation, which concerns itself with mathematically formulated models. (The first type, the token-type relation, depends on similarity, which is not strictly a logical relation, as it depends very much on the usability of the token. For example, is the resemblance good enough for it to be used as a vehicle for educational purposes?) According to them, an isomorphic relationship holds between a particular conceptual model, which is a structure, and a particular mathematical model, also a structure. The representational relation in physical science is based on a relation between two such structures: The idealized conceptual model, referred to as O (an idealized original), modelling the physical phenomenon, W (the real phenomenon in the world). M stands in a representational relation to W, the mathematical model corresponding to the conceptual model. To be understood, the representation of the physical system idealizes reality, stripping away most of the observed features, and focusing on only a few, considered the most salient. A conceptual model is represented by a structure (a token of a type or a model of an original.) A conceptual model is an idealized copy of the actual physical structure. All instances of representational relationships and abstract/concrete structures may be observed in a scientific explanation.

As an example, Debs and Redhead explain how modern physics uses a representational structure regarding X-radiation: "Modern physics models this phenomenon with the concept of an electromagnetic wave. This wave-like behavior may be modeled through the use of orthogonal sinusoidal functions summarized in Maxwell's equations."[23] Thus scientists try to represent a physical phenomenon by introducing a model that allows them to discuss what the phenomenon is and what it isn't, and how it should be understood. In this case, the idealized model of the phenomenon is electromagnetic waves visualized via orthogonal sinusoidal functions. The most salient features observed by the scientist in relation to a physical phenomenon have been represented by a concrete scientific structure, the mathematical model.

96 *The Nature of Scientific Thinking*

As Giere points out, it is of course important to choose the *right* features to include in the representational relation.[24] At least initially, the scientist is unlikely to know which features are "right"; thus, in the real world there are most likely to be a variety of competing models, and the one which wins out – by virtue of being used most frequently – comes to be seen as the one which chose the "right" features. Within the social dimension of representation, the information relation is important. Here, this relation holds between two scientists, as opposed to two structures. A feature of the physical world, W, is represented, via the use of a model or token (including what is referred to as "material culture and apparatus"), by one scientist (as an actor) to another scientist (as an audience).

Of course this is a simplified version of real science, as the authors are quick to point out. The actor and audience roles are interchangeable, each scientist taking turns in either role. The communication of representations is described as a "social chain of media."[25] In this use of the word, "media" refers to the vehicles for understanding presented above, namely particular and abstract structures, in both modes of relation. I have mentioned the social dimension not only to do justice to the structure of Debs and Redhead's theory, but also because I am defending a rhetorical view of explanation and representation.

The most interesting, and challenging, part of the representational theory of Debs and Redhead is their explicit reiteration of the need for isomorphism in scientific representations between the conceptual model and the mathematical model. Since the conceptual model is an *idealized* version of the physical phenomenon, as described earlier, isomorphism cannot hold between the physical phenomenon and the conceptual model. However, Debs and Redhead claim that isomorphism is central to scientific representation of physical phenomena: "Scientific representation ... relies heavily on the notion of resemblance, as expressed in terms of isomorphism between structures."[26] I agree that scientific representation often relies heavily on the notion of resemblance, but I deny that 'resemblance' can be explained as an isomorphism between structures. Moreover, it is even doubtful that 'resemblance' can be clearly defined as an extensional feature of the representans.

As Mauricio Suárez points out, similarity or isomorphism is not a sufficient foundation for representation.[27] Debs and Redhead agree. The relations between its elements characterize a structure. To reiterate, this is a statement concerning an ideal version of representation, and the purpose is to show the logical aspects of this representational relation. If one stipulates that structures are logically determined by their elements standing in specific relations to each other, a correspondence between the two structures can be clearly defined as 'isomorphic.' Suárez presents what I take to be an adequate definition: "Isomorphism is a relation preserving mapping between the domains of two extensional structures, and its existence proves that the relational framework of the structures is the same."[28] As Debs and Redhead themselves note: "One might be tempted to believe that if M represents O then it is isomorphic to O, or alternatively that if M is isomorphic to O, then it represents O. Strictly speaking, neither is the case."[29] Isomorphism is necessary, but not sufficient for a representational relationship.

When considering the nature of representation, the intentions of the scientist are important. On the one hand, the reason for constructing a mathematical model, which corresponds to a conceptual model, is to allow scientists to focus on the structural similarities between them. Only the most salient features of the real-world phenomenon are mirrored by the idealized original. On the other hand, the model must be exactly similar structurally to the original. But the model may use elements not contained in the original, such that they are in a homomorphic relationship. However, this does not exclude any elements of the original from existing in the representation. Also, the model may be part of a larger mathematical structure, making some of it parts supported by the larger system, which is not part of the formal relation between original and model.[30] The boundaries between the model and the mathematical structure of which it is a part might be flexible so that, if needed, more of the supporting structure can be brought in to direct relation with the original.[31]

Given the logical framework erected by Debs and Redhead, their stipulation of the importance of isomorphism between O and M makes sense. But given the level of abstraction employed in their explanation, how much does this really apply to actual science outside of abstract mathematics? Van Fraassen points out that the 'conceptual model' part of the representational relation is problematic.[32] I agree with his criticism of the example of X-radiation mentioned above that: "It is hard to understand how a concept can be isomorphic to a mathematical structure. If it is the sort of thing that can be isomorphic to a mathematical object, then it instantiates Maxwell's equations, which the mathematical structure does equally well — why have both in this account of representation?"[33]

What is the reason for having a conceptual model in between the physical phenomenon and the mathematical model? One might suppose that it is because it would be very hard to defend an isomorphism between a concrete object in the real world and an abstract mathematical model. For a firm believer in logic as a tool for understanding representation, this is not a good sign. Nevertheless, it is also questionable whether isomorphism is a necessary component of all representations. If we consider mathematics as a language, it is difficult to see how it can describe the world only by making an isomorphic mapping of the world. This is not the case with respect to natural language or iconographic symbols. Only the author of *Tractatus*, Ludwig Wittgenstein, defending his picture theory, claimed that any combination of natural sentences consists of a logical structure of atomic sentences and that these atomic sentences stand in a direct relation to the corresponding states of affairs (possible facts), making them isomorphic with the atomic states of affairs they picture. In my opinion, it is just as wrong to claim that a mathematical model forms a logical structure, which is isomorphic with the structure of the world, in order for it to represent reality. There are no such mathematical structures that are logical pictures of possible real structures.

Suárez proposes an entirely different way to approach representation. Instead of trying to find an ideal account of representation, meaning that incomplete or inadequate representations are excluded, he claims to describe only how a vehicle for representation, the representans, coincides with its representandum. The vehicle might be a theory, a model, a token, etc. The same is true for the representandum – it might be a physical phenomenon, a mathematical entity, or something else. As he says, "we shall not require a theory of representation to mark or explain the distinction between accurate and inaccurate representation, or between a reliable and unreliable one, but merely between something that is a representation and something that is not."[34] Because Debs and Redhead's account describes an idealized scientific representation in the language of logic, it is also inclined to be normative with regard to what counts as the representative force of a given physical phenomenon and what does not count as such. Ultimately, the strictness of their account makes it unhelpful in explaining what scientific representation is. Instead, a more fruitful way of looking at scientific models sees them not as pictorial representation, where the model functions as a representans of the representandum, because of conventional relations that have been created for particular purposes. Surely, the point is whether or not a scientist and his audience believe that the convention is adequate and unambiguous given the purpose of their representation (another audience and scientist might of course have different aims). This does not have anything to do with resemblance relations or strictly logical relations.

3 What do laws represent?

It is *prima facie* reasonable to suppose the models "represent" because representation figures essentially in attempts to define "models." But I shall argue that the case with "laws" is somewhat different. At least theoretical laws might be read more as stipulations of the meaning of quantities or instructions to the scientist on how to construct the model in specific situations. Again there is a difference here between theoretical laws and empirical laws. The latter are more "representational" than the former.

Philosophers of science commonly think of a theory as consisting of a set of fundamental law-like statements. Therefore, we can better grasp a theory if we have some understanding of what a law represents. I hold that the law's explanatory role boils down to its representational function. First we need to investigate the various possible ways of expressing laws to see whether they play different representational roles. One way to see such a difference is in terms of the traditional philosophical distinction between *empirical* laws and *theoretical* laws. The contrast is here between expressions concerning observable entities and expressions concerning unobservable or so-called "theoretical entities." The strictly empiricist approach would attempt to reduce all theoretical laws to empirical laws; whereas a more whole-hearted realist approach, accepting the distinction, would attempt to explain empirical laws in terms of theoretical laws. However, I believe that the empiricist distinction between entities that can and cannot be seen by the naked eye is epistemically uninteresting and irrelevant.[35] Whatever "laws of nature" are, the fact is that they happen to be expressed in terms of visible as well as invisible entities.

Nancy Cartwright points to the distinction in physics between *phenomenological* and *fundamental* laws as different from the traditional philosophical distinction.[36] In physics the contrast is not between laws governing observable entities and fundamental entities. Rather, physicists make a distinction between laws that explain and laws that merely describe. As she says, "In modern physics, and I think in other exact sciences as well, phenomenological laws are meant to describe, and they often succeed reasonably well. But fundamental equations are meant to explain, and paradoxically enough the cost of explanatory power is descriptive adequacy. Really powerful explanatory laws of the sort found in theoretical physics do not state the truth."[37] A little later she continues, "I will argue that the falsehood of fundamental laws is a consequence of their great explanatory power."[38] Unfortunately,

Cartwright does not give a more precise characterization of the distinction between "phenomenological laws" and "fundamental laws." But she provides us with some further clues: "The causal story uses highly specific phenomenological laws which tell what happens in concrete situations. But the fundamental laws ... are thoroughly abstract formulae which describe no particular circumstances."[39] Now it is difficult to see how fundamental law statements can, at the same time, be false and yet contain no descriptive content. Assuming that fundamental laws are contingent, they must describe either some particular or universal facts, but then they are either true or false, or they don't describe facts at all, but then they are neither true nor false. Fundamental laws do not describe particular facts, but neither can they describe universal facts, because if they did, they would sometimes be true. But, according to Cartwright, they are always false. Hence they cannot be true either. Nonetheless, she still thinks that these fundamental laws enable us, somehow, to tell a causal story. How can false laws tell a true story? She realizes there is a problem in having no account of how fundamental laws manage to yield such a story.[40] She remains puzzled by how such laws have explanatory force without being true.

Perhaps this puzzlement should be taken as a sign that something has gone astray in her analysis. Beneath her rhetoric, her argument seems to be that all quantitative laws of physics, like the law of gravity, have exceptions. (In this context, Cartwright does not distinguish between fundamental and phenomenological laws, but I am assuming what she has in mind are fundamental laws.) Therefore physical laws are in general false. The law of gravity is only true of a physical system that contains no other forces than gravitation. Genuine examples of such covering laws are scarce because they hold for only very few systems. If physical laws are formulated by including their exceptions instead, then we will get *ceteris paribus* laws. But *ceteris paribus* laws, she claims, are not genuine laws, for they hold only under special and idealized conditions.[41] So without the *ceteris paribus* modifier physical laws are false and with it they are not really laws. Therefore, *ceteris paribus* laws cannot figure in a covering law model of explanation in spite of the fact that they play a fundamental explanatory role in physics.

I think that Cartwright's argument is flawed. Why are *ceteris paribus* laws not real laws? Indeed, if one defines "laws" to be true universal statements that are factually true under every circumstance, then perhaps there are no genuine *ceteris paribus* laws. But is this definition necessary? Cartwright's characterization of *ceteris paribus* laws does not differentiate a fundamental law from a causal law. It might be true that fundamental laws do not describe any real situations, but not true, I would say, of causal laws. One may say that causal laws must in their complete formulation exclude exceptions by including all nomologically relevant circumstances, which, because they leave out other possible causes, must at least be fixed for one particular event to be a cause of another. The "nomologically relevant circumstances" are those conditions that always have to be fulfilled for one kind of event to cause another. No event causes another event in every possible circumstance. A *ceteris paribus* generalization therefore need not, as we shall see, collapse to a Ramsey generalization, i.e. an universal conditional where the antecedent contains all necessary predicates to prevent the expression from not being true in the appropriate circumstances. A Ramsey generalization binds contingent features of a certain situation by claiming that all cases to which the generalization applies are only those exhibiting all these contingent features.

If empirical *ceteris paribus* generalizations should not be regarded as Ramsey generalizations, we have to solve the philosophical puzzle of differentiating between a *ceteris paribus* law having only few instantiations, and a singular causal fact that does not entail a law, i.e., what have been called "accidental generalizations." There are alternative suggestions for solving this dilemma. One proposal distinguishes properties belonging to sorts of entities from properties pertaining only to an individual. On the one hand, if an empirical *ceteris paribus* generalization involves those properties that an entity possesses in virtue of being a certain kind of entity, then it is a genuine law. On the other hand, if an empirical *ceteris paribus* generalization relies only on characteristics that an entity has because it is that particular individual and not because it belongs to a class of entities, then it is not a real law but an accidental generalization. This strategy avoids Ramsey generalizations, but it still seems to be the case that genuine generalizations that rely only on sortal properties must be supplemented with *ceteris paribus* clauses which are either built into the generalization or qualifications explicitly added to them. Take, for instance, Kepler's laws. They are 'true' for all planets of a solar system that is governed by one central force and one planet, but for a solar system that consists of more than one central force, such as a double star-system, they would not be true. So Kepler's laws are 'true' only in certain circumstances.

Another alternative is Giere's proposal that empirical generalizations should be considered simply law statements without embedded *ceteris paribus* clauses. Then they should be understood "as part of the characterization of an abstract (representational) model and thus being strictly

102 The Nature of Scientific Thinking

true of the model."[42] In other words, empirical generalizations refer to abstract models and not to nature. If they report something about nature, they would have to be "definite about something that is decidedly indefinite, and so the resulting package ends up being incomplete. Alternatively, in trying to be indefinite, this approach risks making laws vacuous, claiming, in effect, that it holds except where it does not." I think that Giere is wrong on this for two reasons. First, I believe most scientists intend empirical generalizations to be about concrete objects, not about models; second, Giere seems not to draw the full consequences of his own perspectivism. If it is scientists as agents that do the representing, it is necessary to say that representing is context-dependent. A representation is 'adequate' only with respect to a specific context. It is only because we commonly think of representations (like empirical generalizations) as objective and universal that it is necessary to add *ceteris paribus* clauses to them to express their usefulness for attaining a particular representational goal. However, Giere explicitly denies this picture.

For my part, I maintain that most fundamental laws, in contrast to phenomenological laws, are neither contingently true nor contingently false because in contrast to causal and structural laws, which are descriptive, fundamental laws play a prescriptive role. Because of this prescriptive character, fundamental laws are well qualified to form the semantic basis of empirical generalizations. Moreover, I think science operates with more than one kind of 'fundamental law.' Thus, I propose a threefold typology of natural laws: (i) *principles*, (ii) *theoretical* laws, and (iii) *empirical laws or generalizations*. Principles play mostly a constitutive role, stating the rules for describing nature objectively; theoretical laws play a regulative role by defining basic concepts such as mass, force, charge, energy, momentum, temperature, electric field, magnetic field, etc.; and empirical generalizations describe how nature behaves. None of them fits the inherited view of 'laws' as universally true representations whose truth is determined by universal second-order relations.

These categories of laws contain various subcategories: Among the principles are conservation laws, symmetry principles, and theoretical postulates; and among the fundamental laws we have those that express relations between quantities, which may be called *theoretical* definitions.[43] Empirical laws or generalizations include, for instance, *causal* laws, *structural* laws, and *functional* laws. Here I shall focus mainly on theoretical definitions in contrast to empirical generalizations. The pronounced difference is that theoretical definitions don't describe how physical objects actually behave in their domain, they state how two or more quantities are interrelated, whereas causal laws provide us with an account of what makes physical objects behave as they do. So-called theoretical laws, on the other hand, are the result of definitions for how causal laws must be formulated in a model in order to have any explanatory function.[44]

4 The linguistic approach to theories

Here I shall argue for a conception of "theory" regarded as highly controversial, partly because there is no uniform meaning of the word "theory." In both common and scientific parlance the term signifies certain ideas, a group of assumptions, an axiomatic system, a set of models, a collection of fundamental laws, a general hypothesis, or a paradigm. In my usage "theory" is one component of "scientific practice" and distinct from "model." I use "scientific practice" much more broadly than "theory" because my favoured term includes models, explanations, interpretations, standards, experiments, and methodologies, as well as theories. A theory, as used within this practice, contains a vocabulary and linguistic rules of description as well as theoretical definitions and "correspondence rules."

There is an analogy between natural languages and theories. Any language consists of a system including a vocabulary and a set of syntactic and semantic rules. These rules specify how terms from the vocabulary can be put together in sentences and what such sentences mean within this language. They are neither true nor false. It is only when we actually pick up a correctly formed sentence and use it to *describe* or *state* some concrete state of affairs such that it makes sense to say that this particular sentence is either true or false. The same holds for scientific theories. These provide scientists with a vocabulary and certain syntactic and semantic rules for how this vocabulary can and must be used to describe nature, but they do not have any independent truth-value aside from their use in building models.

Thus, I shall argue *contra* Cartwright that theoretical laws neither express facts, nor have any empirical content, and therefore they lack *ceteris paribus* clauses. However, empirical laws state what are purported to be facts, and in doing so, they always must contain a *ceteris paribus* clause. As definitions of quantities, theoretical laws do not express any such restriction, but they may be applied only in contexts that fulfil certain ideal conditions. Theoretical laws do not describe any particular states of nature, whereas empirical generalizations deal with the causal laws of nature. Examples of theoretical laws are Newton's second law, Maxwell's first law, and Schrödinger's equation.

The quantities that are regarded to be fundamental depend on whether they are directly measurable. Today the international research community has agreed on seven fundamental quantities, namely length, mass, time, electric current, thermodynamic temperature, amount of substance, and luminous intensity. All other quantities can in principle be defined in terms of these seven. But in earlier times there was no such convention.

Several authors besides Cartwright have proposed there are different kinds of laws in science, but the categories remain vague. Kuhn's paradigms, for instance, include what he calls "symbolic generalizations." These may have symbolic form like Newton's second law of motion, $F = ma$, or Ohm's law, $I = V/R$; or they can be expressed in ordinary words: "elements combine in constant proportion by weight," or "action equals reaction". Kuhn contrasts symbolic generalizations with "exemplars" that illustrate the symbolic generalizations. Exemplars of Newton's second would include the case of free fall described by the equation $mg = md^2x/dt^2$; the inclined plane by $md^2x/dt^2 = mg \sin \theta - mg \cos \theta$; or the simple pendulum by $mg \sin \theta = -mld^2\theta/dt^2$. These exemplars correspond to various models in which Newton's second law allows us to derive phenomenological laws. However, symbolic generalizations never refer to concrete situations. According to Kuhn, "they function in part as laws but also in part as definitions of some of the symbols they deploy. Furthermore, the balance between their inseparable legislative and definitional force shifts over time."[45] How can symbolic generalizations be both descriptive and prescriptive? Kuhn admits that this requires a further analysis because our commitments to a law of nature are very different from our commitments to a definition. Laws are corrected little by little, whereas definitions, being tautologies, are not corrigible at all, although the definitions of many terms are revised over time. Kuhn has isolated the symptoms, but has failed to provide a diagnosis.

Kuhn's claim that laws act like definitions flies in the face of the realist opinion. But the claim is not new. Already in 1905 Henri Poincaré argued that with respect to establishing the equality of two forces, Newton's second law of motion "ceases to be regarded as an experimental law, it is now only a definition."[46] And he was not alone. Brian Ellis writes regarding the origin and nature of Newton's laws of motion: "In the tradition that has succeeded Mach, Newton's second law of motion has been widely regarded as a definition of force."[47] In addition, through historical studies Norwood R. Hanson showed that Newton's laws of motion have had various distinct uses. Considering the practices of physics, he said, one will discover that "the law of inertia," "the second law of motion," and "the law of gravitation" all stand for "umbrella-titles." Wittgenstein's idea of a language game underlies Hanson's suggestion; laws form a family of roles, their mathematical counterparts can be taken as definitions, *a priori* statements, heuristic principles, empirical hypotheses, rules of inference, etc.[48] The particular role a law statement actually plays depends on the historical or experimental context. None of these distinct uses could be claim to be the only correct one.

My explanation of the source of these various uses differs from Kuhn and Hanson; moreover, I have doubts about several of the uses of laws. First, I shall argue that theoretical laws are implicit definitions without descriptive content. Secondly, I shall argue that although theoretical laws are definitions, they do not have *a priori* status. In his discussion of Newton's laws, Hanson, like many philosophers, conflates analyticity with *a prioricity*. Thirdly, I shall argue that theoretical laws are neither contingently true nor contingently false.

The basic physical vocabulary of Newtonian physics consists of "force," "mass," "distance," and "time" mathematically represented by "F," "m," "s," and "t."[49] Whereas "distance" and "time" designate *observable* quantities, both easily definable by reference to basic empirical operations with physical objects, neither "force" nor "mass" is definable by direct empirical measurements. So Newton's theory must introduce these other terms as *theoretical* quantities. To define them, Newton's theory contains two theoretical laws, the second law of motion "$F = md^2s/dt^2$" and the universal law of gravitation "$F = m_1 m_2 / s^2$," relating the theoretical quantities to the observable ones. Indeed, for a long time the mass term in the two equations was considered to refer to different quantities, the 'inertial mass' and the 'gravitational mass,' but physicists were unable to discover their distinctness experimentally, and, thanks to Einstein, today we hold that there is no difference that could have made an empirical difference. The only way physicists could have discovered a difference seems to be if the second law was stipulated as a definition of "force," and "mass" referred to an observational quantity, in which case the law of gravitation would be strictly an empirical law. But even though "mass" is considered as a fundamental quantity in the SI-system, whose unit is a kilogram, "mass" does not refer to any observable quantities. "Mass" designates an object's ability to interact with other objects, and such an ability is not observable. As Johansson emphasizes, "The mass unit is given as the mass of the mass prototype, but the dynamical meaning of mass is given by the second law and the law of gravitation."[50] Apart from these two laws, Newton's theory also contains two other laws. Newton's first law is a specific case of the second law, while the third law constitutes a principle that states that the interaction between two systems is reciprocal. With "force," "mass," "distance," and "time" in place, other important terms in classical mechanics, like "velocity," "acceleration," "momentum," represented by "v," "a," and "p" respectively, can then be given a lexicographical definition in terms of the three fundamental terms as "v" $=_{df}$ "ds/dt"; "a" $=_{df}$ "dv/dt" $=_{df}$ "d^2s/dt^2"; "p" $=_{df}$ "mds/dt".

I have three arguments for my claim that Newton's laws of motion and law of gravitation are rule-making definitions, and they are neither literally descriptive sentences stating empirical facts nor contingent states of affairs:

First, I think Brian Ellis is correct when he argues that forces are queer entities because they "are not like other theoretical entities, such as atoms, or genes, since the existence of atoms or genes is not entailed by the existence of the effects they are supposed to produce."[51] Instead "the action of forces," he says, is "supposed to explain certain patterns of behaviour, the occurrence of these patterns is considered a sufficient condition for the existence of the precise force required to produce them."[52] In other words, the existence of forces entails and is entailed by exactly those effects they are supposed to produce, but this means that forces are very different from ordinary causes in the sense that ordinary causes, but not forces, are logically distinct from their effects.

Second, Newton's three laws of motion are not the only useful formulation of classical mechanics. There are alternatives, which avoid introducing mechanical forces, for example, Lagrange's and Hamilton's formulations. Thus the movement of a classical system can be described either in terms of the Newtonian function F, the Lagrangian function L, or the Hamiltonian function H. Two different reactions to these alternative formulations are possible. One might argue if any particular one of the above formulations of the laws of motion had a true descriptive content, that particular function would have different observational consequences. But all three formulations are empirically equivalent. Alternatively, one could argue that they are all equivalent because they share the same descriptive content, since "energy" and "force" are interdefinable.

The basis for these arguments is the belief that the empirical equivalence of all three formulations shows that their descriptive content is in fact the same, not that there is no such content. Such a claim is plausible only if we have some semantic criteria to judge the sameness or difference of their content. But what these criteria are is rather obscure.

Traditionally, positivists argued that theoretical equivalence reduces to observational equivalence (because the theoretical terms get their meaning from the observational terms); thus, since all three formulations are observationally equivalent, they have the same descriptive content. Another possibility is to maintain a holist-theory of meaning, claiming that the meaning of theoretical terms is fixed entirely by the role they play in their respective theories. In this case, two theories are theoretically equivalent, just in case there is an appropriate structural correspondence between the two theories such that either theory can be obtained from the other by a simple term-by-term interchange. Again, the conclusion is that the descriptive content of all three formulations is the same. However, Lawrence Sklar pointed out that both these candidates face a series of problems. The idea of what counts as the "same descriptive content" is not in any way unproblematic.[53] A third possible candidate would be to hold that two expressions are semantically equivalent if, and only if, they are synonymous; that is, if they have the same intension. This view implies the three formulations would *not* be equivalent, since the functions have different dimensions, and therefore they would not have the same descriptive content. So if we make the reasonable claim that intensional identity between quantitative expressions intuitively requires the same dimensionality, we get an altogether different picture.

Thus there seem to be no convincing criteria of semantic equivalence of theoretical terms and, depending on one's choice of a criterion, it turns out that Newton's, Lagrange's, and Hamilton's formulations have or have not the same descriptive content. We have no objective grounds for saying which of them is the correct one in the sense of picking up a real entity. I believe that this is not because the different formulations share the same descriptive content, but because of the bad habit of associating theoretical laws with any descriptive content.

Third, and finally, the mathematical structure of theoretical laws often allows us to talk about possible phenomena that may or may not exist, for examples, negative energies, advanced potentials, supersymmetries, magnetic monopoles, etc. Furthermore, all second-order differential equations have negative solutions, which physicists usually discard as useless, since they do not consider them to give us a literal description of anything. If instead we think of theoretical laws as implicit definitions of the quantities involved, this feature can be easily explained. Like natural language rules, theoretical laws provide us with a wealth of descriptive possibilities that may never be used to state or describe concrete facts. These are the rich possibilities which scientists use when they construct scientific models.

There is no problem with holding theoretical laws are implicit definitions, though all such laws imply counterfactuals. All expressions defined according to rules entail counterfactuals. For instance, "Sunday" is defined as "the day after Saturday," implying the counterfactual statement: "If it had not been Saturday today, it would not be Sunday tomorrow." A definition stipulates a necessary connection between definiens and definiendum; and since this analytic necessity is stronger than counterfactual necessity, the former entails the latter, but not vice versa. Although analytic, theoretical laws like Newton's three laws of motion cannot be known *a priori* to be true, for if they were known *a priori*, it would be inexplicable why the Ancients and the Medievals developed theories of motion which denied them. Also, it would make the later replacement of Newton's theory with Einstein's absurd. Each era, of course, found good evidence for its theory of motion, but the fact that Aristotle's theory of motion was replaced by Newton's theory, and then Newton's theory was replaced by Einstein's theory does not tell us much about their logical status.

Therefore I conclude the most appropriate way of looking at theoretical laws is to see their function as language rules helping scientists to represent a certain domain of phenomena. They are not used to represent anything but are used as instructions to scientists concerning how to describe and speak about certain phenomena. A theory contains a vocabulary and a set of rules or definitions for how this vocabulary of quantities can be put into well-formed sentences based on which we can formulate, for example, dynamical models.

5 Models as focal points of scientific explanation

Models differ from theories. They may be abstract or concrete; they may be mathematical, physical, diagrammatic, or pictorial. Regardless of their constituents, they are constructed to represent phenomena in a particular manner, allowing scientists to ascribe certain properties to them according to specific theories. In doing so they apply their theories to the models in order to produce predictions or hypotheses about the phenomena they want to explain. Models reflect the scientists' understanding of the structure and interrelationships of a concrete system of phenomena and help them formulate statements about the system. Thus models map individual entities, but theories do not. Theories furnish a vocabulary of quantities and define their interconnections by which models represent things and the causal connections between them. We may have models of concrete systems we are studying, or we may have models of data used as evidence of the adequacy of these models. Data models include graphs, diagrams, pictures, or other non-sentential means.

Therefore I propose that a scientific model is a theoretically-structured representation of concrete systems the purpose of which is to be used for describing and explaining a certain empirical domain.[54] A model is not itself a linguistic system, but it consists of a system of abstractly constructed objects describable in a language defined by a theory. A model can still be useful even though the assumptions behind it need not correspond to reality. My approach to the relation between models and theories conflicts with a naïve semantic view of theories that holds a theory consists of a set of models, and among this set of models one should be isomorphic with reality in order to be true. But if models are often built on strongly unrealistic or idealized assumptions, there must still be ways in which their usefulness can be explained and justified.

For example, consider a container with gas in it that we want to heat to a certain temperature T. Prior to doing so we want to know to how much pressure will the sides of the container be subjected in the process. We are not interested in any other particular gaseous properties. Our epistemic interests determine on which ones we focus; so in this context we are not interested in the gas's mechanical properties or its taste, odour, or colour. Given our interests, we model the gas as having only the properties of temperature T, pressure P, and volume V, and we accept that the relation between these quantitative properties is expressed by an empirical generalization also known as Boyle-Mariotte's law: $PV = nRT$. Pressure, volume, and temperature can all be measured. Even though Boyle-Mariotte's law, like all phenomenological laws, is limited by *ceteris paribus* clauses, namely that it only holds within a range of moderate temperatures, it is capable of explaining how high the pressure will be, and one can compare it to the maximum pressure the container can resist.

However, our interests could have been in understanding why the properties of pressure, volume, and temperature interact with each other the way they do, or in how the container reacts to heating in the temperature beyond the range in which Boyle-Mariotte's law holds. In this case we construct a model of the gas as composed of a very large number of invisible particles, and we define the temperature, pressure, and volume in terms of the mechanical properties of these gas molecules such as mass, velocity, elasticity, and the ability to collide with each other and with the walls of the container. Temperature is defined as a function of the average velocity of the many particles. In this model we are not interested in all the properties of particles. We select individual properties and idealize them. Our interests lead us to a model that builds on two assumptions about molecules that we know are false: (1) no force is operative among them; and (2) they have no extension. Thus we have devised a model of an *idealized* gas in the container. On the basis of linguistic rules fixed by this kinetic theory of gases, we are able to derive Avogadro's law and the equation of state of an ideal gas, specifying how such an ideal gas will behave. With these consequences of the theory, we are able to provide an explanation of the actual pressure at any given temperature and volume.

A real gas is in no way an ideal gas. We may discover that the model in which we map particles without structure and extension is insufficiently exact for our purposes. To yield a fair representation, the molecules must be without extension and the distance separating them has to be relatively great. If either of these assumptions fails, we must determine whether the gas in the container is monoatomic or polyatomic. This distinction may still be inadequate for some purposes, so we will need to add more properties and structure to the model. If we take into account the kind of gas in the container, the mutual attraction exerted by the molecules and each individual molecule's volume, the kinetic theory of gas yields van der Waal's equation.

All three equations express *ceteris paribus* laws applying only to ideal gases in a model, never to any actual gas. No single mathematical expression, nor any individual law, could ever describe an actual gas in complete detail. Nonetheless, in most contexts we can treat an ideal gas as representative of certain types of gases; in relatively fewer cases we may need to represent the particular gas by a model that adds further constraints tailored to the concrete situation. Depending on the degree of exactness required for the purpose of answering the particular question posed, we choose the specific level of abstraction for our model, and we use the corresponding law to produce an explanation. The standards for justifying our choice are determined by the epistemic costs of being wrong: if the costs of being wrong are rather small, then our commitments to high standards will be quite low, but if the risks are rather high, our commitments are similarly very high. In the light of such pragmatic considerations the scientist chooses a model that affords the best means of interpreting and representing reality. She applies accepted theories to this idealized structure and develops new hypotheses in relation to it.

Low-level generalizations, constants of nature, and chemical reactions have been found by direct confrontation with and intervention in the world. But causal discourses are often based on theoretical models in which scientists attempt to represent only some aspects of the real world, since these aspects are too complex to be directly manipulated. Models operating with a particular causal mechanism take for granted that there exist nomologically relevant conditions in each and every case of causation, and that different outcomes of one and the same type of event require a difference in at least one relevant condition. But because of the complexity of many systems, it is technically impossible to specify all the relevant circumstances (even if you somehow knew which were the relevant ones, often of course the scientist doesn't) in which events of a particular type invariably give rise to events of another type. In these circumstances, and before we know any exact causal mechanism, we may construct statistical models where we take probabilities as evidence for the existence of causal connections: most often positive correlations are seen as an indication of complex causal structures between kinds of phenomena. Hence, in many cases, we use probabilities only for purely pragmatic reasons.

While theories provide us with the principles stating the relationship between specific quantities, models give an abstract representation of concrete things according to the principles of the theory. A theory does not represent things or objects in the real world, but it defines certain properties in terms of other properties. Theories supply scientists with a lexicon and some exact linguistic rules, which they then use to describe the objects represented in a model.[55] If the properties attributed to the objects in the model are of a kind that is assumed to make the objects of a physical system causally connected, then we can offer a causal explanation based on the model. Naturally, a model can be made more or less exact depending on the extent of details taken into account in describing the system, the physical circumstances applying in the particular case, and the specific epistemic aspirations and interests we might have in explaining some aspect of the behaviour of the system in a particular way. When the theory is applied to the model, the scientist can explain a concrete phenomenon.[56]

Ronald Giere maintains a view of models that I believe is similar to the one presented here and elsewhere. In discussing how models represent reality, he agrees with my argument that scientists are intentional agents with goals and purposes that play an important role in how they choose to represent nature.[57] His understanding of the function of purposes in the representational practices of science is explicated by the following relationship:

S uses X to represent W for purposes P.

Here S can stand for an individual scientist, a group of scientists, or a larger scientific community. Giere points to the fact that X can be many things, but what is important in this context is that X is a model used to represent certain aspects of a concrete system W for the purpose P of explaining a certain phenomenon. Models may also be used for a variety of purposes: scientists may want to learn the structure of a certain system, to know the causal mechanics of the system, or to predict the future behaviour of a system given certain constraints.

Purposes are determined by personal or shared interests of the scientists. Therefore scientists may easily use for different purposes incompatible models to explain different aspects of the causal behaviour of one and the same system. The use of alternative models in the practice of science seems impossible to understand if one accepts Hempel's covering law approach to explanation, or if one wants to provide, as does Wesley Salmon, an ontic account of explanation in terms of causation. Instead, the fact that scientists use mutually inconsistent models shows the need for a pragmatic understanding of scientific explanation, including causal accounts.

An example may help to clarify my point. The atomic nucleus is a quantum system, meaning that its energy, spin, and angular momentum are quantized and specified by different quantum numbers. One simple means of studying the atomic nucleus experimentally is by sending an ensemble of positively charged particles each with the same energy into a target. This causes the atoms in the target to emit gamma particles, which are distributed with a spectrum of different quantized energies, allowing the scientist to infer the nature of the target. Physicists explain the detected energies by means of a causal story of what happens inside the atom from the moment the particle beam hits it until the energy leaves it again. To do this physicists use a model of the nucleus to which to apply their equations. In general, nuclear physicists work with two incompatible models, both borrowed from classical physics. The first is *the shell model*, which represents every nucleon as moving in an orbit determined by the mean power field of the other nucleons. Such a model is especially useful for explaining the effect of the individual particle movements inside the atomic nucleus. Whenever one knows the potential energy governing the nucleon's motion, Schrödinger's wave equation gives all properties of the possible orbits. This allows physicists to calculate the spectrum of the various energies emitted. But the shell model does not permit a precise description of *all* properties of the nucleus. This is not due to the fact that scientists do not know the precise form and magnitude of the potential, but rather that the complex interactions in a nucleus cannot be represented by a single common potential for all nucleons. The second model is *the drop model*. Here the nucleus is considered to act as an ensemble, behaving like a water drop. Like a water drop, the nucleus is assumed to be able to change its shape and exhibit various empirical phenomena taken as caused by what the model represents as vibration and/or rotation of the nucleus. And the drop model is also useful because nuclear matter is assumed to be rather incompressible like most fluids. The drop model focuses on the collective movements of the nucleons, making possible an explanation of phenomena such as nuclear fission.

Neither of these two models yields the one and only correct representation. Indeed, from my perspective, since 'correct' is relative to intention, the expression "correct representation" is meaningless unless a context is specified. Each model helps the scientist to describe different aspects of nuclei depending on their epistemic problems and pragmatic interests. So in a given situation the selection of a particular model is determined by the scientist's purpose in giving a representation by which he can produce the most accurate explanation according to his need. Thus, a causal explanation should not be understood as a two-place relationship between a sentence and the world where the focus is on explanation as a true representation. Since the agents' purpose in representing the world in a particular manner plays a significant role in determining whether or not the explanation is accurate and acceptable, causal explanation must be understood only in terms of representing the world according to a particular epistemic perspective. All the claims I have defended in this chapter imply the conclusion that explanations are both pragmatic and context-dependent in the sense that their value as appropriate explanations depends on the scientists' interests and knowledge.

# 5
Scientific Explanation

Philosophers have tried to grasp the notion of scientific explanation, either by looking for some essential features, such as inherent logical properties that every explanation has to satisfy, or by arguing that a true *scientific* explanation must appeal to some specific kind of factual relationship existing in the world. For instance, Hempel's model of explanation in science was based on clear conviction that all explanations could be characterized as an argument which states that the phenomenon to be explained follows deductively from some general laws. Although he did not deny the pragmatic side of explanation, the model of explanation he presented is abstracted from those contexts in which scientists use explanations. One may therefore wonder whether it is possible to find such a context-free notion of scientific explanation without losing sight of the cognitive purpose of providing explanations?

Explanation is a man-made activity, created for certain communicative purposes. Thus the role of explanation is nothing over and above its pragmatic function, which has both logical and factual characteristics. Instead of abstracting 'explanation' from the context in which it appears to find a common logical or factual factor, we may grasp 'explanation' better by asking questions like: Why do we ask scientists to produce explanations? What is explanation used for in science? What is its purpose? An obvious reply is that we seek explanation because it provides information and imparts understanding. But the kind of information and understanding one looks for depends on one's cognitive situation.

One reason getting a firm grip on the notion of explanation is difficult is because a wide range of distinct kinds of entities "explain." People, facts, events, hypotheses, models, and theories are all regarded as having explanatory power. Behind this plethora lurks an ambiguity. Like other similar words, "explanation" ambiguously designates a process and a product. "Explanation" refers not only to a linguistic or human activity but also to the result of this activity. The sentence "Rutherford *explained* why a few alpha particles did not penetrate a thin film of gold" treats explanation as an act of discourse, whereas a sentence like "The sun's gravitational attraction explains why the planets move in elliptical orbits around the sun" regards explanation as a product of an explaining act. That it is a product implies that there would be no explanation unless somebody had produced it. But can we conclude that it makes no sense to speak of things, facts, events or theories as "explaining" anything, unless the explanation is a result of such an explaining act? This is controversial.

Sometimes we assume that an explanation exists completely independently of whether any person "discovers" it, but at other times we assume that an explanation is a specific linguistic activity, so explanations "exist" only in a linguistic realm. There are good reasons in favour of each view. Most people believe that phenomena in the world happen the way they do regardless of whether human beings exist or not. It is an objective fact that metals expand when heated, and what explains this fact is an objective matter. Therefore explanations of phenomena exist irrespectively of whether anybody has ever attempted to formulate them. From this perspective, "explanation" refers to a kind of objective fact which takes explanation to be a proposition or argument, and thus publicly accessible, open to inspection and evaluation.

It also seems to be true that people do the explaining. Nature by herself explains nothing, nor do theories. Human beings intentionally explain by appealing to what they take to be the explaining fact of the matter. We call upon theories to throw light on a problem when explanations of this problem are needed. Since an explanation responds to a question raised in a particular context, what counts as an explanation therefore depends on the context; there is no objective fact of the matter to determine the explanatory content alone. It is just as much a matter of choosing the appropriate linguistic acts, which involve cognitive interests and background assumptions of the explainer and an explainee. It is difficult to see how both the objective and the contextual characteristics of explanation can leave room for each other.

Another difficulty in grasping "explanation" arises from the fact that different kinds of explanation seem to be the norm in distinct subject matters. In the physical sciences, where we have no direct influence on the course of events, we produce *causal explanations*. We explain, for instance, the destruction of a village on the seacoast by saying the tsunami caused its demolition, or that the tsunami was caused by an undersea earthquake. Even when we can influence physical events, say by building levees, our actions and purposes still must be physically manifested. But in biology we often produce and find *acceptable* explanations in which the existence of a feature or a quality is explained in terms of the *functional role* it plays. Heartbeats are explained by their function of pumping blood through the blood vessels. Furthermore, in cases when human beings are acting and deliberating persons we may give, and be inclined to accept, *intentional explanations*. A man goes shopping because he wants to buy his wife a present. Thus, the intention of buying a present explains why this man went shopping. However, functional and intentional explanations are problematic because they seem to explain a present phenomenon in terms of its future state, which does not yet exist.

A third difficulty arises when we differentiate *scientific* explanation from mundane everyday explanation. Some philosophers argue there are important differences because science provides 'objective' explanations. Others hold that such a demarcation cannot be drawn. Our cognitive interests determine how general or rigorous we want an explanation to be. Thus, sometimes explanations are requested for *singular* events, sometimes for *kinds* of events, or, as in science, sometimes for *general laws*.

In everyday situations usually we want to explain a *singular* fact or event. In science we also want to explain single events such as a certain track on a particular experimental recording, the extinction of the dinosaurs, or the current increase of the global temperature on Earth. The most natural way to explicate what is called for when such an explanation is proposed, is to say that an explanation is needed when we know that E, the effect has happened, but we lack knowledge of C, its cause, normally the necessary and sufficient physical conditions for its occurrence, although in explanations of conscious behaviour, a mental state (motive) may also be a causal condition. In many everyday contexts, why-questions demand purposive answers; but in natural science causes are usually purely physical or mechanical and are regarded as merely proximate triggers when what is requested is the purpose of the *explanandum*. Although everyday explanations seldom refer to laws, some have argued that in *science* any explanation of singular events involves laws. Sometimes in everyday circumstances we are seeking explanations of *kinds* of events, but this is more typical in science. A substantial difference between ordinary explanations and scientific explanations is that, in contrast to everyday life, science wants to explain lower level empirical laws in terms of more general laws, such as "explaining" Kepler's and Galileo's laws in terms of Newton's laws, or the ideal gas laws in terms of the kinetic theory of gasses. Ultimately, fundamental explanations in sciences seem to depend only on *causal laws*; we talk of the force of gravity causing the elliptic movement of the planets, heat causing metal to expand, water causing salt to dissolve, electrons and positrons fired at each other causing their annihilation and the emission of gamma rays, and so forth. If one recalls the distinction between description and explanation, the motivation of searching for causal explanations is quite straightforward. Explaining a phenomenon by the causal mechanism that brought it about ensures that the additional information provided by the explanation is relevant to the phenomena in question. And it seems quite plausible that knowing *why*, as opposed to knowing *that*, the sky is blue or that the Moon always faces its same side to the Earth amounts to being able to identify the causal factors responsible for these facts. However, it does not follow from this that all physical explanations are causal, or that scientific explanations necessarily point to causal factors. In particular, it may be necessary to rely on certain non-causal laws of nature to explain physical phenomena. So let us take a more substantial look at these issues.

1 Nomic explanations

Philosophers of science have long regarded deductive derivation from universal laws and initial conditions as the paradigm of scientific explanation. But, as we shall see, this deductive-nomological explanation is merely one form of scientific explanation. If one of the general statements in the explanans is a statistical law, then the explanans does not entail the explanandum with certainty. Instead the conclusion is stated with a degree of probability, because the information contained in the explanans can merely inductively bestow a certain probability on the explanandum. Hempel called this type of explanation the "inductive-statistical" model and held that such explanations constitute another kind of covering law model. We are still able to explain why certain phenomena occur by an argument that subsumes the phenomena under statistical laws. But if one holds that all explanations involve causal regularities, inductive-statistical explanations cause considerable trouble.

First and foremost, philosophers have seen the regularity account of causation and the covering law model as fitting nicely together. If we assume that causality is nothing but regularity between type events, that regularity can be expressed by a universally quantified conditional. The truth conditions of singular causal statements are simply spelled out in terms of a universal statement. Thus, the proponent of causes as regularities holds a singular event is caused by another event if, and only if, the statement expressing the universal regularity implies a sentence about the occurrence of the events. Since the covering law model explains an effect by proving that a statement describing the fact follows from premises containing at least one universal statement and a singular statement about another fact (the cause), the covering law model can easily be associated with causal explanation.

Inductive-statistical explanations now appear as a form of causal account, where the regularity is not universal but probabilistic. However, leaving inductive-statistical explanation aside as perhaps a dubious causal explanation, there is still not much ground to believe that all universal statements of a theory are causal laws. Many are not, and possibly none are, for causal laws are context-dependent, universal statements are not. Indeed it depends on how precisely the causal laws are formulated. If all the relevant conditions are specified in the antecedent of the law statement, then the resulting conditional statement is true "universally." All of the necessary conditions can be packed into to the formulation of the law itself, or some can be supplied by establishing a context, allowing the law to be expressed with fewer conditions, since the context makes them assumed. But it seems impossible to specify *all* relevant conditions, and even if it were, the resulting universal "law" statement would probably have only a single instance.

Universal statements concerning principles, symmetries, invariance, and conservation are not *causal* statements, but we shall call these universal statements *"fundamental* laws." Physics prefers to describe systems in terms of properties that are *conserved* or *invariant* because these properties are independent of particular observers. A certain quantity of a system is "invariant" if it remains constant through a change of frame of reference. Likewise, conservation laws claim that a certain quantity of a closed physical system is "conserved," if the system is symmetric under a continuous transformation of a parameter with which the quantity is associated (*Noether's theorem*). These are purely *prescriptive* principles ascribing certain fundamental properties to a physical system. In fact, physical processes can be characterized as processes that transmit conserved or invariant quantities.[1] Obviously, on pain of circularity, principles of symmetry, invariance, and conservation cannot themselves contain causal information if these principles determine physical processes.

Elsewhere I have argued that the distinction between "causal" laws and "fundamental" laws should be characterized in term of their facticity.[2] Causal laws are empirical laws restricted by *ceteris paribus* clauses; they describe the interaction and temporal development of systems given certain nomologically relevant circumstances. As we saw in the previous chapter, "fundamental" laws form a heterogeneous group, including theoretical laws, conservation laws, and theoretical postulates. *Theoretical laws* like Newton's laws of motion, Maxwell's electromagnetic laws, Einstein's field equations, or Schrödinger's wave equation, express interdependent relations between certain quantities. They should be regarded as a kind of definition stating how certain quantities are defined in terms of others; they don't express facts, and for that reason, contrary to what Nancy Cartwright has argued, they don't contain *ceteris paribus* clauses.[3] I regard theoretical laws as definitions or rules for formulating empirical laws in a model, generally through expressing them in the form of second-order differential equations.[4]

Thus, as implicit definitions of certain quantities, *theoretical laws* are universal, unconstrained by exceptions, and cannot be used to explain anything other than how some quantities are connected. But scientists construct idealized models of systems that can be described according to these definitions, for instance, the impact of a neutron of certain energy on U^{235}. Theoretical laws provide scientists with an abstract vocabulary by which they can talk about their models, which are constructed such that the objects have exactly those properties that can be described by the predicates of the laws. Based on experiments and the predictions of theoretical models, scientists may gather information about nomologically relevant circumstances to establish *causal laws*. In contrast to theoretical laws, causal laws are the kind of empirical laws to which scientists appeal whenever they give a causal nomological explanation of certain events.

Geert Keil has argued by that there cannot be *ceteris paribus* laws.[5] If we assume, as most philosophers do, that a law is a universally quantified conditional and that the *ceteris paribus* clause should be understood as other things being equal, then, because of the logical form, there cannot be *ceteris paribus* laws. The idea is that the restriction with a *ceteris paribus* clause makes a demonstrative reference to particular circumstances, whereas universal statements are not controlled by such circumstances. Accepting this objection means one must give up on the thought that causal generalizations can be explicated as universally quantified conditionals. But take the actual circumstances in which a particular c causes a particular e. If this single instance of causation can be generalized, it implies that we have to say that the relevant circumstances also have to be generalized, and we can at least say that given the same circumstances (i.e., other things remain unaltered) another c would cause another e. Moreover, I think it makes sense to separate these circumstances under which c causes e into those that are nomologically relevant and those that are nomologically irrelevant for the outcome.[6] Here scientists' models and experiments help to determine how to specify these circumstances and how they can be divided. Therefore in principle we can think of a causal law as a quantification over those particular circumstances which existing theories and background knowledge make nomologically relevant for asserting that C causes E. We then arrive at something like the following:

Ceteris paribus, C causes E' is a law of nature if, and only if, $(x)(y)(C(x)$ & $K_J(y)$ causes $E(x))$, where y runs over all nomologically relevant circumstances K_J.[7]

Such generalizations are not true unconditionally, but true only so far as causally relevant circumstances hold.

Conservation laws comprise another kind of fundamental law. I agree with Lars-Göran Johansson's view that conservation principles can be explained from an epistemological perspective as a consequence of the objectivity requirements scientists put on the descriptions of nature.[8] They are claimed to hold for every observer. But conservation laws can enter into the explanation of singular facts. Consider the following reaction: $\mu^- \rightarrow e^- + \nu_e + \nu_\mu$, where a muon decays into an electron, an anti-electron-neutrino, and a muon-neutrino. If a student asks why there is one charged particle in its decay product, the answer is that conservation of electric charge implies the total net charge remains constant for all observers. Thus, all particle decay processes must have the same total charge in the initial as well as in the final product. Since the decaying muon has a negative charge and charge is conserved, one, and only one, surplus charged particle turns up in the decay product.

Not only conservation laws yield nomic but non-causal explanations. Other fundamental laws, theoretical postulates, assign certain fundamental properties to a physical system. Such postulates include the quantization of action, the constancy of the velocity of light, Pauli's exclusion rule, etc. The Michelson-Morley experiment to establish the existence of the ether was designed so that two light beams coming from the same source were transmitted perpendicular to each other along two arms of the apparatus, until they were reflected by mirrors back to an interferometer. The underlying idea was that the velocity of light would depend upon the variation in Earth's velocity with respect to the ether. Thus the two rays were expected to create an interference pattern due to a phase change caused by their different velocities; and a subsequent 90° rotation of the instrument was expected to produce the opposite phase difference, resulting in a displacement of the interference pattern. But the expected effect did not appear. The outcome was completely negative.

Nevertheless, H. A. Lorentz gave a causal explanation of this unexpected result. He assumed that among the molecules in the two arms there were molecular forces acting like electromagnetic forces, and that the molecules would behave as charged particles through attraction and repulsion when the arms moved with respect to the ether. From these assumptions he calculated that the length of an arm parallel with Earth's direction of movement is shorter compared to its length when it is rotated 90°, and the contraction with a magnitude $(1-v^2/c^2)^{1/2}$ would explain the result of the Michelson-Morley experiment by the now well-known Lorentz-FitzGerald contraction. Ten years later Einstein proposed the special theory of relativity, which did not offer a causal explanation of the null result but deduced it from the first principles of the theory. Thus the theoretical postulates of the special theory permit a nomic, but non-causal, explanation of the null result. This kind of explanation by principles is sometimes, given the context, considered superior to causal explanations.[9]

In these examples the explanandum is explained by the rules for describing the phenomena in question. In the language of science we often call such constitutive rules "laws" or "principles." Given these rules for applying certain descriptions to the world, it is true that a negative charged muon decays into particles with a total of one negative charge, and that there will be no variation in the interference pattern in the Michelson-Morley experiment. These particular facts are simply the consequences of the way nature is described according to the principles of scientific theories. But there is more to it than that.

The same appeal to descriptive norms appears in the explanation of empirical laws. A question like "Why do the planets obey Kepler's laws?" also cannot be given an appropriate causal answer. The question asked is about why certain descriptions of the planets' behaviour are true. An adequate response to that question is to point out that Kepler's laws are a result of applying Newton's laws to the model of a central body system, but this explanation is not causal, since Newton's laws in no way cause Kepler's laws. From Newton's laws and some suitable additional premises we can infer Kepler's laws; thus we can say that the descriptions of the movement of the planets around the Sun in terms of Kepler's laws are true as a consequence of Newton's theoretical laws. Therefore theoretical unification offers no factual understanding in itself. Rather it furnishes us with a grasp of the linguistic rules which prescribe how to use the quantitative terms in a uniform description of the world. Causal understanding is the other factual part of scientific comprehension. This latter understanding is, however, completely expressed in quantitative terms in accord with the theoretical system. So when scientists change the rules of language used to describe the phenomena, their causal understanding may change too.

Depending on the meaning and the context of the question, we therefore have to operate with at least two kinds of nomological responses to a why-question in the natural sciences. These different sorts of accounts could be labelled *analytic* explanations (which appeal only to definitions) and *synthetic* explanations, i.e., those based on empirical laws. Which one you are interested in depends on the kind of problem you have. Most often the different kinds of problem are well separated as when, on the one hand, a particular phenomenon is explained in terms of its cause and, on the other hand, the truth of a certain description is explained by a principle acting as a rule for applying that description to the phenomenon. But sometimes they seem more or less to coincide, and therefore are easily conflated, as when the subsumption under a law is taken to explain causally the occurrence of a singular phenomenon. Nevertheless, we can also here differentiate between the synthetic account and the analytic account. The first account explains why something occurs, and it could in principle have been expressed in terms of singular events. The truth of that kind of explanation depends on the world itself. The second kind explains why the description *simpliciter*, i.e., the description in which the causal relation is stated, is true in virtue of pointing out that it is in accordance with certain fundamental descriptive principles. The truth of 'analytic explanations' is verbal; the truth of 'synthetic explanations' is factual.

2 The Hempelian view

Do these different forms of nomic explanation fit Hempel's view, or "the received view" as Salmon called it? Hempel didn't require that the law covering the explanandum is a causal law. His model is often mistakenly viewed as causal, but in fact it is nomological because he adopts Hume's empiricist view of causation.

Other philosophers have argued that explanations in science directly or indirectly must appeal to general statements that express a causal law under which the particular fact to be explained is covered. If one believes that we can separate scientific explanation from everyday explanation in terms of content, and not in terms of the degree of evidential support, it is reasonable that scientific explanation appeals to a causal law of nature. However, Hempel himself did not reduce all explanations to causal explanations nor did he believe that only causal laws provide an appropriate scientific account. He distinguished between "laws of succession" and "laws of co-existence," which may also figure in explanations.[10]

We should interpret Russell's notorious remark that "The law of causality...is a relic of a bygone age, surviving like the monarchy, only because it is erroneously supposed to do no harm" in a similar vein.[11] He held it is not causal issues, but mathematical functions, that provide genuine models of explanation in science. In either case it is only subsumption under a general law that is essential to the notion of scientific explanation. It is not causal explanations as such, but nomic explanations in general that serve the aim of science.

In their classical paper, Hempel and Oppenheim identified three logical conditions of explanatory adequacy and one empirical condition. First, *the entailment requirement* demands that the explanandum must follow validly from the explanans. Second, the explanans must contain at least one general law, which is necessary for deducing the explanandum. Third, the explanans must have empirical content in the sense it is possible, at least in principle, to test it by experiment and observation. The fourth empirical condition stipulates that a "sound" or "correct" explanation requires the sentences of the explanans to be true. Later Hempel amended this fourth condition by distinguishing between *true* explanations, more or less strongly confirmed explanations, and *potential* explanations.[12] Thus a *potential* explanation fulfils only the logical conditions, whereas a *true* explanation must additionally meet the empirical condition.

Since Aristotle, it has been known that deduction from premises containing an accidental generalization does not give explanation. Accordingly, to have a scientific explanation of empirical phenomena, at least one premise will have to be a non-accidentally true general statement, whereas others are about particular circumstances. So according to the D-N model of Hempel and Oppenheim, explanation in science can be conceived as a deductive argument where the description of the phenomenon to be explained is validly deduced from statements of universal laws and particular circumstances. (Eventually Hempel gave up the idea that all explanations are *simple* derivations when he attempted to broaden up his notion of explanation to include inductive-statistical arguments. But let us set this side of the issue aside for the moment.) Before we proceed, we should notice a number of things about this deductive-nomological model of scientific explanation. The main features of Hempel's theory can be summarized in a few points: (1) Scientific explanations are answers to why-questions; (2) scientific explanations are arguments to the effect that the event being explained was to be rationally expected; (3) scientific explanations subsume the event to be explained under laws of nature; and (4) there is no logical or structural difference between explanations and predictions. Although Hempel and Oppenheim opened their seminal essay by stating that scientific explanations are answers to why-questions, they never asked themselves whether their account left out some informal but essential aspects of explanation, which also were necessary to understand explanation. They believed an explanation should be capable of being expressed as a formal argument in which the conclusion is a deductively valid consequence of premises stating certain antecedent conditions and general laws.

Hempel held what are deduced are not the phenomena themselves but *sentences* describing these phenomena. What are deduced are sentences (propositions) but what are explained are phenomena, events, or sentential facts that are described in *explanandum*-sentences. From that deduction the phenomenon is regarded as explained.[13] The objects of explanation are *descriptions* assigning properties to the phenomena and not the phenomena *per se*. So explanations are a matter of formal logic and not a matter of facts. The law statements in the *explanans* need not express causal laws or reflect any form of physical necessity. Their unlimited scope of evidence and augmentation distinguishes law-like statements from statements of accidental generalizations.

Finally, in Hempel's account there is no logical difference between explanation and prediction. It depends on the viewpoint of those who make the argument. If the occurrence of the phenomenon, as stated in the consequence, has not yet been observed, the argument functions as a "prediction" (or "retrodiction"), but if the phenomenon has been observed, the argument is an "explanation." Because of this logical symmetry and the deductive character of explanation, the phenomenon to be explained is nothing but what was to be expected given the nomic situation in which it appears and the initial conditions determined by observations.

Hempel imposed two requirements on "scientific explanation." The first is *explanatory relevance*: "the explanatory information adduced affords good grounds for believing that the phenomenon to be explained did, or does, indeed occur."[14] The other is *testability*: "the statements constituting a scientific explanation must be capable of empirical test."[15] By invoking these requirements Hempel hoped to exclude all kinds of phoney claims from having explanatory force. He seemed to have thought the demand of explanatory relevance implied that the appropriate scientific explanation has to be an argument. Unfortunately, there are well-known examples fulfilling both requirements (although the facts of the matter are considered irrelevant,) like the man who doesn't become pregnant because he regularly takes his wife's birth control pills; or the salt that dissolves in water because it is hexed. Hempel's two requirements, as explicated here, do not exclude every form of irrelevant explanation. Also a phenomenon may be subsumed under a law that in most cases would be relevant as an explanation, but may still be irrelevant for this particular phenomenon. Effects of the same type can have different types of causes and therefore can be subsumed under different laws. A shot in a person's heart "affords good grounds for believing that the person's death to be explained did, or does, indeed occur," but this particular person died, say, because of cancer. The logical structure of an argument cannot ensure that the *explanans* is relevant in the circumstances for this *explanandum*. It applies as well in cases where we possess inductively established reasons to believe the law cited in the *explanans* often can be associated with the phenomena mentioned in the *explanandum*.

3 The received criticism

We should not forget that logical empiricism provides the philosophical motive and background for Hempel's view of explanation. Hempel didn't just happen to explicate "explanation" in purely logical terms or just accidentally (through oversight) ignore pragmatic considerations and context. It was a natural consequence of logical empiricism (and its historical background with Frege and Russell's logic and Hume's empiricism) that Hempel took this course. Indeed, he presents his deductive-nomological model as just a formalization of what had been implicitly assumed by the positivists (especially Carnap) well before World War II. His approach was determined by his positivistic conception of knowledge: except for matters of pure logic (and mathematics), there is no knowledge except that provided by scientific explanations and the observations and theories required for those explanations. As a good empiricist, he reduced causal knowledge to what could be expressed in terms of empirical regularities, which may elucidate why he hoped to reduce explanation to a logical relation between propositions.[16] Also Hempel was strong on the use of the deductive-nomological model univocally outside as well as inside the natural sciences because of the background of the *verstehen*-controversy over the social sciences, which positivists wanted to combat. One should understand his approach in this light.

Standard objections against Hempel's covering law model have established themselves over the years. One is that subsuming under a law according to the covering law account is not sufficient for having explanation. Sylvain Bromberger's famous flag pole example shows we can explain the length of the shadow of the flagpole on a flat level ground by deducing it from the height of the pole and the angle of the sun above the horizon together the principles of Euclidean geometry. But it seems surely wrong to suggest, as the deductive nomological model does, that it is possible to "explain" the height of the flagpole by the length of the shadow. This asymmetry of explanation lies in our strong belief that the height of the flagpole *causes* the length of its shadow, but not vice versa. Similarly, we can explain the red shift of the light spectrum from a galaxy by deducing the antecedent conditions containing the law of the Doppler effect and the tangential velocity of the galaxy, but it doesn't seem reasonable to say that its velocity can be *explained* by deducing it from the antecedent conditions containing the same law and its red shift. Again the reason seems to be that the *explanans* states what acts as the cause of the *explanandum*. The direction of explanation is determined by the direction of causation, because evidence cannot explain that for which it is supposed to be evidence.

But is this objection always valid? Isn't it possible that we might want to explain the height of the flagpole or the velocity of a star? If so, the explanatory direction depends on the context of explanation rather than corresponding to the causal direction. Recall Bas van Fraassen's fictive story of the tower.[17] But real life examples also show that the explanatory arrow need not point in the same direction as the causal arrow. More to the point: neither the causal nor the explanatory arrow needs point in the direction of the logical arrow. The logical arrow, so to speak, shows the direction of what we know; while it doesn't explain the height of the pole, it can very well be taken as an explanation of how we know how high the pole is. The causal arrow is metaphysical. And the explanatory arrow falls somewhere in between.

In the experimental sciences we often see that explanation runs opposite to causation. A textbook experiment asks the students: "Explain the motion of Barnard's Star that is displayed in Figure 10.1."[18] The information available for the students is a superimposed photograph taken of the star at six-month intervals and a spectrogram with a set of comparison lines and an absorption feature of the stellar spectrum. Based on these experimental data and assuming the physical principles of the Doppler shift, it is possible for the student to deduce the answer to the question. This is a fair example of a real-life explanation we might request from an astronomy student. Indeed, according to the covering law model, an answer to this question would not count as a scientific explanation. But experimental practice shows there is no reason to dismiss the answer as unscientific. The student will back up her explanation with information about the reliability of data, the estimate of measurement uncertainties, the apparatus efficiencies, etc., just like a working astronomer would do. A question like this shows that in a concrete experimental context many different types of explanation-seeking questions may appear side by side, and each particular response to them may support one another in a network of explanations. The received view cannot make sense of such an explanatory network, but it seems to be undeniable that within ordinary scientific practice explanation can go both ways. The example also shows that explanatory direction depends on the particular epistemic context in which the *explanandum* appears. Since the deductive-nomological model permits the explanation to run opposite to the causal direction, it is undeniable that the received view tacitly presupposes that explanations are context-dependent.

Hempel denied that all deductive-nomological explanations of singular events are causal explanations, but what about the opposite claim: Can all causal explanations of singular events be captured by the deductive nomological model? One may consider another standard objection, first due to Michael Scriven. There are cases (if not most cases,) he says, where the explanandum is not explained by being subsumed under a covering law. Instead, an explanation of one particular fact consists in a reference to other singular facts. Only if this explanation is put into question, says Scriven, do we appeal to a generalization for justifying it. In other words, an explanation does not demand the explanans refers to a law; rather it is *the context* of the explanation that determines whether an appeal to a law is needed. This may be true in everyday situations, but does it hold in science too? To a certain extent the answer is definitely affirmative. As David Hull has emphasized, statements of particular circumstances are equally necessary in the covering law model, and in historical sciences like geology and biology these particular circumstances ought to have the explanatory attention.[19]

For example, the palaeontologist Christopher Bennett explained why the fossils of *Pteranodon* fall into two distinct groups depending on the size of their leg-bones, finger-bones, etc. The larger boned fossils also have very big crests and narrow pelvises, while the smaller ones have small crests and wider pelvises, and the smaller pteranodons were twice as numerous as the larger. Bennett proposed the two groups were female and male pteranodons respectively, living together in herds. Thus the males had competed over the females using their big crests as mating displays like deer antlers. In such a case we have a genuine explanation with no need to invoke an unspoken covering law to complete the explanation. A non-vacuous causal law may not even exist.

Cartwright agrees: "Many phenomena which have perfectly good scientific explanations are not covered by any laws."[20] She argues such phenomena are covered by *ceteris paribus* generalizations, holding only under certain ideal conditions. Without the *ceteris paribus* clauses these generalizations would usually be false, and with them the generalizations may be true for a few ideal cases where the conditions hold. But Cartwright seems ambiguous. On the one hand, she holds *ceteris paribus* laws are not true laws and "what happens on most occasions is dictated by no laws at all"; on the other hand, she maintains these laws are nevertheless generalizations. I suspect Cartwright does not really wish to deny that *ceteris paribus* generalization are causal laws, but wants to say that covering laws as general statements about regularities between kinds of events have no role in science. Her conclusion seems to be based on the observation that whether or not one considers a specific instance as falling under a covering law depends on how approximate or exact a prediction/retrodiction one expects. As long as a phenomenon of a certain type occurs regularly under some general circumstances, it seems to appear according to a covering law. But the generality of the circumstances is arbitrary. If one allows as a "covering law" any generalization, no matter how complex and specific it might be, any event can be subsumed under such a "law." Thus it becomes tautologous to claim that all phenomena can be explained by subsuming the explanandum under a "covering law." Any general statement that contains so many constraints and exceptions that it applies to perhaps only one instance is not reasonable as a candidate for a genuine law.

How can non-causal genuine laws of nature be restricted? I hold that *theoretical* laws, like Newton's laws of motion, "don't express facts, and therefore they don't contain *ceteris paribus* clauses, whereas causal laws state facts, and by doing so they always contain a *ceteris paribus* operator. Thus, law statements without exceptions are the theoretical laws; they don't describe any law of nature, whereas law statements with exceptions deal with the real laws of nature."[21] Since theoretical laws stipulate definitions, they explain nothing other than our linguistic behaviour. However *causal* statements are highly context-dependent, true only given *ceteris paribus* clauses. No event can be characterized as a cause of a certain effect without the insertion "in the circumstances." In different circumstances the same event might not have caused the same effect. Thus when we generalize from singular connections to causal laws holding between similar kinds of events, the causally relevant circumstances have to be stipulated as nomologically relevant in terms of a *ceteris paribus* modifier.

Cartwright is right to argue that there may be particular phenomenon due to a combination of so many different more or less well-defined processes that no single causal law would cover them. However, her conclusion that laws play no role is false. What her discussion shows is that the context of a given explanation determines which causal laws are needed. All causal laws apply only to some degree of exactitude, which is determined by the context, because a host of other factors are ignored in any particular explanation. Nevertheless, they do have tiny influences and so make the observed values of measured quantities very slightly different from those deduced from laws and initial conditions. Surely it is misleading to say this inexactness implies we do not apply the laws to real phenomena to explain them scientifically.

Our current explanation of the cosmic background radiation provides an example. Here the signal discovered in 1965 by Arno Penzias and Robert W. Wilson implied a small amount of energy coming from everywhere in space with a uniform temperature a few degrees above absolute zero. This phenomenon is explained as the residual energy of the Big Bang, the singular event that began the Universe. Indeed, this explanation presupposes *ceteris paribus* clauses, physical principles, and assumptions about the nature of the early Universe, but causal laws are also used in calculating the temperature and frequency of the radiation. The explanation is the causal story that can be told about how the background radiation came about. It is hardly very reasonable to hold there is a "covering law" generalizing that all universes come about exactly the same way.[22]

In nature phenomena will be affected by a host of uncontrolled causal factors, and "covering laws" do not apply simply or directly to them. Thus experimental scientists often create phenomena that have never occurred before in nature, so that they actually apply theories to models representing these highly controlled phenomena of the laboratory.[23] These phenomena are artefacts of experimentation. An experimental phenomenon is a regularity of a certain type of event or process that can be re-observed, reproduced, or remanufactured under definite circumstances. But whether artefactual or natural, they can often be explained in terms of a theory. But this does not imply that this explanation of a phenomenon involves a certain covering law. As Hacking suggests, the experimenter may explain phenomena according to a few rules of thumb.

Hans Christian Ørsted discovered the electromagnetic effect one day when he realized that when a compass needle was placed near to a closed galvanic circuit, parallel to the wire, the needle started to move. He observed that the deflection of the needle was correlated with the electricity in the wire. Further experiments confirmed this observation. On this basis he could justify holding a causal connection existed between the electricity and the movement of the needle. On this level he already had an acceptable explanation of the behaviour of the needle; he could start to think that it happens because of the voltaic current in the wire. Thus he could formulate a general empirical statement, i.e., a causal law, saying that whenever a compass needle is placed parallel and close to an electric circuit, it will be deflected. Ørsted never conceived the quantitative laws from which he could derive the electromagnetic effect, but this hardly means that he could not explain the movement of the compass needle. He could, and did, give a causal account of this effect.

Of course I do not deny that science often appeals to all sorts of causal laws in explaining phenomena, but not always. The context includes the scientist's interests and background knowledge of potentially applicable covering laws The stock examples Hempel and others give of the deductive nomological explanation are misleading if taken to represent allegedly context-free scientific explanation. If we accept that causal explanations are context-dependent, it does not make sense to maintain that they are explanatory only because they fit into the deductive-nomological model. In general, this model fails because it ignores that causal explanations are about relations of dependence in the world, not about deductive relations between sentences.

Apart from these problems with the deductive-nomological model, there are also difficulties with Hempel's inductive statistical model. Hempel required that the probability involved in the inductive-statistical account should be near-certainty or very high. The *explanans* bestows high probability on the *explanandum*, and this makes it relevant to the *explanandum*. In some cases, however, this requirement of high probability cannot be satisfied. The probability for persons with untreated syphilis to catch paresis is quite low. Nevertheless, nobody catches paresis unless they suffer from untreated syphilis. The inductive-statistical account can be used to predict the chances for such an individual to develop paresis, but the probabilistic law can hardly explain why a certain individual caught the disease. What explains the paresis in a given case is the fact that the particular victim had previously contracted syphilis. In this context we do not have to increase the probability by adding unknown genetic and physiological factors in order to regard paresis as explained by reference to syphilis.

Whether or not subsumption under a statistical generalization is considered 'explanatory' depends on context. If one lays down a general definition of "explanation" such that subsumption under a statistical generalization counts as an explanation and then, as Hempel does, specifies that the definition is dependent on a high numerical value of the correlation, it *makes statistical explanations highly contextual*. A very high correlation, say 99 per cent, seems plausibly explanatory: if I ask, why did Jones die of lung cancer, and you reply because 99 per cent of two-pack-a-day smokers die of lung cancer, and Jones smoked that much, it seems reasonable to count your answer as explanatory. But if I ask, why did Smith get sickle cell anaemia, and you answer because 0.2 per cent of black men get the disease and Smith is a black man, then if someone wants to say that is an "explanation," at least it is a very weak one. After all, why do 99.8 per cent escape the condition? The point is that when the statistical correlation is very high, we automatically assume there is a very direct causal connection between the two correlates even though we might have no knowledge of the mechanism (as is often the case in medicine). So relying on a statistical correlation is not really getting away from a causal connection in explanation. In effect the high correlation acts as a surrogate causal law. But when the correlation is a low number, we naturally would not be very much impressed by such an "explanation" because we assume whatever the connection is between the correlates, it is not a very direct one, but requires a host of other factors as well. Again the absence of causal connection or the rather scant evidence that such a connection exists leads us to think rather poorly of this "explanation."

The basis of the criticism of the context-free covering law model is that a causal story may explain the phenomenon without our having to assume the existence of certain causal laws. Even if such laws exist, they need not be explicitly mentioned to have a scientific explanation (and they can only be expressed as *ceteris paribus* generalizations). For example, suppose we explain the Cretaceous extinctions by a story involving the impact of an extra-terrestrial body with the Earth. This story is explanatory only if we presuppose dynamic laws governing what happens when a body of a specified mass collides with another body at a particular velocity, etc. Calculations using these laws showed that the particular impact off the Yucatan had the characteristics necessary to cause this event in a way consistent with all available evidence. Scientists take it for granted that their audience will already be familiar with a huge background of the sort of causal relations at work in a collision of this sort, which enables them to "tell" the "story" of the collision and its effects without specific reference to the causal laws involved. They do not talk about these causal laws because it is not them that are on trial here; it is the explanation, which makes use of them, that is at issue in this context. So in science we may explain particular events as the causal results of other particular events. We do not need to appeal to a covering law to have a satisfactory explanation. Of course there may be underlying laws, but we do not have to know them, and if we do, we do not have to involve them. The context of the explanation determines whether or not they are needed for this or that particular account given to this or that particular audience.

One may challenge the covering law model beyond pragmatic reasons by arguing that we sometimes have causal explanation without causal *laws* as long as we can point to causal *processes*. The causal laws that might underlie the Cretaceous extinctions were not without exceptions.[24] Dinosaurs and pterosaurs were wiped out but some birds and some mammals survived (perhaps because they nested or lived underground.) In this case no universal law or a set of universal laws were in place. But does it still make sense to insist that these extinctions were a result of the manifestation of a manifold of causal laws and we have to refer to them in order to have a complete and genuine explanation? The explanatory situation complicates because the impact of the bolide was apparently not the only causing event. There were many other causal factors such as the Indian Dacca Traps that may have significantly contributed to the extinctions. In this case the explanation is a causal story appealing to a long series of many different causal mechanisms.[25] Acting together in the circumstances, these mechanisms caused the extinctions of dinosaurs and pterosaurs but not all birds or all mammals. But how these mechanisms played together and how they enhance one another was partly accidental and not something that can be described by a nomological abstraction. The recognition of the many causal processes that participated in the extinctions gives us the causal but complicated story. So in this particular case the causal explanation does not get its credibility from following the covering law model but from our ability to point to causal mechanisms. The deductive-nomological model of explanation seems to be useful only in those cases where we can consider the explanandum-event as being part of a very isolated, well defined, and idealized situation.

A third objection raised many times against the covering law model is that there is no symmetry between prediction and explanation as the model implies. Scriven's two classical examples show that we may either have prediction without explanation or have explanation without prediction. A sharp drop in the barometer reading may invariably allow us to predict that a storm is building up, but we don't say that the drop explains the storm. What we wish to maintain is that the low atmospheric pressure explains both the storm and the drop in the barometer reading. On the other hand, syphilis explains paresis, simply because it is the only cause of this illness, but only very few people who contract syphilis will develop paresis. Thus, given a case of syphilis it is not possible to predict paresis with absolute certainty.

Hempel and proponents of the covering law mode responded to these criticisms by accepting the counterintuitive view that explanation goes both ways; that is, from the height of the flagpole to the length of the shadow, and vice versa. Hempel also held that the examples where we appear to explain something without introducing a covering law always rely on the presence of a suppressed law.[26] And, finally, Hempel argues that since paresis doesn't follow invariably from syphilis, there is no deductive explanation and hence no deterministic predictability.[27] In the reversed case Hempel seems to admit that one cannot explain the storm by the barometer reading, but only because not every drop in the barometer reading is followed by a storm.

However, Hempel's rejoinders are unsatisfactory. His response to the example of paresis provides us with a convincing answer to the question of whether scientific explanation always needs a law. He is right that a probabilistic law, even one of very low probability, permits probabilistic predictions. But since the general statement in the so-called *explanans* ascribes a low probability to the *explanandum*, on Hempel's criteria the argument fails to qualify as an "explanation" even though we know that there is a causal relationship between particular cases of untreated syphilis and paresis.

There are also good reasons for scepticism about associating 'understanding' and 'rational expectation,' since they are distinct notions.

Several critics have pointed out that one can predict that a particular phenomenon will occur under certain conditions without having any understanding why it occurred. Such predictions may use so-called "indication laws"; for instance, the drop of the barometer indicates the rise of a storm, and Koplick spots indicate the present of measles. Quantum mechanics provides precise predictions but little understanding; for example, radon 222 has a half-life of 3.8 days. From this decay law we can rationally expect that after 3.8 days in a particular sample of such an isotope half of the atoms will have decayed into isotopes of other atoms. But this expectation, while a deductive consequence of the decay law, provides no understanding of the phenomenon in question. Thus we have rational expectation without understanding.

In fact, the opposite seems to be the case as well; we can understand a phenomenon even though it was not rational to expect it. For example, the probability for a certain incident to take place can be very low, and so it would not be reasonable to expect it, but we still are able to understand why it happened when it does. Only 10 per cent of smokers get lung cancer, but when a doctor diagnoses a smoker with lung cancer, he can justifiably say he understands why this phenomenon occurred. The conclusion is quite clear: understanding is not identical with rational expectation.

Thus the widely accepted deductive nomological model of scientific explanation that sprung from Hempel and Oppenheim's original hope has met with important challenges. In my opinion one of the most important reasons it fails is that it proposes such a narrow definition of explanation – so narrow that it unproblematically applies to only part of physics. The logisistic orientation of his times led Hempel to assume both that explanations are clumps of propositions and that explanation reduces to the purely deductive relationship between the *explanans* and the *explanandum* statements. The death of positivism shows such a meagre diet was insufficient to sustain philosophy. But with it that whole *modus operandi* of doing philosophy of science died, and a new climate, which took a pragmatic, contextual approach to explanation, arose to replace it.

By making laws essential to explanation, the covering law model automatically associated scientific explanation especially with the *natural* sciences, where reference to laws is undoubtedly quite common. But I think a theory of explanation that cannot account for research practices in the social sciences and the humanities is entirely inadequate. Even in the natural sciences, if we want to characterize explanatory practice correctly, it is a misrepresentation to think that explanation merely subsumes phenomena under laws. Therefore it is reasonable to demand that a theory of explanation gives an account of our explanatory practices in the natural sciences, the social sciences, and the humanities, as well as outside of science. No argument has ever proved that the logic of explanation in everyday life differs from that of explanation in science.

6
Causal Explanations

Our daily actions rely on causal connections so instinctively that we hardly notice that we must presuppose these connections in order to get a successful result. The tacit presupposition of the causal connectedness of phenomena takes place beneath the level of our immediate consciousness. But when observed phenomena deviate from normal, we may reflect on why things were different and express this in causal statements. So in our daily way of thinking causal understanding allows us to see the world as well-ordered and structured. Causal discourse expresses this understanding allowing us to describe what we see as non-accidental relations in both science and everyday life. Moreover, causal claims made in science do not differ in kind from those made in our daily lives, but in science causal discourses require more extensive and detailed theoretical commitments and are justified by more sophisticated means of observation than in daily life.

We search for causal understanding because our grasp of the world through causal concepts allows us to experience the world as structured by causal connections between phenomena. This causal scheme of thinking developed over eons of evolutionary history, going back to a time long before our progenitors became human. Together with other cognitive schemata 'causality' was induced into higher organisms by cognitive adaptation to their environment. Hence, the capacity of causal thinking is not learned by each individual. Rather, this capacity for causal thinking is inherited, acting as a cognitive vehicle that is stored in our neuro-body genetics carried over from our forefathers. The reflective mind then adds on the top of our adapted causal understanding its own conceptual contribution in order to make a sophisticated distinction between regular co-variations (correlations) and regular causations. In Kantian terms the notion of causation is *a posteriori* for the species but *a priori* for the individual.

Today we acquire causal beliefs directly by observing singular causal facts: just as we perceive particular things and events as having certain properties, so also we see these things and events as causally connected and participating in causal processes. Thus I hold, *contra* Hume, that causal connections are directly experienced. We obtain causal beliefs immediately when something acts directly on oneself. However, during human history a more sophisticated notion of causation has been abstracted from the regular succession of our own actions and from our ability to intervene predictably in observed regularities in our surroundings. The innate causal schema leads us to gain causal beliefs whenever we experience daily things in the right circumstances.[1] But contingent regularities, in the form of contiguity and co-variance, do not exclude accidental correlations. When we perceive cases of causation, the phenomena produce in us a causal belief due to the causal notion that was originally obtained by our direct awareness of our own action and agency. But, in contrast to ordinary knowledge, most scientific knowledge about causal processes and causal mechanisms is not something picked up by the naked eye or by instruments, but is based on a vast amount of data and model-based inferences. Here the scientific mind adds further constraints to those of regularity, contiguity, and co-variance.

We could not act in a world where we did not know – or had no quite reliable expectations about – how things would behave in this or that situation. Knowledge of causes allows us to plan actions whose result we wish to bring about or avoid. Knowing causal processes operating in experimental apparatus is a precondition of successful experimentation and the manufacture of technical devices used in experiments. Thus, as a matter of fact, most causal beliefs can claim to be genuine knowledge about the world because when they are employed, especially in the design of experiments, they have been extensively confirmed by everyday experiences.

1 From embodiment to modal reflections

If we accept that the naturalistic notion that causation had its origin in the biological evolution of higher organisms and that our concept of causation arises from an innate cognitive schema due to our ancestors' interaction with their environment, it follows that the adapted sense of causation preceded the much later acquired conscious ability to apply causal schema effectively. So the sense of cause and effect came into the biological world millions of years before *Homo sapiens* developed science and advanced technology. The *reflective* sense of causation developed only when human beings began to impose their innate schema on causal processes completely alienated from the human body and its sensory interaction with the environment. This has happened with understanding natural powers or supernatural powers, seasonal changes, the movements of the planets, or if we look into the recent history of mankind, in connection with scientifically described phenomena. A naturalist account of causation has implications for a proper understanding of causation. The general notion of causation goes far beyond any particular science. It is a cognitive schema for understanding all phenomena that fulfil some common domain-independent criteria that do not depend on realizing a certain type of processes. It makes no sense to claim that causation has a particular physical meaning, biological meaning, or economic meaning that assumes that the notion of causality refers to one particular kind of processes in these fields. Causation can have many different manifestations depending on the context in which a causal judgment is framed. The only requirement is that there exists a relation between facts, events, or phenomena that obeys the criteria necessary for applying the embodied notion of causation.

As Hume observed, any causal process is a series of contiguous events, but 'causation' also entails the series of these types of events also has to be regular; that is, the same type of events has to succeed the same type of events. When our experience reveals nature behaving repeatedly in the same pattern of regular succession of events in like circumstances, when these circumstances prevail, we come to expect the latter event as soon as we experience the former, which Hume made the empirical basis of forming the belief (or making the judgment) that these events are causally connected. The same behaviour is found among cognitively developed animals; for example, the case of the ravens' hunt of red crossbills mentioned by Marzluff.[2] An analysis of the various steps in the ravens' behaviour reveals that their notion of causation consists of more than the simple idea of regular succession among types. Their notion rests on the following criteria: (1) a certain type of action brings about a certain type of effect in the proper circumstances, i.e., a specific type of action is effective only with respect to a particular type of environment; (2) a certain type of action prevents a certain type of effects to occur in the proper circumstances; and (3) certain types of events have causal priority over other types of events. The ravens' particular actions were causally successful in this particular context only because there was a corridor between two buildings and one of them had a huge window pane at the end of the corridor. The male's chase of the crossbills caused them to attempt escape through the corridor towards the window, and the females' blocking their escape route made the crossbills turn left into the window pane instead of right to freedom. The action of the male and the female were not only both necessary but together sufficient for killing the crossbills. The ravens had learned a causal connection, so they could bring about a desired effect. Without a robust concept of causation, we could not explain the ravens' ability to adopt a stratagem guided by their intentions and execute this strategy in a very particular environment. They could foresee what was sufficient to reach their goal in the particular circumstances, and how their actions could realize the possibilities they wanted to happen.

The evolutionary story explaining causal schema tells that through variation, selection, and retention, vertebrates' interaction with their environment developed innate cognitive schemata of understanding. Initially vertebrates automatically interacted with the world without any form of understanding. Eventually natural selection favoured the capacity to make a behavioural distinction between rewarding actions and unrewarding ones. The final steps seem to be the selection of organisms that were able to reinforce the behavioural tendency to rewarding actions by classifying them as a type and connecting them causally with a type of goal. Naturalists adopt a functional, non-mentalistic definition of a "concept," according to which we can attribute concepts to animals – even invertebrates – as long as they can distinguish types from tokens.

The embodiment of causal information in individuals depends on both of two things: (1) the kind of environment to which the individuals are adapted; and (2) their ability to learn from other individuals acting in their environment. Hence the individuals learn which type of action is efficient in producing a certain type of effect in a certain kind of circumstances by activating their innate causal schemata, instantiated in each individual as bodily experienced notion acquired via learning by doing. The schema itself functions as a pre-conscious organizational schema of understanding that the individuals have inherited as members of a particular species.

Indeed, higher and very complex organisms like ravens will also frequently experience an accidental, non-causal series of events. How did organisms originally acquire the ability to distinguish between accidental and non-accidental regularities? To understand causal relationships they had to develop a mental capacity of abstraction and construction. In order to be able to distinguish causal relations in their environment, which seems to require a sense of modality, the organism had to be able to subtract features, and then to add other features. Abstraction requires mentally removing those features from their behaviour and environment that tie them to the actual circumstances. Having acquired the ability of subtracting features from actual actions and events, an organism acquired the notion of sorts, i.e., it was able to identify its actual action and effect to be of a type similar to previous actions and effects.

Yet to know what is causally dependent on what, in order to be able to distinguish between accidental and non-accidental regularities, the organism had to be able to add further features to the actual events that would make a difference between them in possible but not actual situations. An organism must be able to remember situations where its actions were successful in producing a desired effect and compare them to situations in which they were not successful. So to develop a useful concept of causation, an organism had to be able cognitively to remove causally irrelevant features of the earlier circumstances, and be able to imagine the circumstances necessary to carry out similar actions in the future. An organism's action would appear non-accidentally, i.e. causally related to an effect if, and only if, it knows that in similar circumstances its action could bring about the desired effect and that abstaining from acting would not produce the desired effect.[3]

To obtain a functional concept of causality an organism must be able to grasp that actions and events similar to the actual ones exist at other places and times than here and now. Causal dependency implies more than mere succession and contiguity of action and its outcome, but no organism directly observes the modal features essential to the embodied notion of causal dependency. An organism cannot experience that an action would bring about a certain effect before it occurs. The modal belief that a possible action *could* bring about a certain kind of effect in the appropriate circumstances is not empirically accessible in any immediate way. The expectation (or belief) that in given circumstances a certain event will follow another event has been reached by induction from what an organism remembers about similar actions in other situations. The organism acquires the modal expectation that a certain event *necessarily* follows a certain action because it experienced what happened in relevantly similar circumstances at different places and times in which it controlled similar events or intervened in their succession. Thus, the abstracted and constructed features disclosed to animals through control and manipulation of their environment became the modal features which humans identify with understanding of causes, because even though the non-accidental patterns were observed only for past events, they are generalized to apply to any present or future cases of causation under relevantly similar circumstances.

The embodied notion to make causal connections is a functional notion, so the criteria for an organism to experience its action as causal are functional criteria. A certain kind of bodily movements can act as a cause of a certain kind of effect if, by carrying out an action of this kind in the proper circumstances, the organism is usually successful in making its intended effect occur. As we just saw, a direct line runs from the functional criteria of causation to a modal characterization of causation. Bodily manipulation and intervention with objects in the environment, together with mental abstraction and construction, allowed organisms to form a useful concept of causation. The same criteria that led cognitive evolution to embody a notion of causation as a cognitive schema of understanding allow us to determine whether the concept of causation applies to processes alienated from our bodily actions or perceptually unobservable. After our predecessors developed a reflective consciousness, the already embodied notion of causation began to be consciously applied beyond our own actions and our immediate environment, ultimately covering all kinds of things on order to gain understanding through experiential coherence and predictability. Thus, the reflective and self-aware consciousness automatically draws causal connections in what it experiences because evolution induced this form of comprehension is in our cognitive apparatus as a basic schemata of understanding. When we can ascertain that the events satisfy the criteria of being causes and effects, we cannot but ascribe causation to that connection.[4]

I believe that when people appeal to causal intuitions, they are hinting at this embodied understanding of causation. But philosophical reflection and deliberation also add some features to our conception of causation: Consider once again the ravens' hunt of the crossbills, but from *our* reflective perspective. We regard the kind of action that a raven uses to bring about a desired outcome as that which, in the circumstances, necessitates the effect, and the action existing independently of the outcome is claimed to be necessary for the effect in the circumstances. Designating an actual set of actions by c and a particular crossbill's hitting-the window pane by e, we take "c *necessitates* e" in a strong sense which says in the actual circumstances this particular e would not occur if the ravens did not do c. We believe this because of our observation that under similar circumstances these ravens reached their goal whenever they performed an action of a type like c. But we also believe c is *necessary* for e in the strong sense defined above. Indeed, this *necessity* is an unobservable modal feature we constructed from observations of similar cases where an event of the same kind as e did not occur unless an action of the same kind as c occurred. If we shot the ravens, no crossbill would have hit the glass. Such observations of similar but numerically different cases eventually have become embedded in our reflective notion of causation as the modal features of an actual, non-accidental succession of events. Hence we readily ascribe counterfactual implication to causal beliefs as part of our reflective understanding of our experience with similar types of events other than the actual events under consideration.

2 Causal explanation as reflective understanding

Causation is not the same as explanation, but it is the ontological counterpart to causal explanation. A causal explanation expresses *our* understanding of particular causal relationships in nature. Nevertheless, our explanatory interests determine what we consider a "cause" as long as the *explanandum* fits some partly innate criteria of causation. As a schema of understanding the notion of causation gives direction to cognition without determining its content. The content always must fulfil the criteria constituting the embodied notion of causation. However, these criteria are only partly due to the biological organism's interaction with its environment. They combine the organism's interactional behaviour with its ability to think abstractly.

When our predecessors developed a reflective consciousness, the embodied notion of causation came to be consciously applied to other things than the organism's own actions. After they acquired a language for expressing their thoughts, they could use their causal notion of understanding to explain explicitly what happened in the world by an appeal to causal connections. Thus "causality" came to refer to objective processes in the world; it denotes an extensional relation between events holding regardless of how the events are described. While the concept of causation stands for a natural relation holding between particular circumstances, events, and facts, the concept of causal explanation is concerned with a person's beliefs about causal circumstances, events, and facts. Hence "causal explanation" signifies an intentional relation. Thus, in contrast to cases of causation, causal explanations are sensitive to how we think about and describe the world. Therefore they may include pragmatic or subjective elements, but they are meant to give us understanding of what we take to be objective causal connections in the world.

A causal explanation answers why a phenomenon, X, occurs (or a similar question such as a what-caused-X-question) by appealing to a causing event, C, that brought about X. Although Hempel believed not all explanations fitting his model were causal explanations, as he put it: "causal explanation is a special type of deductive nomological explanation," but not every nomological explanation is causal.[5] As we have seen, scientists often tell causal stories about singular phenomena without reference to causal laws that would turn the stories into deductive nomological arguments. Hempel would certainly respond that such causal stories should not be regarded as *complete* explanations, but are incomplete or partial explanations.[6] But even if we believe, for the sake of argument, that causes are law-governed, causal laws always have *ceteris paribus* provisos. Since it is impossible to know every relevant condition, an explanation has to be judged as "complete" with respect to the particular context in which the explanation is produced and not with regard to some ideal demands. The Hempelian account is not merely an ideal; abstracted from a context, it is strictly speaking an unattainable ideal.

Scientists use theories and models in their explanations, but these references to abstract structures and relations should never be considered as statements expressing causal relations. Outside the hard-core of the natural sciences, in the social sciences and humanities, singular causal explanations are more the rule than the exception. Proclaiming that there must be a law that always can be generalized from the singular cases must be considered an empty gesture. Although we must accept singular causes as a genuine element in the scientific explanatory practice, it partially contradicts the received view. Nevertheless, there is a rationale for such a practice, for not only is our acquaintance with singular causation epistemically prior to that of any empirical generalization but also any appeal to causation usually provides an explanatory connection in the account of the events. In fact, the explanatory power of singular causal explanations seems to give us all the features we believe explanation should have. We immediately assume the explanatory connection reflects the causal connection between the facts themselves. We also get a highly desirable *asymmetry* of explanation: causes determine their effects; effects fail to determine their causes. Pointing to a cause gives us the explanatory information we need. And, finally, offered causal explanation is indubitably *relevant* to the explanation-seeking why-question. The cause seems always explanatory relevant for its effect, since the presence and absence of a cause makes a difference in the way the world is: the effect would not have occurred unless the cause had occurred, and an explanatory question addresses the effect as the *explanandum* for the existence of which an *explanans* is requested. There are always many causes, some immediate and some more distant. What we regard as "the" cause depends on the context in which the question is raised and the nature of the inquiry.

Causal issues are first and foremost the immediate objects of investigation in science. But based on claims about particular causal connections, science also aspires to general causal claims by determining the recognizable and recurrent circumstances under which one type of phenomena causes another; that is, whether or not the relationship between singular phenomena obtains in general. Phenomena do not appear in isolation but are habitually assumed to occur always in relation to other phenomena. Some of these seem to co-occur accidentally, but others do not. What is usually called our "world-view" makes *most* cases of accidental correlations immediately obvious as accidental. It doesn't take a trained scientific mind to see that "all the coins in my pocket are copper" is an accident that could be otherwise. But this shows that, strictly speaking, the accidental/lawful generalization distinction cannot be made outside of a presumed ontological context.

It is common to regard causality as essential to differentiating between accidental and non-accidental happenings, and between the actual and the possible. Not only do we use it whenever we try to grasp why something new and unexpected happens, but also our immediate experience in daily life is conceptualized in terms of causality. Habitually we assume tacitly that thousands of causal relations hold between our bodies and the world and among objects in the world. No wonder that so many scientists and philosophers associate explanation with causation. When we can show that one event causes another event, we can use this fact to explain why the caused event happened, and this explanation is certainly relevant because the cause is defined as that which *brings about* the effect. Thus, "explanatory relevance" becomes equivalent to "causal dependency," which defines causes as necessary and sufficient – in the specified circumstances – for their effects.

A final feature of causal explanation should be mentioned here – a feature that seems to characterize explanation as such – is that it is an answer to a question like "Why e instead of f?" rather than simply "Why e?" A causal explanation attempts to eliminate alternatives to e by implying that *only* e could have occurred given the cause c and the circumstances O.[7] To be sure, I have no intention to deny that science most often provides us with causal explanations because there are objective causal processes. But I hold that causal statements and their substitutes are partly context-dependent and therefore that our use of causal explanations depends on our interests as much as on objective matters. The contextual nature of causal explanations implies that the deductive-nomological model of explanation does not serve them well.

This chapter should aim to solve two puzzles. The first is to specify what characterizes causal explanation, and the second is to say why reference to a cause explains its effect. These two puzzles are not independent; an answer to the second question partly determines the answer to the first. The short response to the second question is to say that causal explanations give us causal understanding. I have claimed that the cognitive capacity of causal understanding is not limited to humans but also belongs to higher animals. It is so essential for predicting, preparing for, and preventing what might happen, for humans as well as for some of our social fellow creatures, that sometimes it becomes important to share this comprehension with others of our species. This seems to tell us that causal explanation expresses causal understanding.

However, this short answer is not the whole answer. It raises two more challenges: one is what is 'causal understanding' in science? I have argued that it consists at least of a grasp of the regularity between types of events so that if one is present, we expect the other will follow. But it may also involve a weak notion of co-variation between types of events, but that requires a rather strong conception of co-variation to distinguish accidental correlations from genuine causal regularities. The second challenge is that if causal understanding is nothing but a comprehension of regularity between cause and effect, then causal explanation in science seems to involve much more than mere embodied causal understanding. Any causal explanation may involve many other causal beliefs that are part of a complicated view of the world. Assuming that causal *understanding* is embodied, but causal *explanation* is part of our reflective cognition, and precisely because of this, causal explanation may add something further to the embodied notion of causation. The part reflective cognition adds is important in accounting for causal explanations in science. When we understand and accept a causal explanation, we become *reflectively aware* of *the* causal principles that are explicitly invoked, but we're also *pre-reflectively assuming* numerous other causal connections – all those pertaining to the tacit *ceteris paribus* assumptions. Indeed since we can have causal understanding only relative to our world-view and what it permits about the sorts of things that can be the causes of other things, the huge lower part of the iceberg of causal assumptions involved in any causal explanation exist only at a pre-reflective level.

Various philosophers have suggested different characterizations of a causal explanation. We have already scrutinized the account of causal explanation within the standard covering law model. But there are several other proposals. John Mackie argues that an appeal to 'c' causally explains 'e' in case c and e belong to event-types C and E, and C is an inus-condition for E.[8] David Lewis holds that 'c' causally explains 'e' if c participates in the causal history of e, and that a causal explanation provides information about the causal networks in which both c and e take place.[9] Wesley Salmon thinks that 'c' causally explains 'e' if c is connected with e though a continuous causal process, where a "continuous causal process" is one capable of transmitting a mark or carrying a conserved quantity.[10] In earlier works I have argued that 'c' causally explains 'e' if c produces e in the circumstances, and only if c has causal priority to e in the actual specified circumstances. Our causal knowledge of c's efficacy to produce e comes from our discovery that 'effects' (e-type events) are manipulable via their 'causes' (c-type events).[11] Hugh Mellor holds that 'c' causally explains 'e' if the circumstances make e more likely to occur given c than without c, and that chances, expressed by single case probabilities, are real features of the world, so single causes are probabilistic rather than deterministic.[12] James Woodward claims that 'c' causally explains 'e', if in cases when c were manipulated by intervention, e would change too.[13] And finally, Stathis Psillos maintains that 'c' causally explains 'e' if c and e fit into the nomological structure of the world, which is determined according to the Mill-Ramsey-Lewis view of laws as those regularities which are members of a coherent system that can be represented by an ideal deductive axiomatic system fairly balanced between 'simplicity' and 'strength.'[14] These suggestions all have in common that there are always certain external constraints acting as standards for the correctness of a causal explanation.

I do not intend to discuss these proposals in greater detail. Some of them are quite general, applying to all forms of causation; others are oriented towards physics in particular. But all attribute some mind-independent feature to causation, other than regularity, which a causal explanation has to grasp. Some proposals consider causal understanding to be a form of apprehension that can be expressed by a co-variant relation among event-types; others avoid any introduction of counterfactual reasoning in order to generalize causal understanding to physical processes or causal mechanisms. However, each proposal has its virtues and vices depending on the explanatory context. 'Causality' originates as a category in daily life experiences, which has no 'natural' scientific interpretation. What scientists consider characteristic of a causal explanation varies from subject to subject, and what scientists regard as the objective content of causal relation depends on the particular case they want to explain. By representing things in terms of 'types' and 'tokens', causal explanation expresses an embodied type of notion, which can be tokened in various ways depending on our interests. The specific formulation we give of our causal beliefs is context-dependent, and a causal explanation rests on this formulation. But this doesn't imply that our general causal understanding expressed by that particular causal explanation does not as share formal features as a type of schema with other causal explanations.

Thus, I see no reason why any of these cogent proposals should not describe what a group of scientists, working with a particular problem, might have in mind when they provide us with causal explanations. Scientists working in the context of the discovery of the Higgs particle will generally be looking for interaction processes and conserved properties as sufficient to pinpoint a particular causal explanation, whereas climatologists will attempt to give a causal story in which they connect many different sorts of causal mechanisms. For particle physicists 'causality' is often identical with the interpretation of scattering processes and the application of conservation principles, but for climatologists 'causality' is identical with the trillions and trillions of processes that interact with each other in the atmospheric system from the transportation of energy from the Sun down through the atmosphere to the Earth's surface, processes involving all sorts of damping and amplifying forces. Let us see what else we can learn by elaborating on causal explanation in physics and biology.

3 Causation in physics

Causal explanations in science reflect the way in which different scientists use their causal understanding and how they think we can determine the existence of causal relations. Describing causes and causal processes depends on the vocabulary of the particular science and the phenomena under study. As long as a certain set of phenomena obeys a relationship that fits our cognitive criteria of causation, many different formulations may express a causal relationship. These cognitive criteria are also those that on a practical level have led to advanced methods of finding causal relations involving statistics and probability measures.

In physics, four different suggestions for identifying a causal process have usually been put forward; it can be identified with (1) the transfer of (positive) energy; (2) the conservation of physical quantities like charge and linear or angular momentum; (3) the interaction of forces; or (4) microscopic interactions in the framework of quantum field theory. Although I have defended (1) in attempting to understand backwards causation, I hold that none of these four proposals characterizes the true nature of causation in physics.[15] Which one is more to the point than the others can only be determined in a particular context of discussion. If we distinguish between 'causes' and 'causal processes,' these four suggestions don't even exhaust the possible understanding of causation in physics. A cause may initiate or trigger a continuous process but not itself be part of the process it starts. And an effect is often the termination of a continuous process.

We often distinguish 'causes' from 'causal processes' for analytic reasons. Causes are distinct from their effects, so their relationship may give rise to laws. In contrast, causal processes involve a dynamical replacement of some entities in space and time. This example from quantum mechanics illustrates this distinction: When an electron in a hydrogen atom is in an excited state, there is a finite probability that it transitions from a higher energy level to a lower one within a very short interval, emitting a photon. It starts a process, so we can say that the transition of the electron causes the atom to emit a photon with a specific frequency. But none of the four proposals describe the actual transition and therefore the actual cause. The transition starts a causal process transferring energy from the atom to the radiative field. Yet, it is still a causal law that electrons "jumping" from a higher to a lower state emit photons with a wavelength determined by the principal quantum numbers.

Another example from classical mechanics is the Moon's motion around the Earth, which is due to the gravitational force of the Earth. Without the presence of that force the Moon would move in a straight line. Thus in this context it makes sense to say that the Earth's gravitation causes the Moon to orbit it, because it is the gravitational force that makes a difference: the Moon would move uniformly in a straight line, if it were not attracted by the Earth. The process is the spatial-temporal motion of the Moon, but the cause is the deflection of the course from a straight line. It is questionable to reduce the mechanical description of the Moon's movement to a process of interaction. Processes may change over time, but changes are what causes do.

The two examples have little in common. Transition in an atom and gravitational force are different kinds of things, but both count as causes because both make a difference. Thus, I agree with Nancy Cartwright when she claims "there is a great variety of different kinds of causes and that even causes of the same kind can operate in different ways."[16] We both have learned from Elisabeth Anscombe that "the word 'cause' itself is highly general...I mean: the word 'cause' can be *added to* a language in which are already represented many causal concepts."[17] The problem is not that various philosophical accounts of causation fail to account for some cases, but that among supporters of one there is a hegemonic tendency to extend their particular theory of causation to the explanation of *all* phenomena. However, I am less enthusiastic about another claim Cartwright makes:

> The problem is not that there are no such things as causal laws; the world is rife with them. The problem is rather that there is no single thing of much detail that they all have in common, something they share that makes them all *causal* laws. These investigations support a two-fold conclusion (1) There is a variety of different kinds of causal laws that operate in a variety of different ways and a variety of different kinds of causal questions that we can ask; (2) Each of these can have its own characteristic markers; but there are no interesting features that they all share in common.[18]

I agree with her that causal laws are many and various, and that they are used in many different ways, but I am more dubious of her claim that there are no interesting features they have in common. The above examples show that they have some functional features in common, perhaps not many, but still enough to prevent causal explanations from being equivocal. The transition of an electron and the gravitational force of the Earth share in common that they bring about changes, and their discovery helps us at least to understand nature. The use of causal terms obeys certain minimal criteria implying that we are ready to commit to a counterfactual discourse. Materially causal laws may be of many very different kinds, but functionally they are very alike.

4 Causation in biology

Also what counts as causation in biology changes with the context. The kind of analysis biologists find explanatory depends on epistemic and pragmatic considerations. They choose the form of explanation depending of their research interests, and then they find the causal relationships appropriate to these interests, i.e., they find the relationship among the phenomena under study that fulfils the criteria of causation and serves their research interests. These interests may concern

150 *The Nature of Scientific Thinking*

genetic, evolutionary, physiological, neutral, or behavioural issues. For some biologists analyses at the level of genes will tell many things about the causes in which they are interested. But this does not imply that an account solely on this level would be sufficient for understanding more complex causal relations. Other biologists might see organisms as complex self-organizing systems to which 'agency,' functional part-whole relations, historicity, etc. can be attributed. These scientists argue correctly that much about organisms and their environment can only be understood on this level. Such an approach indicates an organismic or mereological point of view. I agree with John Dupré that on the *ontological* level methodological reductionism is not confirmed by actual scientific practice in the biological disciplines.[19] It is merely one out of many fruitful approaches but not "the only game in town." So let us illustrate the last claim with some examples of alternative approaches.

Recently debates about explanation and causation in biology have been related to questions over descriptions of *mechanisms* as the prototype of modern biological explanations. The so-called New Mechanistic Philosophy is a reformulation of biological explanations in terms of mechanisms, but without the reductionist tendencies of earlier accounts of mechanism. An important contribution is the MDC-model formulated by Peter Machamer, Lindley Darden and Carl Craver, which suggests that mechanisms are entities and activities organized such that they produce regular changes from start or set-up to finish or termination conditions.[20]

The MDC-model makes entities and activities interdependent; thus, it focuses on both properties of entities and their activities and processes. Causation results from entities carrying out activities. This characterization of causation in biology is in stark contrast to earlier formulations in terms of causal laws, such as that of Stuart Glennan who earlier defined a mechanism as "a complex system which produces that behaviour by virtue of the interaction of a number of parts according to direct causal laws."[21] But the MDC-model claims that such "direct causal laws" are rarely found in biology, and the intelligibility of mechanisms is not directly reducible to their regularity but involves higher-level productive activities in inter-level explanatory models.[22] Nobody would deny that this model provides some rewarding insights into causation in biology, but it is doubtful that it can cover all forms of causation from genes to organelles to cells to tissues to organs to whole organisms. The problems are at least twofold: Are there always definite entities involved whenever biologists point to a cause of a specific effect? Are the activities always of a kind that can be accounted for by inter-level descriptions?

Concerning the first question, Dupré argues that many biological entities, like a gene, are not stable enough to carry a mechanism. Genes do not exist as ontologically persistent material entities. A given protein cannot be equated with a fixed part of a DNA-sequence, and therefore in many contexts genes are now functionally rather than structurally defined. The difficulties of establishing criteria for identifying entities challenges a view of causation focusing, even only partly, on properties of entities. Concerning the second question, we observe that finding causally relevant entities that carry out activities also turns out to more complicated than previously thought, because there are many cases in biology where the environment must be taken into account if one wants to understand the activity of a system of which the mechanism is part.

Also the MDC-model does not clearly distinguish between causes and causal processes. Sometimes such a distinction is not explanatorily relevant, but from a conceptual and ontological point of view it allows us to introduce other explanatory approaches. Hans Reichenbach talked about processes, like a beam of light, as a series of genidentical events supported by the same objects that do not change their properties over time. In contrast the MDC-model sees the process as being "productive of regular changes," and I take this to mean that these changes are not merely change of spatial and temporal locus. But this conception implies that there is no room for a clear distinction between a process and a cause that changes the process. A process has to have temporal thickness, i.e., it lasts over a duration of time, whereas a cause simply needs to occur at a point in time at the beginning of the duration over which the process lasts, or in other words a cause is always external to the process. It belongs to the environment of the process, so that any change of a particular process (apart from its spatial replacement) is caused by factors external to that process. Indeed such a clear distinction is important when we want to understand the causal interaction, say, between an organism and its environment as in evolutionary biology, etiology, and theories of infections, or between a specific process and the changes that may have happened to such a process caused by other processes external to it. Only within a discourse that separates causes from causal processes does it make sense to talk about biological laws. I am far from claiming every scientific investigation aims to find causal laws, but I am saying that sometimes this kind of *ceteris paribus* generalization offers valuable understanding of regular changes of those objects it covers. Moreover, it seems that much of our knowledge of organisms and their behaviour stems from finding such causal regularities.

There are other approaches different from the MDC-model. It was assumed for years that genes caused the phylogenetic features of organisms, but it then becomes a problem with that, for instance, the same genes are present in almost all cells of an organism even though they develop very differently according to their position in the organism. The Developmental Systems Theory (DST) responds to this conundrum by holding that genes do not have causal priority, and there is no singular relationship between genes and the organism's phylogenetic development. We now understand that heredity and development are caused by genetic, environmental, and epigenetic factors. This approach holds that there are many equally important factors to consider when analysing developmental processes, instead of focusing on genes alone.[23] We must take the whole developmental matrix into consideration. A gene is an explanatory resource of developmental biology, but it is not a "resource" of the organism. In addition to genes, membranes, organelles, methylation patterns, cellular chemistry, and behavioural patterns are inherited from one generation to the next. Understanding inheritance and development requires taking into account the whole system including the environment.

Systems Biology (SB) is a related but even more general view, which emphasizes the non-linear relationships between biological variables and the importance of *organization* for understanding living systems. A central method in SB is large-scale mathematical and computational modelling, paired with an increased focus on quantitative analysis. Robert Rosen, a central figure in systems biology, suggests that instead of focusing on material causation of behaviour (properties of entities) we should be looking for principles governing the *organization* of phenomena.[24] He defined his new approach as *relational biology*, emphasizing the need to transcend the backward-looking paradigm of mechanistic analogies and understand biological organisms "quite independently of their physical or chemical constitution."[25] So systems biology moves away from entities and concentrates on dynamical processes. Progress in understanding biological systems requires choosing *functions* and not *structures* as units of analysis. Systems biology marks a reaction against earlier reductive tendencies in biology, and one of its aims is to discover and to describe the function of emergent properties of the systems.

The SB-model assumes that the properties of any level depend *both* on the properties of the parts "beneath" them and the properties of the whole into which they are assembled. Thus the SB-model introduces emergent properties of the system, leading some scientists to talk about "downward causation." But then how can we have efficient downward causation from the whole to the parts? It is difficult to see how this downward causation can fulfil the criteria of causation. The whole only comes into being by putting the parts together, but it seems to be logically impossible to bring about the whole unless its parts already exist. A related problem is that an event acts as cause only in given circumstances – but under which circumstances can the whole act as a cause? We can avoid such problems by defending only a "medium" version of downward causation, where higher-level entities are *constraining conditions* for the activity of lower levels, but there is no efficient causation from higher to lower levels.[26] Consider a huge flock of starlings dancing back and forth in the air. The form, density, and actual movement of the flock seem to constrain the flight of each individual in the sense that each individual moves according to the movement of the flock. Nevertheless, it has been established by observation that in fact starlings (and other birds, fish, and mammals) causally interact with only their six or seven closest neighbours. Thus models of flocking behaviour are described by three simple constraints which apply to the individual bird: (1) *separation*: the focal bird avoids crowding neighbours; (2) *alignment*: the focal bird steers towards average heading of neighbours; and (3) *cohesion*: the focal bird steers towards average position of neighbours. With these three rules scientists are able to simulate flock behaviour in an extremely realistic way, creating complex motion and interaction. The function of this behaviour is indeed to protect against predators acting on the system of starlings. In my opinion such a view of the part-whole relationship explains why biologists in many cases concede to functional descriptions where the actual effect is used to account for the function of the cause. In the following chapter I shall argue that functional explanations are high-level explanations of lower-level phenomena.

Considering causality at the level of human beings, one often encounters neuro-physical reductive explanations attributing to the brain properties that are definable only by considering the whole living organism. For example, it makes no sense to claim that the brain "makes decisions" and then look for a place in the brain where the decision-making faculty is located. Reasoning this way commits the fallacy of division. We cannot identify an activity in the brain with making a certain decision unless we already have a clear concept of what a decision is independently of any neurological structure. A decision is a mental commitment to one of several possibilities of action, resulting from deliberating which of them to pursue. Human beings, not brains, make decisions; indeed we could not make them without our brains, but this does not imply that it is the brain that makes decisions. When scientists understand *Homo sapiens* (and other higher animals) as a psychologically and socially influenced creature, they look for forms of causal influences at the level of societal and culturally induced beliefs. Here we have recourse to intentional explanation where we understand behaviour and agency according to biologically inherent intentions, but it is fallacious to infer from this fact that there is a locus somewhere in the organism where we might find these intentions. Generally one may say that the lower the level of organization scientists consider in a biological system, the closer they get to forms of causal explanations we find in physics. However, when they analyse higher-level relations between organs and the organism, they attempt to understand them as functional systems. Whenever they understand animals as looking for food, making decisions, etc., they get closer to the kinds of explanation in terms of which we understand human intentions and behaviour.

5 Causal beliefs about causal facts

Causal understanding allows us to produce causal explanations, which are often responses to why-questions. We ask for a causal explanation when we are ignorant of why X occurs and hope to find information that explains its occurrence. Because we consider causes as explaining the occurrence of their effects, usually, we will cite a particular *cause* as the relevant explanation. This creates a problem, for some philosophers argue that only *facts* can explain facts.[27] Taken literally this claim implies that the relata of an explanation are causal facts, thus ontologically speaking, the world explains itself, but in my opinion explanation is an epistemic notion, aimed at understanding. Causes do their job independently of how they are described, but a causal explanation can only relate things described or conceptualized in certain ways. To achieve causal understanding the description must pick out the right events or circumstances.

Once John Austin and Peter Strawson debated whether facts are nothing but true statements. Strawson held that facts are linguistically defined entities, while Austin argued that a fact is not the same as a true statement but that which makes a sentence true. I agree with Austin and those who see facts as non-linguistic entities. So here "true statements" (and "false statements" too) refer to *linguistic* entities, while "facts" refer to states of affairs *in the world*. The same fact can be described by many different statements; thus we must be able to identify the sentence and the corresponding fact independently of each other. No fact can work as a truth maker unless we are able to identify it independently of our language. This is possible, only if we have empirical access to non-linguistically determined facts.[28]

Irrespective of which of these two alternative notions of fact one adopts, I claim facts cannot be the relata of explanation. Indeed Davidson was correct in arguing that a *causal explanation* should be considered distinct from *causation* because the same event or thing can be identified by different descriptions. It seems to me that Salmon did not clearly separate causation and explanation when he claimed that the *explanans* consists of "particular events that constitute causes of the explanandum, causal processes that connect the causes to their effects, and causal regularities that govern the causal mechanisms involved in the explanans."[29] But accepting Austin's position does not imply that Salmon's attempt to provide a general account of explanation in terms of causation is sound if we assume that facts are the relata of causation. Quite the opposite. Causal explanations are fallible, but causation is not. As with other epistemic claims, causal explanations may be highly justified, given what we know, even though they may, given further experience, prove to be false. Recall global warming; in spite of all the evidence to the contrary, it may turn out that the present increase in the global mean temperature is not caused by burning fossil fuel but rather by increasing solar activity and/or the invasion of fewer cosmic rays into the atmosphere. Causal explanation is fallible only if the relata of explanation are not facts but causal *beliefs*. Facts might play the role of the explanatory relata only if causal explanations could convey infallible, demonstrative knowledge, but since causal claims, as expressions of empirical knowledge, are inductively justified, they cannot be infallible. In other words, if there were no human (or other intelligent) beings there would be no causal explanation, but there would still be causality and causal relations between facts in the objective sense.

A typical simplified situation of causal explanation in science can be described briefly as follows. A climatologist believes a particular fact, say, that the present global mean temperature is increasing, and now she wants to know what causes this phenomenon. As a climatologist, she asks herself whether it could be burning fossil fuel, which has grown rapidly in the past century or so. As part of her background knowledge, she already knows that carbon dioxide is a greenhouse gas produced in burning fossil fuels. She also knows that since the industrial revolution energy consumption has doubled many times. Her background knowledge also suggests her hypothesis that the emission of carbon dioxide might cause the average temperature on Earth to increase. Therefore she looks for further evidence by comparing temperature curves and measurements of carbon dioxide levels in the atmosphere to see whether there is a significant correlation between them. She finds out that there is; she may then look for a common cause and eventually, if she is unable to find a common cause, conclude her belief is true.

A causal explanation expresses the agent's actual causal beliefs, but at the same time causal explanation must be linked in some way to real causes and causal processes. We can do this is by introducing a kind of five-level model to describe the process of causal cognition: we have (1) causal relations in the world; then (2) an evolutionary acquired schema of how facts in general have to be related in order to count as causes and effects; this innate schema of how facts in general have to be related has become (3) inherent in our conscious thinking because of being educated in a certain world-view; this capacity of understanding may therefore give rise to (4) particular beliefs about *actual* causal connections between certain facts, if these facts are thought to obey our inherent criteria for causal connectedness, and finally we produce (5) causal explanations, i.e., statements expressing our beliefs about the links among actual facts (but not the causal relata themselves).

How does this causal account function as an explanation? First scientific reflection connects two beliefs according to the causal schema, one about the increase of the mean temperature, which we may call the "*explanandum* belief," and the other about the human emission of carbon dioxide, the "*explanans* belief," both assumed to be true. Bringing these two beliefs together under the causal schema gives her a causal understanding, which may or may not represent a genuine causal connection in the world. She can express this by statements that articulate her causal understanding. The explanation is true if her understanding is correct, that is if the organization of her two beliefs about those two facts represents correctly what is causally connected in the world. Thus, the causal schema helps create an *epistemic connection* in the mind of the climatologist between two first-order beliefs, and it may, if her understanding is vindicated, give rise to a second-order belief concerning a causal connection in the world. It is not the two first-order beliefs that stand in a mutual causal relation themselves; it is the facts which make these beliefs true that are supposed to be causally connected with each other. Therefore, a causal explanation is fallible in two ways: the explanans-belief or the explanandum-belief may be false, and therefore the causal belief would be false. Or the explanans-belief and the explanandum-belief may be true even though the belief that they are related as cause and effect could be false in case there were no causal connection between the two. Causal understanding is about the causal structure of the world but engenders epistemic connections in the mind.

6 Context-dependent relations and context-dependent descriptions

As we have seen, causal explanations are context-dependent, but so are causes themselves. An event is causally potent only if it appears in the right circumstances. For instance, climate change depends on variations in the greenhouse gas concentrations in the atmosphere. The abundance of greenhouse gases acts as radiative forcing. Strictly speaking, forcing is the change in incoming radiation with respect to time. An increase in greenhouse gases may be a source of radiative forcing. The heating and cooling of the Earth's surface affect the release of CO_2 and other greenhouse gases from the oceans or of CH_4 from the tundra. When the global temperature becomes warmer, more carbon dioxide gets into the atmosphere, and when it gets colder, carbon dioxide is reduced and restored to the oceans. So changes in the Earth's orbit (or the density of the intervening material between Earth and Sun) can trigger an interglacial period releasing greenhouse gases into the atmosphere, which will increase the global temperature. The reverse process reduces the amount of carbon dioxide forcing the temperature to cool down. A study of ice cores indicates a strong correlation between carbon dioxide concentrations and variations in the global temperature during the last 600,000 years.[30] But other factors such as burning fossil fuels may also increase the amount of greenhouse gases in the atmosphere. However, the conditions necessary for this human activity to have a causal influence on the climate include the radiation of the Sun, the gravitation of the Earth, the thickness of the atmosphere, and the present orbit of Earth around the Sun, just to mentioned few possible factors.

Pointing *to one* salient feature as *the* causing factor in the present case is very much influenced by one's interests and cognitive orientation. We say that the global average temperature is increasing because burning fossil fuels pumps a lot of greenhouse gases into the Earth's atmosphere. So in these circumstances, burning fossil fuels is counterfactually necessary for the rise in temperature, but so are many other events or states of affairs. As John Mackie observed, a cause only produces its effect in a causal field of other necessary circumstances.[31]

Mackie's observation accords with our earlier recognition that causal statements are generally true only under *ceteris paribus* conditions. A singular causal sentence like "c causes e" is true if, and only if, c and e

occur and certain factual circumstances are realized so that c causes e. Therefore, riders of various sorts seem to be parts of the truth conditions of all causal statements. Some of them can be specified in terms of positive existential claims, as we have just seen; namely claims concerning things and states of affairs that have to be present in order for a particular event to produce a specific effect. Others consist of negative existential claims, i.e., all those claims which specify things that must be absent for "*the* cause" to produce a specific effect, most of which are not specifiable at all, or at least taken as tacitly understood. For instance, other factors like the intensity of the solar activity, variations of the Earth's orbit, the cloud cover, the amount of precipitation, the rate of evaporation, the ocean currents, deforestation, and volcanic activity – only to mention the most conspicuous conditions – are also necessary for the temperature increase in the right circumstances. Their absence or their non-influential presence is required if only the burning of fossil fuel causes the heating of the Earth. The actual circumstances must satisfy the *ceteris paribus* provisos, which demand or exclude other causal factors, in order for burning fossil fuels to make the average temperature rise. But in those cases where these factors are not absent or their presence cannot be deemed insignificant, it is us who single out the human production of greenhouse gases, or possibly something else, as *the one* among the necessary factors that should be regarded as *the* causal fact.[32] Usually, we select what we take as the most salient feature, the most unique event, among all the necessary factors as the event to be pronounced *the* cause. We point to burning fossil fuels as the cause of global warming, only because we believe this to be the timeliest of all the factors in question. Some factors, like the presence of gravitation to keep the atmosphere in place, are deemed not to be causes at all, but seen as standing conditions, rather than events, which by themselves cannot trigger the causal process.

Although this view is common, a causal explanation need not always reflect the way everybody sees it. For instance, forests function as carbon sinks, thus, the reduction of forests for cultivation over the last centuries may have devastated nature's capacity to bind carbon via photosynthesis. So today's wisdom that burning fossil fuel has caused the present increase in the global mean temperature is true only under the proviso that everything else remain unaltered; for instance, that the forested areas on Earth are kept constant. So an environmentalist might argue instead that deforestation has caused global warming by heavily diminishing the amount of carbon being locked up in forested land and by burning wood which releases carbon dioxide. Although there may not yet exist enough quantitative data to support such an explanatory suggestion – it would require a strong correlation between destroying forests and the increase of carbon dioxide since the beginning of industrialization – awareness of the significant impact deforestation may have on global warming has grown among the great majority of environmental scientists. My main point with this example is that when we seek a causal explanation, it is possible to *redescribe* the actual set of facts in such a way that another cause is singled out as *the* explaining factor.

Thus, the distinction between "the causing event" and "the circumstances in which causation takes place" is both epistemic and pragmatic. Among all the events in question, which one is honoured by being elevated to *the* cause depends on our interest in that particular context. The use we want to make of an explanation determines our choice of the particular feature as unique for the entire situation. Nature herself contains nothing like a distinction between what we regard as 'causes' and what are regarded as 'background conditions.' Often the factor that we consider as "the" cause we are looking for is the one we can control. We cannot do anything about the fact that carbon dioxide is a greenhouse gas or that combustion emits carbon dioxide, but we can control the amount of wood and fossil fuel we burn.

The fact that causal statements have certain contextual elements built into their truth conditions does not mean that causation is not an objective relation, or that causal sentences do not have an objective truth-value. Although we slice it according to our interests, the cake itself is not a product of our subjectivity. In the continuum of processes leading to an effect, it is us who determine which of these events to which we wish to pay special attention as causes. As we have seen, the selection of a candidate for *the* cause is limited to those necessary factors complying with the requirements of causal connectedness and causal priority. Fulfilling these requirements gives us an objective description of causation. If we can prove in principle that whatever event we pick out as *the* cause meets the epistemic criteria for a cause, we are justified in choosing this event as the cause. Whether or not the event we select *does meet* them is something that is the case independently of our actual investigation. This is the closest that we can come to what we mean by "objectivity."

The distinction between causation and causal explanation, and the fact that all causal laws are *ceteris paribus* generalizations, seems to imply that explanations appealing to causal laws must be *ceteris paribus* explanations. Nevertheless, I distinguish between a *ceteris paribus* explanation and an explanation utilizing causal laws that are *ceteris paribus* generalizations.

In the pragmatic view of explanation I am defending there are no *ceteris paribus* explanations.[33] Consider the following example: Sometimes citing causal laws is the proper explanatory response to a question like "Why did *this* lump of gold dissolve in *aqua regia*?" So we respond, "This lump of gold dissolved in *aqua regia* because it is a causal law that gold reacts with a mixture of nitric acid and hydrochloric acid found in *aqua regia*." But when pragmatic considerations are taken into account, citing a law of nature is an acceptable response only when the necessary criteria for that regularity to take place are already met. For instance, if it is a law that *C-type events* cause *E-type events* in the appropriate circumstances, and we want to explain why this particular case of *e* was caused by this particular case of *c*, then both must fulfil the necessary criteria, including the contextual criteria of being considered as cases of *c* and *e* fitting into the types *C* and *E*, such that they can enter into the causal relationship: This case of *c* caused this case of *e*. Thus, a question like "Why did this lump of gold dissolve in *aqua regia*?" is not explained by a causal law (facts are not literally explained by facts); rather, the causal law statement is part of the explanation of why this particular lump of gold dissolved in this particular *aqua regia*. This lump of gold and this particular *aqua regia* have to fulfil specific criteria of being gold and *aqua regia* for the causal process to occur. Pragmatic considerations would also include the problem that led to the explanation-seeking question. A law abstracted from the context is disconnected from the question, i.e., from the actual explanatory situation itself.

In no way am I suggesting that laws of nature do not play a role in explanations; however, laws by themselves are not explanations. Thus, there might be aspects of *ceteris paribus* clauses in some explanations, especially from physics or chemistry. It seems *always* to be the case that when we say "All things being equal" we know that is not in fact the case; there are of course "inequalities" (i.e., differences) from one case to the next, but what we are really saying is that the differences involved have such small causal influences on the effect in question that they can be "safely" ignored. Scientists often use the expression "for all practical purposes" when they say the influence of this or that possible cause can be ignored, where the word "practical" surely indicates the consideration is pragmatic. Pragmatic considerations override our background knowledge that strictly speaking the *ceteris paribus* conditions are not being met but that the best we can get is to approximate those conditions as closely as is *practical* for our present purposes. In that case, there is no such thing as "all else being equal" in the strict sense of a *ceteris paribus* clause – pragmatic considerations remove (or preclude) this generalized jump. Pragmatic considerations may very well indicate that differences between exigencies are irrelevant. In such cases, the qualification "*ceteris paribus*" does not apply to the explanation as a whole, but only to those factors deemed irrelevant via pragmatic considerations. Even in this regard, I think we get past the use of *ceteris paribus* explanations by understanding explanations pragmatically.

But before turning to the pragmatic theories of explanation we shall consider in the next chapter other forms of explanation, such as structural and functional explanations, to see why non-causal explanations are also informative, how they relate to causal explanations, and what their scientific status is.

7
Other Types of Explanations

Outside the domain of physics explanatory discourse as practised in natural sciences such as biology, physiology, evolutionary psychology, and cognitive science involves other kinds of explanations than what physics permits. Because the objects these sciences study are different from those of physics, scientists in these fields often use *functional* or *intentional* explanations. Daniel Dennett calls different commitments to various types of explanation different "stances": the *physical* (causal) stance is adopted in physics and chemistry, the *design* (functional) stance in biology and engineering, and the *intentional* stance in software and minds.[1] Dennett sees them as reflecting varying levels of abstraction – the physical stance being the most concrete and the intentional stance the most abstract. He also holds that the stance we choose in a particular situation depends on how successful that stance is in explaining objects. I defend exactly this pragmatic, context-dependent view of different kinds of explanation and prediction. Although Dennett's motives were different, nothing prevents realists from holding the same position and explaining the success of a certain kind of explanation, say, intentional explanation, by appealing to emergent mental properties.

Although it sometimes seems that the natural sciences are concerned only with general laws and regular behaviour, we should not think that this is all these sciences do. Objects of scientific explanation include both singular events such as the outbreak of bird flu in East Asia or phase transitions in the early Universe, and general facts such as that planets move in ellipses. One might object that it is impossible for the natural sciences to explain singular events in their uniqueness, isolated from the kinds they exemplify. We can only understand this outbreak of bird flu as an instance of a kind of disease outbreak pattern. But, of course, the fact is that natural scientists do explain events in their particularity just as much as do historians. The Cretaceous extinction is a particular mass extinction, but we don't understand it as merely just a mass extinction event like all the others. This one hit the dinosaurs and was caused, it is argued, by a particular bolide, and we have little understanding of other mass extinctions that may have had wholly different causes.

From an overall explanatory viewpoint, I hold that the various sciences are not substantially different despite the long tradition within philosophy of science of distinguishing sharply between kinds of explanations offered by different sciences. I call Dennett's 'physical stance,' the *causal stance* and his 'design stance,' the *functional stance* because all three stances may occur side by side in a particular discipline. Science provides us with several kinds of explanations as the appropriate response to different types of why-questions. Other than a strong reductionism, there is no reason to narrow down science to physics and chemistry. Any philosophical position should be able to account for the diversity of explanations in different sciences. Indeed, the *content* of explanation varies depending on the subject matter at issue. The explanation of a falling body is different in content from the explanation of a pendulum coming to rest. But both explanations have the same form, and were selected to be of the same type, because in both cases we want the same type of understanding. So the *form* of explanation is also sensitive to the object of scientific inquiry since we normally have different cognitive interests in different objects. The form may vary according to the kind of objects a particular field of research treats and according to the context in which an explanation is requested, mechanical, moral, or intentional, etc. I dare to believe that a satisfactory theory of scientific explanation must be able to do justice, in one form or another, to the many varieties of explanation-seeking questions.

1 Types of explanations

Many scientific explanations are answers to why-questions seeking causal understanding of the phenomena involved, but there are many other different types of answers to why-questions which make no direct references to causes. Therefore, Salmon's and others' attempts to establish a general account of scientific explanation in terms of causal mechanism is doomed to fail not only outside physics but even inside.[2] Apart from mechanical, statistical, and genetic explanations, science also produces *structural*, *functional*, and *intentional* explanations, depending on the subject matter and the stance of the scientists, shaped by their beliefs, background knowledge, assumptions, and mutual interests.

Structural explanations explain the existence of a particular phenomenon – whether it is a certain property of a particular entity or its behaviour – by referring to the structure of the system to which the phenomenon belongs. For example, to account for the behaviour of a particular individual in a group of mountain gorillas, an explanation will typically appeal to the hierarchy of the group and the individual's position in this hierarchy. This structure is upheld by intimate causal interactions (e.g. the near celibacy of young males) between the members of the group because the positions of the animals in the social order also determine the kind of interactions taking place. In chemistry we find many similar examples; for example, the benzene molecule consists of six carbon atoms organized hexagonally with three double bonds, allowing two forms of resonance. These different forms can be explained by the overall structure of benzene, because the six electrons creating the double bonds do not belong to any particular carbon atom but are delocalized. Whenever we are concerned with systemic properties that the components of a system have only because they are members of the system, scientists will appeal to the structure of the whole system to explain features of the components.

In the sciences of living nature we often find *teleological* accounts, another large group of "non-causal" explanations. Causal considerations reflect on the succession of events and explain a future state of a system in terms of its past and present state. Even in physics there are non-casual nomological explanations involving conservation principles, symmetry principles, or gauge invariances that do not appeal to any past or present state of a system as causing its future state. Moreover, in evolutionary biology, for example, we do explain the development of certain traits in an organism by reference to a future state. Such *teleological explanations* occur when biologists say that the giraffe has developed a long neck because that gives it an advantage in picking leaves in treetops. The point here is that a past or a present trait and its development is explained by referring to a future state, *i.e.* by pointing to the advantageous ability to pick leaves from high treetops. However, I contend the future state appearing in the *explanans* need not be taken as a 'teleological' cause guiding, dragging, or even determining a past or present state; such teleological explanations are used when information given by the *explanans* does not come from the cause, but the effect.

Functional explanations are a subgroup of teleological explanations, which explain traits and qualities in terms of the functional role they play. Teleological explanations in general refer to the goal of the *explanandum*; in functional explanations the goal is the functional role played by a given phenomenon. Thus, the circulation of the blood can be explained in terms of the function of providing the body with nutrients and oxygen. Also the explanation of the giraffe's long neck can be given in functional terms. So all functional explanations are teleological, but not all teleological explanations are functional.

We should notice that although functional explanations have had a dubious reputation among many scientists and philosophers of science, nowadays at least some admit them into the scientific fold. Larry Wright and others initiated a major breakthrough in this direction, which I will discuss in a while. Equally important for the renewed interest in the irreducibility of teleological explanations has been the status of functional explanations in neuroscience, cognitive science, and philosophy of mind, where many see mental states as nothing but functional states. Moreover, functional explanations also find their natural place in archaeology, anthropology, sociology, and linguistics.

The behavioural sciences make use of another subclass of teleological explanations, those that use *intentional* terms like "beliefs" and "desires" to explain purposeful behaviour. For example, if asked why John went to the drugstore, one might answer because he *desired* to buy a newspaper and *believed* the drugstore sold them. Intentional explanations refer to an intended goal as the reason for action, whereas in functional explanations the goal need not be intended. Thus, intentional explanations are invoked only when dealing with things that can be said to have intentions, i.e., conscious beings. This implies a further difference because an intended end state does not need to be realized, or even realizable, in order to appear in an explanation. But when we appeal to functions, we presuppose that the function is realizable in the light of the present state of the organism or system. I have looked at intentional explanation in the philosophy of humanities elsewhere, so I shall not discuss it here.[3]

Sometimes explanations personify 'Nature' and give it intentions, e.g. "Nature does nothing in vain," "Nature abhors a vacuum," or "Nature always takes the path of least action." We do actually say these sorts of things intending to explain phenomena, and I hold they can work in scientific explanations, as long as they meet standards of scientific rationality like testability.

"Teleological" explanations differ from "mechanical," "structural," "genetic," etc., but are these types of explanation really distinct from "causal" explanation? I think so. However, I see no reason why structures, functions, intentions, or historical processes should not be considered as "causes," simply not *mechanical* causes. Low rank in a social order, or the advantage given by a certain adaptation, or the hope in a man's mind, all can have real causal powers, and so there is no reason they are less real than the stone which broke the window. For example, social status clearly is a real cause, not to mention other human intentions, for a great deal of human as well as animal behaviour. There is no point in denying such things causal efficacy other than hewing to a reductionism born of a prior commitment to mechanistic materialism. For a strong materialist, the man's belief that his social status is low is reducible to a certain biochemical state of his physiology, and that is the real cause of his behaviour, the intention is but an epiphenomenon. But even so, he is not proposing a non-causal explanation.

Furthermore, there is a long tradition going back at least to Plato in which logical implication is assimilated to 'cause,' and 'explanation' of a phenomenon is identified with validly deducing the occurrence of that phenomenon from other propositions previously accepted. In a less empiricist age the association of causation with logical implication was literally taken for granted. This may have been a mistake, but the opposite view seems to me to be taking its denial for granted. What can be considered a "cause" and what cannot be considered as such is a matter of one's *metaphysics*, but we need not agree at a metaphysical level in order to agree on a theory of explanation. A good theory of explanation can be metaphysically agnostic as to the specific sort of being that causes have.

Nevertheless, the various types of explanation are distinct because they express different kinds of understanding. Even though structural or functional explanations may presuppose a causal structure of the world, causal explanations are not able to exhaust the grasp we want to have of this structure.

Our particular cognitive goal within a certain domain determines how we use each of these particular explanations in response to an appropriate why-question, and our cognitive goal is indeed partially determined by what we believe about the domain in question. If we believe it is true of nature that phenomena are causally connected, it will naturally be our cognitive goal for science to answer why-questions in terms of causes. Similarly, if we think that the specific feature of conscious beings is their intentionality, then it becomes a cognitive goal for the behavioural sciences and the humanities to answer why-questions by referring to intentions, motives, wishes, or meanings behind human behaviour and linguistic actions, simply because we take intentions, motives, wishes, and meanings to be the fact of the matter. And finally, if we think that the characteristic feature of some of the products of human actions consists of symbolic meaning, it is our cognitive goal to answer why-questions about, say, a work of art by an interpretation that explains its symbolic meaning, so we want the interpretation to say something true about this artefact. In each case the sort of why-explanation we choose to ask depends on what we want to know; similarly, this partially depends on what we actually believe to be true about the *explanandum* phenomena or circumstances and therefore on the kind of question we want answered.

In each case we explain a fact by relating it to another fact that we take to be relevant for understanding what we do not initially understand. If this relation were arbitrary or accidental, such an account would not work as an explanation. In causal explanations we point to the causal nexus because we take causes to be relevant as explanation of their effects. In an example of a broken window, the explainee accepts a reference to a stone as the cause, since it is part of his background knowledge that a stone can break a windowpane. So the explanation just relates the effect to its cause. But if knowledge of the causal connection alluded to by the respondent is not part of the questioner's background understanding, the respondent must offer an extended narrative story about the circumstances before she has successfully given an explanation.

Imagine that the pane broke because a fly hit it. No matter how true the response that the window broke because a fly hit it might be, usually it would not count as an explanation until more information had been added. Before it is explained, in order to accord with the questioner's background understanding, the fact must be put into a broader factual context by providing a story leading the interlocutor to understand how these facts can be causally connected in the given circumstances. This narrative discourse transforms the response into an explanation by telling that the windowpane was so fragile that the momentum of *this* fly was enough to break it, or that the window already had some cracks, or that the glass was frozen, etc.

Other non-causal responses to why-questions also become explanations by being embedded in a narrative discourse; either implicitly – if it already fits a background understanding shared mutually by the interlocutor and respondent – or explicitly in terms of a whole story – if the interlocutor and respondent are not cognitively on a par. Intentional explanations, for instance, answer why-question by pointing to the *intended* effect in order to explain a certain action. Unless the interlocutor already has the appropriate background knowledge, an intentional explanation accounts for why a person engages in a certain action (rather than another) by pointing out that this action was chosen because the agent believed it was the most effective way to reach his or her goals. Similarly, *interpretative* explanations respond to why-questions by giving an account of, say, a mathematical sign in terms of its symbolic content, relating it to certain rules or conventions. The respondent may explain her choice of a particular symbolic system instead of another by pointing to her understanding of the problem or to the general rules of a theory, which give the formulation its meaning. She explains why the mathematical formalism is as it is by saying that it presents an effective means for expressing the physical meaning of what she wants to say.

There is some controversy over whether all these examples represent genuine scientific explanations. Some people might not wish to count psychological explanations, for instance, among scientific explanations, because they claim that psychology does not meet the objective standards of proper science. Many people also have difficulties with explanations referring to end states, because these can be difficult to test empirically. And, as mentioned, some people will claim that interpretations lying at the heart of the humanities disqualify them from being 'scientific.' There are, however, good reasons to speak against such restrictive views. Interestingly enough, even physics employs teleological explanations, for a physics of indeterminism, attractors, equilibriums, and chaotic phenomena once again introduces an appeal to final states. Furthermore, the selection of a particular type of explanation by the working scientist or scholar depends on the kind of explanation she (and her scientific community) believes delivers the desired sort of understanding of the phenomenon to be explained. There is no other point of view from which we can judge the 'correctness' of an explanation; only such pragmatic considerations allow us to evaluate the diversity of scientific explanations and the scope of theories concerning these explanations.

2 Structural explanations

Structural explanations do not focus on particular individuals but on collections of them because a certain feature of an individual is explained by its membership in a class. An individual may have structural properties; that is properties that it has because of membership in a coalition or a certain class of objects. Structural properties form overall patterns in a population, which can explain facts on their own level.[4] A social position as a judge or a farmer is a structural property, defined in relation to other social positions. Many social or behavioural phenomena are explained as a result of such structural properties. A judge has the right to enter the courthouse in which he is employed, and being a judge, and not a courthouse clerk, explains why he can sentence another person to prison. We also saw how the social position of a male mountain gorilla affects his mating behaviour.

Neither the position as a judge nor as an alpha male mountain gorilla explains a *particular* behaviour, but it explains the *type* of behaviour in which any individual with the same position can engage. When a judge declares a certain individual to be guilty in a particular crime, she does so because she finds that that person is accountable for the crime. Her position as a judge does not explain why she finds the defendant guilty or the sentence she declares, but her position explains her capacity because her job gives her legal rights to sentence people to prison. Similarly with the alpha male mountain gorilla; he can mate with females in the group because of his position in the hierarchy. His position gives him a social ability to mate, although the actual behaviour is caused by the rise of sexual desire at a moment shortly earlier. Both are examples of properties, which clearly could not be reduced to a structure of possessed (non-dispositional) properties. We are attributing to the judge and the male gorilla characteristics that will be exhibited only in specific contexts where they bear certain relations to other objects in the social structure. It is possible no one will ask the judge to take the seat in the courtroom, and thus these dispositions will never be realized.

Structural explanations are found not only in the behavioural sciences; they are also found in any explanations that appeal to dispositional characteristics. In chemistry many compounds, called "isomers," have the same molecular formula but different structural formulae. Some are structural isomers; some are stereoisomers, and among them are enantiomers whose mirror-images are not superimposable. These different structures have different structural properties. For instance, one enantiomer of the compound $C_{10}H_{24}N_2O_2$ causes blindness, the other enantiomer, etambutol, is used to treat tuberculosis. The toxic feature is explained as a disposition with respect to an organism, not reducible to some possessed property of $C_{10}H_{24}N_2O_2$. Again, the reason why litmus paper turns red when it touches acid is because of its disposition to react to acid in this manner. In both cases we explain the disposition of a certain stuff in relation to something that is not part of this stuff. We often explain a dispositional property, or a certain state of affair, in terms of the causal powers which the underlying entities, the bearers of the disposition, have in relation to something else.

I believe it is futile for the explanatory reductionist to hope for more. Another example may illustrate why this optimism is illusory. Consider the explanation of many physical and chemical properties of the basic elements. Scientists describe the organization of the basic elements in the periodic system on the basis of the structure of the various kinds of atoms and the configuration of electrons around their nuclei. Electrons orbiting the nucleus within a particular range of energy make up a shell, organized in the system by the atomic number and various quantum numbers. The order in which electrons build up these shells depends on certain structural principles such as Bohr's rules, the Pauli exclusion rule, the Madelung rule, and Hund's rule. We may think of these rules as explanatory because physicists use them to explain how the atomic structure is organized. Structural rules like these "explain" how the various atoms get their chemical properties, but they don't "explain" why atomic structure must obey these rules.

The number of valence electrons in the atom's outermost shell explains many physical and chemical properties. Consider a question like: Why does helium not react with other elements? The structural explanation holds that noble gases like helium have a full valence shell, and chemical bonds usually occur when atoms can share two unpaired valence electrons. Here we may say that the fact that He is inert is caused by the fact that its valence shell (in an unexcited state) is filled, thereby according to the valence bond model, preventing other atoms from forming molecular structures with He atoms. Similarly "being a judge" is a structural property, and if I ask "Why does she have the power to sentence criminals?" you answer "Because she's a judge." So her being a judge is the "cause" of (i.e., a necessary and sufficient condition for) her having this power to sentence people. Ordinary people don't have this power, but because she's a judge, she does.

Structural explanations of chemical properties show that the use of inconsistent models is quite acceptable and therefore, like all other explanations, structural explanations are context-dependent. An alternative to the valence bond model, the molecular orbital model, assigns electrons not to individual bonds between atoms, but represents them as if moving under the influence of the nuclei in the whole molecules. For some purposes this model gives more accurate explanations of spectroscopic, ionization, and magnetic properties of the molecules, whereas the valence bond model is much more useful for calculating the bond energy. Again, a reasonable understanding of these explanatory practices in chemistry seems to call for a contextual and pragmatic account of explanation.

3 Functional explanations or evolutionary explanations?

Humans often create objects for particular purposes, endowing them with particular functions. Therefore it seems unproblematic to explain both the invention of an artefact and even more obviously, its design (the structure), in terms of its functions. But when human intentions are not directly involved in designing an object, teleological explanations seem quite problematic to the contemporary scientific outlook and reminiscent of a time when nature was seen as God's creation. In general, physics is based on a strong rejection of goal-oriented explanations. Functional explanations do not make much sense in a world in which the past appears to determine the future. It is not merely puzzling, it seems, but outright false to refer to things not yet existing in order to explain the existence of present things. Nevertheless biologists routinely talk about functions and regard them as having explanatory weight. The job of the heart is pumping blood, the function of transpiration is the regulation of body temperature, the purpose of camouflage is to protect from predators, and antlers serve as courtship signal. A question like why are gulls' fledglings grey with black patches is answered by saying that looking like pebbles protects them against predators. As an example of a functional explanation, this is a kind of teleological explanation.

How can an appeal to the function of mimicry explain its existence? The real challenge behind the use of functional explanations is that the function comes after what it is a function of. I shall argue that seeing explanation as a practice of answering information-seeking questions makes functional explanations legitimate as long as they provide relevant information. We need such explanations for more than one reason. They furnish a simple and overall insight into a certain relationship between an organism and its environment, where understanding of the causal mechanism might not provide the same kind of information; they may inform us about how a certain item is able to persist or prevail. But more importantly they are useful where no appeal to a causal mechanism is available.

One classical response to the challenge is to say that the function of any property can be regarded as a "final cause" in the Aristotelian sense; however, the notion of a final cause is ambiguous. In case the function of X for a certain organism O is Y, the phenomenon Y is taken to be explanatory because it causes the earlier phenomenon X to exist. The final cause seems to be a peculiar kind of efficient cause acting backwards in time. Nothing is conceptually impossible with a notion of reversed causation, but nothing indicates that phenomena explained by functional explanations require causes acting backwards in time.

Another response is to deny that functional explanations are genuine explanations. Hempel did not regard functional explanations as having sufficient explanatory import. He called them "functional *analyses*" because they do not qualify as *bona fide* explanations, but are effective only as heuristic devices. His main objection against functional explanations is a logical point sometimes referred to as the problem of functional equivalents. He compares functional analysis with the covering law model to see how functional analysis measures up to his ideal of scientific explanation. He notes that the kind of phenomena that functional analysis is invoked to explain are typically some recurrent activity or behaviour pattern in a group or an individual, examples of which include psychological mechanisms, neurotic traits, cultural patterns, or social institutions. The objective of functional analysis is to show how this behaviour pattern contributes to the overall preservation or development of the group or individual, which exhibits this behaviour pattern. The behaviour patterns are thus understood as playing a role in keeping a system in proper working order.

The basic problem with such analyses is best illustrated with an example. Consider the statement:

(i) The heartbeat in vertebrates has the function of circulating blood through the organism.[5]

What does this statement mean? In particular, what does the word "function" mean in this example? One might think it signifies something like 'effect', but this is mistaken because there are many effects of the heartbeat that we would not consider to be functions of the heart. For instance, the beating of the heart has the effect of making heartbeat sounds, which is hardly a function of the heartbeat. But it is crucial that the function actually contributes to the health or survival of the organism. Thus, we may try to reformulate (i) as:

(ii) The heartbeat has the effect of circulating blood, and this ensures the satisfaction of certain conditions (supply nutrients and remove waste), which are necessary for the proper working order of the organism.[6]

Certain conditions in the surrounding environment need to be met for the heart to have this function. The aorta must be intact so that

Other Types of Explanations 173

circulation of the blood is possible, oxygen can only be transported if the lungs are able to provide it, and the kidneys must work so that they can dispose of the waste, etc. Abstracting from all this, Hempel arrives at the following schema:

(iii) *Basic pattern of a functional analysis*: The object of the analysis is some "item" i, which is relatively persistent trait or disposition (e.g., the beating of the heart) occurring in a system s (e.g., the body of a living vertebrate); and the analysis aims to show that s is in a state, or internal condition, c_i and in an environment representing certain external conditions c_e such that under conditions c_i and c_e (jointly to be referred to as c) the trait i has effects which satisfy some "need" or "functional requirement" of s, i.e., a condition n which is necessary for the system's remaining in adequate, or effective, or proper working order.[7]

Hempel cannot accept this pattern because he holds that explanations are arguments. For him to explain the presence of the item i, say, the heartbeat, means to deduce i from premises about the working order and conditions of the organism, but this requirement creates a problem for the functional analysis. Suppose we wish to explain the presence of i (the heartbeat) in a system s at a certain time t. Then Hempel offered the following functional analysis of i:

(a) At t, s functions adequately in a setting of kind c (characterized by specific internal and external conditions);
(b) s functions adequately in a setting of kind c only if certain necessary conditions, n, are satisfied;
(c) If trait i were present in s then, as an effect, condition n would be satisfied;
(d) (Hence), at t, trait i is present in s.[8]

The question is whether, in this functional analysis, the conclusion (d) is a logical consequence from the premises (a)–(c).

The answer of course is negative because the argument commits the fallacy of affirming the consequent with premise (c). The conclusion (d) *could* be validly inferred if (c) claimed that *only* the presence of i could effect the satisfaction of the condition n. But as the argument stands, we can only infer that the condition n must be satisfied in one way or another (this is given by (b)), and therefore we cannot conclude that other alternatives would not suffice to satisfy requirement n, so we have not explained why *i* is present in *s* at *t* instead of an alternative. Put differently, we can assert (a) that *if i* is present in the system and the system is in normal conditions *c*, *then* the system will function normally. We observe (b) that *n* in fact obtains; that is, the system is working properly in the conditions *c*. Now, if we try to deduce the presence of *i* from (a) and (b), we commit the fallacy of affirming the consequent. Given *i* (the heart beats), we can infer *n*, that the body works normally, but given *n*, we cannot infer *i* because other items might be responsible for the orderly workings of the body.[9]

This predicament is the problem of "functional equivalents"; we identify a given 'item' as having a certain function and recognize that it is *sufficient* to produce a given result (e.g., the working of a mechanism), but in most cases we cannot also claim that the trait is *necessary* for that result because we cannot rule out the possibility of other 'items' having the same effect. Salmon gives an example to underline this point: consider the function of cooling off in hot conditions that humans and many animals have.[10] The effect of cooling can be realized in different mechanisms. Some animals have large ears from which the heat radiates (for example species of rabbits and foxes), humans perspire, and dogs pant. All are different mechanisms for obtaining the same goal, and given an animal living in a hot region, the only thing one can conclude is that it must have some mechanism of heat radiation.

How can we salvage functional explanations from Hempel's argument? His point is roughly that given the proper functioning of a system in a certain setting or environment, we can conclude only that there is some trait that has the function of satisfying the conditions needed for the system or organism to work properly. We cannot infer that only this specific trait has the function. But what if we are able to specify the settings and perhaps the trait more accurately?

Consider the example with animals' ability to cool off. Looking closer we see that there might be a very good reason why, say, dogs pant, whereas foxes living the warm regions radiate heat from their large ears. Since most dogs have fur, perspiration would make this fur wet and thus very heavy to move, giving the dog a considerable disadvantage. Therefore it is quite understandable that it should pant and not respire like humans do. However, a panting fox in a desert would lose too much precious water. Therefore desert foxes do not pant.

Similar selective constraints can be found in the evolution of rabbits. They avoid predators by moving very quietly; if they were to pant, it would make so much noise that they would be easy targets for predators. They also have fur; thus the only available cooling method seems to be through their ears. If we specify the constraints even further, we are able to explain more traits and abilities. Rabbits, or similar animals, which live in desert regions, have larger ears not only because they need to cool off more, but also because in a virtually barren desert, sound can travel over large distances and large ears allow them to hear predators far away. In forests where the climate is cooler, a rabbit (or a fox, or a mouse) need not cool off so much and sound travels shorter distances, because it is absorbed by trees, and larger ears in a dense wood would slow down the rabbit's movement, etc.; therefore, the rabbit's ears in such an environment are smaller. Thus, we seem to be able to avoid Hempel's predicament by emphasizing that *the context of the problem* restricts functional explanations in the same way as it limits causal ones. Leaving everything unchanged in the circumstances, the increased size of the ears is causally necessary for cooling off foxes and rabbits living in a desert.

The crucial point seems to be that when we explain that the rabbit living in a dry and hot habitat has large ears, we do not refer to the future advantage this will give the rabbit but by reference to the advantage it has given earlier individuals of that species. Thus, sometime in the past an animal had larger than average ears which resulted in a selective advantage in escaping predators. The presence of large ears is due to the adaptations of earlier animals, not because it will benefit them in the future. This explanation invokes a *causal* aspect; namely that the early specimen's large ears gave it an advantage (not that the future benefits will cause the larger ear). Another important point is that we cannot *predict* future development of a species (which is probably also why Hempel is wary of these explanations) because the available variations on which natural selection operates are, for all we know, random.

These considerations illustrate the view that today has received nearly paradigmatic status: functional explanations are informative because, appropriately construed, they involve references to *antecedent* causes. If the function of X is Y for a certain organism O, it is there because its ancestors did Y, and the present instance of Y is a causal result of X being there. The phenomenon Y is regarded as explaining the existence of X because its functional role is due to a complex set of historical mechanisms that sustains the fact that Y is part of an efficient cause forward in time.

However, I will pursue a third proposal that neither rejects functional accounts nor attempts to reduce them to a complex of forward causes. It holds that if the function of a property or a behaviour X is Y, the phenomenon Y is explanatory not because it is part of a *cause* of X's existence, but because it is an *actual effect* of X's existence. Accordingly, the explanatory work is not done by an efficient cause; *Y does not explain why the property X exists, but Y explains X's particular function, and therefore why Y is helping the object having property X to maintain its existence, i.e. to persist.* The point is that searching for an explanation of why X exists requires a causal account, but in searching for an explanation of the particular role X plays and how this role helps preserving the object carrying X, we are not interested in the causal history leading up to X. The lack of interest is not due to a fact that there are no underlying causal processes involved, but that they do not respond successfully to our explanatory request.

Philosophers of biology agree that a functional explanation is not informative because of the existence of a final cause. The present agreement is based on a proposal by Larry Wright, a suggestion later improved by Ruth Millikan. Wright argued that a statement of the form "the function of X is Y" means:

(1) X is there because it does Y
(2) Y is a consequence (or result) of X's being there.[11]

It may seem paradoxical that (2) tells us that Y is an effect, and not a cause of X, while (1) states that X exists because it gives rise to Y. The latter clause displays the teleological character of the explanation but, according to Wright, "because," in (1), "is to be taken in its ordinary, conventional, causal-explanatory sense."[12] His "etiological" theory of explanation proposes that the fact that Y is an effect of X is among the antecedent efficient causes of X, and therefore provides us with an efficient causal explanation of the existence of X in terms of its producing Y.

There is a certain ring of circularity in suggesting that Y's being an effect of X causes X to exist, and X causes Y to exist. But how can the fact that Y results from X have any causal efficacy in bringing about X before X brings about Y? Here we must distinguish between types and tokens. Moreover, there are counterexamples showing that Wright's analysis seems to include cases, which do not invite a functional explanation. Christopher Boorse suggested the following example playing on a similarity of explaining why X is there in terms of natural selection:[13] Assume that a scientist builds a laser. He uses a rubber hose to connect it to a source of gaseous chlorine. After turning on the machine, he notices a break in the hose, but before he can stop it, he inhales the leaking gas and is knocked unconscious. Apparently, we can describe the situation such that the release of the gas (Y) is a result of the break of the hose (X), and the break is there (continues to be there) because it releases the gas. Otherwise the scientist would have corrected it.

Ruth Millikan showed that both problems are avoidable if certain restrictions are put on the instances of Y that can explain X's existence.[14] She argued that

- X has a function to do Y only as belonging to a "reproductively established family."
- The function of a member X of a reproductively established family is to do Y just in case *ancestors* of X did Y and their doing Y causally contributed to their family's having greater reproductive success than competing reproductively established families.

Thus, a statement that the function of X is Y means:

(1*) X is there because its ancestors did Y.
(2*) Y is a causal result of X.

While Wright's analysis focused on those instances of Y that were among the current effects of X, which made it seem circular, Millikan's analysis points only to those instances among the causes of X. It understands a statement that ascribes to X the function of producing Y as an explanation of X's existence strictly in terms of preceding efficacious causes. The crux of her account is that the function of a trait is the way the trait has contributed to fitness in the past. Thereby no circular reasoning seems to contaminate this account. In addition, her proposal does not run into Boorse's counterexample since the break in this particular hose is not a copy of earlier breaks in hoses that had the same effect on scientists.

Millikan's analysis also has some further advantages. It successfully distinguishes functional effects from accidental effects. Functional effects, such as mimicry in a butterfly, has helped an organism's ancestors to survive and reproduce, and thus led to reproducing mimicry, whereas accidental effects have no influence on a successful reproduction rate. It would seem that "accidental" here would be defined as an adaptation that came along with some genetic "package" but did not itself have any influence on reproductivity, so this would be analytically true on her account.

Finally, it explains how teleology can exist in a world of efficient mechanical causes. Modifications result from random recombination or genetic mutations, producing a small variation in some trait of a certain population. If a modification in an individual gives it a certain advantage in competing with the other members of the population in surviving and successfully reproducing, this trait will be passed on to the succeeding generations and will eventually characterize the entire population.

Nonetheless, Millikan's suggested explanation in terms of causal mechanisms and reproductivity are not the same as a functional explanation, but are more aptly called "*evolutionary* explanations." They originate in biology and may be broadly construed as a species of causal explanation, but they have specific features making it reasonable to give them a separate heading in this context. Sometimes both an evolutionary and a functional explanation can be given of the same phenomenon, but this does not imply that one can replace the other. In a termite colony there is a neat balance between soldiers and workers. So if the colony is attacked by an anteater and loses many of its defenders, the system will be marked by a shortage of soldiers. The queen will thus start to reproduce extra soldiers, resulting in a later state of the system marked by more soldiers. The mechanistic explanation of this phenomenon is that the queen's food is brought into the nest by passing through the mouths of both workers and soldiers. When there is a shortage of soldiers, the normal mixture of salivation in her food is changed and the queen reacts to this imbalance by producing more soldiers. The evolutionary explanation appeals to the fact that earlier generations of this termite queen had a higher chance of continuous reproduction because their queen reacted to the change in salivation and therefore this present queen is caused to do the same thing. However, the functional explanation removes the causal history by shifting the explanatory work to the notion of a functional requirement for maintenance of a particular system. Although the queen's behaviour is required to produce more soldiers, the system as such will not be able to maintain itself unless she does so. The functional explanation requires that the talk of systems as a whole is correct by thinking of them as having properties other than those of their parts.

There is no reason to doubt the causal-mechanistic or evolutionary account. It effectively explains in causal terms why a certain phenomenon exists at all. But functional explanation is not reducible to mechanistic or evolutionary explanations. In order to explain why a particular person sweats, we may point to the actual cause or the actual effect. The actual cause may be the hot weather, and the actual effect is the cooling of the body. So sweating stops overheating and eventually death. In contrast to a mechanistic account which cites a cause C as the explaining factor of a phenomenon X, a functional account appeals to the actual effect E to give us another form of explanation of the phenomenon X. In biology a functional phenomenon can be said to be propitious for the maintenance of the organism or its reproduction. For example, some instances of mimicry help the individual to survive; other instances help the individual to produce offspring. A functional phenomenon may be beneficial for either the persistence of an organism or its reproductive success. Sometimes the benefices go hand in hand, but at other times a trait may have a positive effect only on the organism's survival but not on its reproduction or vice versa.

Since every genetic mutation is subject to natural selection with respect to its effect on the ability to reproduce, a question like why do individuals in a certain population have the property X is properly answered by saying that property X enhances the ability of an *individual* organism to reproduce more than all other nearby alternatives to X, by being a local maximum for its reproduction capacity.[15] Nothing is gained in explanatory understanding by producing a causal explanation in terms of feedback mechanisms. Property X is a result of an accidental mutation in one individual, which subsequently spreads to every descendant in the entire population. A random event cannot by itself explain why an individual with X initially became reproductively more successful. We must know more about the entire environment in which the individual organism lives before we can explain why X gave this individual a better reproductive rate. The effects of most mutations would be irrelevant to reproductive success (of the individual or the group), or may even have a negative effect. Moreover, X could be reproductively advantageous for a particular individual but may turn out to be disadvantageous for the population as a whole.

But neither Wright's etiological analysis nor Millikan's selected-effect analysis are able to handle functions which seem to occur without evolutionary evidence.[16] Even in biology there are clear-cut examples where a functional attribute or characteristic is selected but is irrelevant to reproductive success. These examples count cases where a trait is assigned a function but it is either known not to have been selected or without evidence of having been selected. An illustration of the former case is exaptation: ears arising as hearing organs and co-opted for heat regulation; feathers selected as insulators and co-opted for flight, etc. In combination with other analyses, the selected-effect analysis may account for unselected functions, but all these analyses attempt to justify is why on a naturalistic perspective it is legitimate to use functional explanations in biology. None of the analyses suggested here can cope with functions outside biology.

I contend that a theory of functional explanation should be able to deal with the questions that we have already touched upon as well as others.[17]

A theory must permit that appealing to a thing's function does not necessarily explain why the functional trait is there. On this issue functions in biology differ from functions in social institutions and functions of deliberately made artefacts like tools and equipment. A theory must distinguish between accidental effects and functional effects of certain characteristics. It must be open to the fact that functions can be highly contextual. It must allow some properties, like mimicry, to have different functions (production or pollination). It should permit that a function may be advantageous for one individual but not for the species as a whole or vice versa; as with altruistic behaviour. Moreover, a theory should allow that functions sometimes can be used to make predictions (moths). It should describe a trait as having a function only with respect to the system including the bearer of that trait. It should recognize that functional explanations can be responses to different types of questions: *What* is this trait for? *Why* does this trait (type) persist? and *How* does this trait type contribute in general to the fitness of its bearers? Functional explanations in biology are special cases of functional explanation involving contributions to fitness with respect to an environment. A theory should also countenance both selected and unselected functions in biology. And finally, it should cover questions concerning non-biological functions in other sciences or about artefacts. It should account for functional similarities and differences between the items involved. Biological items have evolutionary functions only when there are reproductions with similar effects, whereas artificial items can have functions without being reproductions at all. All these demands call for a pragmatic theory of explanation.

Therefore I propose that functional explanations are answers to why-questions about the functional role of something, which are contrastive, context-dependent and effect-oriented. An effect plays a functional role if it helps to maintain a system as a whole. The context determines what kind of explanation is appropriate to a question like "Why X (rather than Z)?" If the question addresses a functional problem, it is functionally explained by an appeal to Y (because of Y) given the following conditions:

(1) Y is an effect of X.
(2) Y is beneficial for O with respect to E.
(3) Y is unintended by O which manifests or carries out X.
(4) Y is explanatory for S in a context C.

Here O may stand for a system, an organism, or a group of individuals, E for the environment of O, and S for a subject. The big ear (X) of the fennec fox (O) serves to get rid of the heat (Y_1) and to hear the movement of prey (Y_2) during the night because in the desert (E) it is beneficial for the fennec fox to scatter heat quietly (therefore it does not pant (Z)) and to hear its preys over long distances (there are fewer of them). It only explains the functional task of X, but does not require answering what caused X.

Although sceptical about the explanatory status of functional explanations, Hempel and Oppenheim noticed their widespread use and popularity, but attributed this to their heuristic utility and the idea that teleological explanations give us a certain feeling of understanding.[18] We feel that we understand a phenomenon if it is explained to us in terms of ends and means with which we are familiar from our own experience of human behaviour. But for Hempel and Oppenheim such a feeling does not meet objective scientific standards. In science 'understanding' is a matter of knowing the basic nomic structure of a phenomenon, not a question of feeling familiarity with the phenomenon.

In contrast, I hold that functional explanations give us real understanding because these explanations contribute to bringing together causal and holistic considerations. According to the pragmatic approach I am pursuing here, whether or not some bit of discourse is considered an 'explanation' depends on whether those asking the question that stimulates the putatively explanatory response feel that as a result of hearing this response they now understand the object of the question better than before. Thus, a critic might say, "This feeling of understanding is not *justified* because this teleological explanation doesn't meet the standards of scientific objectivity" (perhaps the information conveyed in the explanatory discourse was false), but he cannot very well deny such a feeling might exist. If an individual feels that he understands something better in virtue of having heard a certain discourse, then that discourse is *ipso facto* "explanatory-for-that-individual," whether or not anyone else finds it explanatory. In effect a pragmatic theory must point to a notion of "Y is explanatory for a subject S in context C" as the only form of explanation that can fulfil the goal of an account of functional explanations as context-dependent.[19] However, such a notion of "subjective understanding" could be made more intersubjective and more objective by adding that all the claims made in the discourse must meet some common form of understanding and must be known to be true. Functional explanations reflect our understanding of how the property of being part of a whole system partakes in goal-oriented behaviour in order to sustain the whole. Therefore such explanations are part of explanatory practice in biology and the behavioural sciences.

What we feel as bringing functional understanding depends on what we take to be essential for explaining the phenomenon in question, and this may vary from person to person and from context to context, but yet be considered very informative, because the feeling itself rises from the way functional schemata organize the explanandum-belief into our epistemic system.

8
The Pragmatics of Explanation

In the preceding chapters we have seen how both the formal-deductive approach and the ontic approach faced certain problems by not paying due attention to the contextual aspects of explanations. We have seen how the context of an explanation makes a difference that cannot be ignored, and especially how ideal notions of scientific explanation run into difficulty. Any evaluation of the relevance of an explanation depends on both the content of the *explanans* and pragmatic factors concerning what is tacitly or implicitly understood in presenting a particular explanation in a particular situation. Since all epistemic norms govern the practice of explanation, simply being true will not suffice to make it an appropriate explanation. An appropriate or relevant explanation will depend on the truth of the explanation (or so it seems) and on whether the speaker satisfies the epistemic norms of relevance. This chapter focuses on some of the main theories that try to incorporate contextual aspects into the notion of explanation, thereby making pragmatics an essential part of explanations.

A primary lesson our analysis of these theories of explanation has taught us is that it seems arbitrary to draw a sharp line between scientific explanations and more mundane everyday explanations, and most pragmatists agree. They treat explanation as an independent object of study and whether or not a proposed "explanation" is a *scientific* explanation is determined by the specific context, not semantic content or logical character. Because establishing necessary and sufficient conditions for an explanation to be scientific is problematic, pragmatists allowed matters other than logical form, causal content, and empirical testability decide the status of a given explanation.

Some have claimed that theories of causal explanation should be considered pragmatic theories.[1] This seems a bit puzzling, but some philosophers, like Michael Scriven, have harmonized a pragmatic view of explanation with a causal account because the truth conditions of the causal sentences in the *explanans* are considered context-dependent. But others, like Salmon, hoped to establish an objective and counterfactual-free notion of causation, i.e., that causal statements have objective truth-conditions, regardless of the context in which they are put forward. Making 'explanation' dependent on reference to the interests and beliefs of the audience and speaker does not rule out reference to 'causes,' which will be understood against the background worldview of the audience and so will reflect that context.

Pragmatic theories of explanation characteristically focus on how concepts and language are actually *used* rather than the syntactical or semantic structures of the concepts and language involved. Of course an explanation also has syntactical and semantic features, but pragmatists hold that these features are subordinated to the context in which the explanation is given. This pragmatic orientation will enable us to see how the act of explaining functions as means of communicating understanding and that all explanations aim to increase the understanding of a particular audience.[2] However, before turning directly to the pragmatic theories, I shall look at a related approach in which explanation is considered from a question-answer perspective, where the focus is on the presuppositions made by the questioner when she requests an explanation. Bas van Fraassen is famous for advocating that to understand how an explanation should answer the question in a satisfactory manner, the essential element is the situation in which the explanation-seeking question is posed.

Understanding how contexts and pragmatics are related requires clarifying these terms. "Context" refers to the fact that certain external, non-semantic factors exert an influence on the use of a concept, which in turn implies that a word's meaning depends on the situation in which it is employed. In linguistics "pragmatics" refers to the study of language *use*. Pragmatics is conceived as a special approach to linguistic phenomena that examines how language functions in diverse situations. Language works differently in different situations because these different contexts influence the way language is correctly used. Thus "pragmatics" and "contexts" are closely related terms referring to the overall use of language. A term, phrase, or concept is "pragmatic" if its meaning is partly determined by the speaker or the audience, and is "contextual" if its meaning is partly influenced by the situation in which it is used. But there is an important difference in their scope. Any linguistic factor referring to the speaker or the audience is a *pragmatic* factor, whereas any linguistic factor determined by the situation in which the concrete linguistic action takes place is a *contextual* factor. A context may, for instance, involve personal interests, habits, and beliefs that do not belong to pragmatics even though such things can directly influence how a person actually uses and understands language. In general, however, the truth conditions of a sentence involving a contextual term depend on situational matters, such as the topic to which the term is being applied and the person employing it. Thus, in the case of explanation, whether a description counts as an explanation varies according to what and to whom the description is being applied.

However, reference to the speaker and audience can be only a *necessary* condition for pragmatics since there are concepts whose semantic explications require reference to a speaker and an audience as well. Indexicals like "I," "you," "here," and "now" are terms whose meaning changes with the context though we would not call them pragmatic. Here the semantics of indexicals includes a reference to the speaker, place, or time. The difference is that in pragmatics, depending on the situation, the speaker selects between alternative sets of truth conditions that can, in principle, be described by a declarative sentence, whereas in the semantic of indexicals a reference to the speaker is already part of the truth conditions of the sentences.

The various forms of pragmatic theories depend on which logic of questions and answers they presuppose as the underlying structure of explanation. Van Fraassen's approach, the "reduction-to-answer" view, takes an explanation-seeking question to be essentially equivalent to its set of answers, holding that every such question has direct answers, and these direct answers are statements.[3] Understanding a question requires knowing what counts as a direct answer to the question, and any statement that implies a direct answer completely answers the question, but any statement, which a direct answer implies, is presupposed by the question and only partially answers it.

Another approach, the "reduction-to-intention" view, holds that to understand a question is to know what the questioner wants with her question, i.e., her intention. Indeed, not all questions "signal" linguistic responses. "Do you have a pencil?" does not request an answer (something linguistic) but a pencil (something physical). An explanation-seeking question does request a linguistic response. Every such question is assumed to have a presupposition, a desideratum, and a direct answer. A conclusive answer satisfies the desideratum by fulfilling certain relevance and conclusiveness conditions. Between the direct-answer and the intentional approaches, there is scant agreement about the logical framework in which to describe questions. In this chapter I shall look

186 *The Nature of Scientific Thinking*

more closely at van Fraassen's line of thought, but at the end I shall reject it as a satisfactory theory of explanation. Later I shall deal with epistemic theories of explanation, offering an alternative to van Fraassen's.

1 Why is explanation a matter of pragmatics?

There are a number of good reasons for changing from a formal-logical to a naturalistic-pragmatic focus on explanation; some of them may help summarize our earlier discussion.

First, we must recognize that even within the natural sciences scientists regard many different types of accounts as explanatory. Though nomic accounts seem to fit the requirements of the formal-logical approach reasonably well, in previous chapters I have shown the deficiencies of this approach. Empirical generalizations generally contain *ceteris paribus* clauses, making them true only if certain idealizations are fulfilled, but such idealized conditions are never met in this world. Moreover, no fundamental law statement applies directly to the world; rather, if law statements are true at all, they are true of an idealized model. Statements of laws, such as Newton's laws, never refer to any particular object; they express merely how various properties are interrelated. Models represent concrete systems by picturing them according to certain interpretative standards. Causal explanations are usually carried through such models, which need not have deductive connection with a theory.

Models help produce causal explanations; fundamental laws do not. Nancy Cartwright has offered a fine illustration of this.[4] In quantum mechanics the phenomenon called radiative damping produces a broadening of the spectral lines. The atom is represented in a model with the nucleus surrounded by electrons in various energy levels. It decays spontaneously from an excited state, emitting a quantum of energy into the radiation field, which then may be reabsorbed by the atom. The reaction of the field on the atom both provides the line width and causes a shift of the line called the Lamb shift, but the line broadening can be represented mathematically six different ways, using three different equations. None has priority, and none gives us the correct covering law; these different approaches are useful for different purposes. While the causal explanation remains the same, the theoretical treatment may differ depending on which mathematical technique is used. That technique is chosen which works best (or easiest or most economically, etc.) in the particular context of that calculation, which again depends on the physicist's ability to find a good approximation fitting the problem. This and similar examples show that, in general, physicists select a particular covering law on pragmatic criteria. No theoretical description applies directly (i.e., without interpretation) to the world. Hence, these pragmatic criteria cannot be regarded as irrelevant but are essential for grasping the notion of scientific explanation.

Second, a pragmatic and naturalistic treatment of explanation focuses on human understanding. I see no reason why the social sciences and the humanities should be excluded from such a treatment. The naturalistic approach must include all kinds of explanatory accounts, especially including accounts depending on information about either motives or meanings. The practices of natural science implicitly contain these kinds of explanations, although they often play a methodological or meta-scientific role in explaining how to give, say, nomic or causal explanations. Naturalists approach this practice as an expression of the explainer's understanding of not only the content of the explanatory request, but also of his or her own understanding of the facts that provide the desired information. Making full sense of the entire scientific enterprise, including theory development, requires making room for interpretive explanations as well.

Third, the meaning of a why-question does not alone determine the relevance of an answer. Pragmatic elements concerning the presentation of the question also play a significant role. Consider a question like: "Why do birds migrate to Africa?" The response relevant to this explanation-seeking question depends on which words the inquirer emphasizes. Bengt Hansson was the first to point out that the two utterances like "Why do *birds* migrate to Africa" and "Why do birds migrate to *Africa*," where the emphasis varies, express the same sentence but nevertheless have different meanings.[5] Unless we take into account such differences in utterance meaning, it is unclear what the inquirer asks to have explained. Does the questioner want to know why *birds* migrate to Africa, or does she want to know why it is *Africa* that birds migrate to? Since the formal-logical approach treats propositions only non-contextually, it will not be able to capture this difference. Hansson proposed constructing a "reference class" for the explanation-seeking why-question. The idea was to capture in precise terms the nuances in meaning that could only be conveyed by style, emphasis, stress, italics, or the relative position of the words. Consider

(1) Why did Adam eat the apple?

While this seems to be an explanation-seeking question, it is unclear what it is that the questioner wants to have explained. Does she want to know why it was *Adam* who ate the apple or why he ate the *apple?* Hansson came up with a solution that permits him to distinguish the difference in meaning between the following questions:

(a) *Why* did Adam eat the apple?
(b) Why did *Adam* eat the apple?
(c) Why did Adam *eat* the apple?
(d) Why did Adam eat the *apple?*

These four alternative interpretations (using italics) of the meaning of the ambiguous question (with no italics) express differences that can be captured only with italics or the relative position of the words. None are independent of the tacitly understood reference class containing things that might have been different and therefore not mentioned by the question. It varies in each of the four cases. For instance, the reference class for "Why did Adam eat the *apple?*" includes not only apples, bananas, grapes, figs, dates, and other edible vegetables in the garden of Eden, but also the snake and other edible animals. Thus a relevant answer should give the reason why Adam chose to eat the apple instead of a banana, a fig, a date, etc. In general, a response to the question is explanatory if it informs the questioner about why instead of another of the many relevant possibilities.

A why-question out of context is ambiguous, but it can be given a fixed meaning by a reference class that is indefinitely large. We may never come to know all the members, but although we do not have the capacity to imagine all the members of this class, we can understand the question if we can, for any given possibility, decide whether or not it belongs to the reference class. Here the context of the question becomes important for narrowing down the alternative possibilities. In the Garden of Eden story, we are interested in knowing why it was the *apple* that Adam ate, and not some other fruit, because God had allowed Adam and Eve to eat all kinds of fruit except the apples, which they were explicitly forbidden to eat from the tree of knowledge of good and evil. But Adam ate the apple because Eve convinced him to do so after the snake had tempted her to taste it. They ate the apple, rather than any other unforbidden fruit, because by eating it they would gain wisdom.

It might be argued that the formal-logical model could handle the difference in emphasis by abstracting the logical features from the pragmatic context differently for each way of emphasis. Indeed, the semantics has *some* role in determining a relevant answer, but it cannot do the entire work alone because the context is also *necessary* for determining it. In other words, *without* a specified context, one cannot know what sort of answer would be relevant.

Fourth, John Searle has correctly argued that the meaning of every indicative sentence is context-dependent.[6] He does not deny that many sentences have a literal meaning, traditionally seen as the context-independent semantic content of a sentence, but he holds that we can understand the meaning of such sentences only "against a set of background assumptions about the context in which the sentence could be appropriately uttered." Thus, the background does not merely determine the utterance meaning whose context dependence may, like indexicals, already be realized in the semantic content of the sentences uttered. Also an extensive background of assumptions, practices, habits, institutions, traditions, and so on determines the literal meaning of sentences. Searle maintains this background network of assumptions cannot be made entirely explicit as a determinate part of the truth conditions of the sentence. Therefore the truth conditions of the sentence will vary with different background assumptions, which cannot be turned into objective implications of the sentences in question, and therefore cannot form part of its semantic content.Searle uses the iconic example: "The cat is on the mat." Asserting this sentence logically implies the presence of a cat and the existence of a mat. A basic assumption, not implied but tacitly presupposed, is that there is a gravitational field defining up and down, or at least a world view in which up and down are reasonably definable. The sentence is true only if the mat supports the weight of the cat. A space-cat, travelling under weightless conditions, would be no more on the mat than the mat would be on the cat. But the existence of gravity does not follow logically from the notion of "being-on." (Even a person without any knowledge of gravitation would have no problem of understanding the meaning of the sentence.) Searle's point is that any attempt to incorporate these basic assumptions into the explicit content of the sentence would not help, for such an explication could be extended endlessly, since logically there is no point to stop. For instance, the solidity of the cat, the mat lying horizontally, and the firm ground supporting the mat, are also among the assumptions, which then have to be included. In fact, the sentence may be true in some contexts without any of these conditions being satisfied.

This example focuses on the context dependence of 'on', but the words 'cat' and 'mat' are even more, context dependence. Thus the sentence does not have a set of truth conditions that uniquely specifies a truth-value in every particular situation of utterance, unless the conditions include features whose existence is not logically implied by the sentence.[7] These features may or may not belong to a determinate set of truth conditions of the sentence, therefore they are not part of its semantic content, but belong to our background knowledge of a non-linguistic kind.

If this analysis is correct, then the meaning of scientific statements contains features over and above their semantic content. Hence, any attempt to establish an objective model of explanation by abstraction from the pragmatic context, as Hempel proposed, is doomed to fail. Even though understood literally, the meanings of the *explanans* and *explanandum* are always relative to a set of background assumptions, not fixed by a determinate and invariant set of truth conditions. Their meaning presupposes conditions that can be determined only with respect to the actual situation in which the explanation takes place and which therefore cannot be taken into account by a formal-logical model. Such conditions are often buried in *ceteris paribus* clauses. Generally, they reflect the explainer's background, her interests, beliefs, assumptions, practices, habits, institutions, and traditions.

Fifth, many explanations take the form of stories. Arthur Danto has argued that what we want to explain is always a change of some sort.[8] When a change occurs, the situation before the change and the one after it are different, and the explanation connects these two situations. This is the story. We have a beginning, a middle, and an end. Indeed, this model of explanation includes not only complex historical-intentional explanations, but also causal explanations fit in as well. A wooden farmhouse lies on the plain; when lightning strikes the house, flames consume the house. The change is the cause that explains why a new situation follows the original; the new situation is the effect of the change. However, stories are not logical arguments. They are told from a certain perspective, determined by the interests and background knowledge of the explainer.

Sixth, a change always takes place in a complex causal field of circumstances each of which is necessary for its occurrence. Writers like P.W. Bridgman, Norwood Russell Hanson, John Mackie, and Bas van Fraassen have all correctly argued that events are enmeshed in a causal network and that it is the salient factors mentioned in an explanation that constitute the causes of that event. For instance, oxygen needs to be present in a certain critical amount for the farmhouse to catch flames, but normally we cite the lightning as the cause. Most of these standing conditions, or other necessary factors, do not interest us, nor do they need to be covered by the answer in order to provide a causal explanation; some may not be explicable. The speaker's interests determine which of these necessary factors he picks up as the cause. The farmer himself may regard the bolt of lightning as the real cause, while the insurance company may consider a defective lightning rod as the real cause. As van Fraassen aptly sums up, having quoted Russell Hanson with approval:

> In other words, the salient feature picked out as 'the cause' in that complex process, is salient to a given person because of his orientation, his interests, and various other peculiarities in the way he approaches or comes to know the problem-contextual factors.[9]

Our tendency to select the salient or the most perspicuous event, say, the bolt of lightning, as the explanatory cause reflects our interests in how we want the world to be, but it also hinges on our general background knowledge of how the world is. Imagine, for instance, that a lightning rod normally protects the farmhouse, but has been taken down for replacement. Assume, also, that you know that the plains are a high-risk lightning area and that lightning had struck the farmhouse many times before, but that nothing else had happened because of the old lightning rod. In this case, in virtue of the known facts, you may point to the absence of the lightning rod as the real cause rather than the lightning itself.

Nothing said here implies that causal explanations are merely subjective. The appropriateness of the answer has a necessary reference to such subjective states such as how the question is 'understood' or 'interpreted' (or perhaps 'misinterpreted') by the listener. But, of course, the context is determined by objective facts of who, when, etc. Moreover, the causal field as such exists objectively, regardless of whether all the necessary factors are entirely explicable. All this tells us that there is not only one correct (or incorrect) way of explaining things, since any correct (or incorrect) explanation still reflects the speaker's interests and background knowledge.

A further point to this analysis of the context-dependence of causal explanations is that any counterfactual analysis of causation makes causation very contextual. David Lewis and Robert Stalnaker have proposed theories for evaluating counterfactuals in terms of similarity between possible worlds. Any appeal to a similarity relation between such worlds is not a purely objective matter because the standards of similarity are selected on a partly subjective basis, since they depend on the conversational purposes for which we assert these counterfactual sentences. Personally I would prefer to say that the idea of causation is a primitive notion and therefore cannot be fully grasped in counterfactual terms, but this would not solve the problem of the contextuality of causal expressions discussed in Chapter 6. Causal statements still logically imply counterfactuals. Therefore whatever is contextual about asserting the corresponding counterfactuals will reappear as contextual elements in causal explanations.

Seventh, the level of explanation also depends on our communicative interest. *Levels of explanation* must be distinguished from *context of explanation*. 'Levels of explanation' suggests a hierarchy from simple, non-technical to highly technical, requiring professional expertise to understand, like a high school text compared to a professional one. Each different *epistemological* level can be said to be *adequate* if it fulfils its purpose of providing sufficient information for the inquirer to have the degree of understanding appropriate to the question which has been posed and appropriate for the audience to which the explanation is presented. In selecting the explanatory level that provides and adequate degree of understanding, the norm is: Don't make an explanation more complicated than the cognitive situation demands. An adequate explanatory response matches the intentions behind a particular explanation-seeking question. Thus, an explanation may be relevant and yet inadequate if it adopts an explanatory level that is not appropriate for the level of the inquirer's cognitive interests and background knowledge. But there are also levels of explanations that presumably reflect *ontological* hierarchies. As I understand context, different contexts could call for different explanations, but they could very well all be on an elementary or all on a professional level. In science an appropriate nomic or causal account can be given on the basis of different explanatory levels, and the particular level one selects as informative depends very much on the rhetorical purposes. If a toxicologist tells the jury in a courtroom that the victim died because she was poisoned by strychnine, he gives the explanation most relevant for this particular purpose. He chooses an epistemic level of explanation appropriate for judiciary evidence and suitable for the audience's background understanding. Had he chosen a chemically more accurate and detailed explanation, telling how the molecules of strychnine interacted with the cells of the body, the explanation would be on a different epistemic level perhaps relevant for other toxicologists. But the toxicologists could also change context (and thereby ontological level), if he no longer focused on the causal mechanism of the poison on a living body but on the effects of the molecules of strychnine on the individual body cells. A physicist might provide a causal explanation on even a lower level trying to give an account of the process on the atomic level, a possibility relevant only to other physicists.

However, the question remains whether all these explanations at various ontological levels and in different contexts are independent accounts, or whether every macroscopic explanation is in principle reducible to a microscopic explanation. Ultimately, extreme reductionism implies the atomic explanation, or the subatomic explanation, is basically the proper scientific account, since every other explanation can be reduced to it. Although it may fare well with the formal-logical view, this reductionism fails because in many contexts we want to keep a scientific explanation of what killed the person in the normal context of explaining a homicide. It does not even make sense to describe a dead body in terms of molecules, atoms, quarks, or superstrings. This seems to imply that every ontological level of explanation is relevant with respect to certain epistemic contexts but not with respect to others. There is neither one correct nor one adequate explanation. The communicative situation, including the audience's interests and the descriptive level of the e*xplanandum*, determines the appropriate *explanans*, and the communicative situation changes all the time, just as the pragmatic theory of explanation maintains.

Eighth, scientific models are empirically underdetermined by data. In principle it is always possible to develop competing models explaining things differently. Therefore, it is impossible to set up a crucial experiment to show which model yields the 'correct' account of the data available. The Bohmian theory of quantum mechanics is empirically equivalent, for all we know, with orthodox non-relativistic quantum mechanics, although each gives a very different picture of the quantum world. The former is a deterministic theory and explains quantum phenomena in terms of well-defined trajectories, whereas the latter is an indeterministic theory that explains everything probabilistically in terms of observables. These various reasons characterize explanation as a communicative practice of science, where explaining a phenomenon amounts to answering a certain question about the phenomenon. The formal-logical approach completely ignores such an interrogative perspective because it fails to realize that even scientific explanations rest on other than scientific conditions.

2 The contrastive dimension of why-questions

The pragmatic nature of explanation becomes evident from the fact that the appropriate explanatory response is very often relevant to an implicit contrastive question indicating the kind of knowledge appropriate for the questioner's interest. Causal explanation is an appropriate explanatory response whenever we believe a causal dependency explains why the *explanandum* fact is the case. Not all causal information is relevant for such an explanation, as we have seen; the selection of germane information is determined by what we believe makes a difference in the particular situation we are addressing. Thus causal explanations are contrastive, whereas causal relations do not have that kind of structure.

How does an explanation-seeking question become contrastive? Often an explanation-seeking question, "Why X rather than Y?" is expressed as "Why X?" leaving the contrast implicit. In these cases the discursive context of the question usually makes the contrast quite obvious. For instance, by asking "Why do we experience global warming?" what we want is an explanation that tells us why the average global temperature is *increasing*, instead of cooling down or remaining constant. The presupposition is that it is a part of our background knowledge of such a question that a temperature may increase, decrease, or stay the same. We are not very interested in an answer which merely informs us that global warming is the case, something we already know; this question requests what causes global warming and at the same time excludes any of the alternative possibilities. But even if we have no knowledge of alternatives, we know any contingent fact might not have been the case. So the understood consequence of the question "Why X?" is always something like "Why X rather than not-X?" (as part of the conversational form of a why-question), and the intention of the person raising such a question "Why X?" is to get to know why X occurs rather than does not occur. Causal explanations are very good for such purposes because, apart from pointing to what determines the effect to occur, they eliminate competing factors which in other circumstances might have made Y or not-X occur rather than X.

Causal explanation generates causal understanding of a belief concerning the occurrence of particular fact X by relating the belief about X to other beliefs about particular facts necessary for X to occur and by *epistemically* excluding a class of other beliefs about alternatives that might have been the case, so that those facts necessary for X are believed to be causally sufficient in the circumstances. This contrastive feature of explanation-seeking questions endows causal responses with their explanatory power or relevance.

But other answers than causal responses to why-questions can increase our level of understanding not only by accounting for X but by excluding Y at the same time. This is true for functional explanations. Why do desert foxes cool off by their long ears rather than by panting? Because heat radiation is quiet whereas panting is not. Earlier I pointed out that the anthropic principle could be seen as a kind of functional explanation.[10] Some cosmologists explain why this or that quality has the value it has by claiming that without this particular value there would be no human beings. The existence of human beings is an effect of, say, the smallness of the cosmological constant, but the cosmological constant could not have been much bigger because we are here. In principle I have no objection against such a kind of explanation because they help at least some cosmologist to get a better comprehension of the universe. Like other functional explanations, it does not inform us about what caused the cosmological constant to be small, but it informs us about its function in terms of an important effect. It should also be noticed that the question is (implicitly) contrastive and that the answer therefore should exclude almost every other *possible* quantity. Further speculations in cosmology include eternal inflation, landscapes of vacuum, supersymmetry, and string theory, and these speculations give rise to multiverse scenarios. But like all explanations the contrastive feature of anthropic explanations is fixed by pure conversational implicatures. Anthropic explanations can be used regardless of whether or not the alternative quantities are realized in their own real world.

In the succeeding chapter, I shall argue that why-questions are not the only appropriate requests for scientific explanation nor are they the only contrastive form of explanation-seeking questions. But before we get to this discussion, I want to present Bas van Fraassen's pragmatic view according to which an explanation corresponds to a class of possible but distinct answers.

3 van Fraassen on explanation

In his seminal book *The Scientific Image* (1980) Bas van Fraassen presented his theory of explanation, agreeing with Hempel and others that explanations are answers to why-questions, but he believes explanations are *nothing but* answers. They are neither propositions nor a list of propositions nor arguments, but simply answers to why-questions. Therefore van Fraassen holds that a theory of explanation must be a theory of why-questions.[11] He also disagrees with Hempel and other formalist philosophers by rejecting the view that explanations are merely two-term relations between theory and facts in the world. Instead he treats explanation as a three-term relation between theory, fact, and context, thus making context become an essential part of explanation.[12]

However, van Fraassen distinguishes between the epistemic and pragmatic dimensions of science, and holds that explanation is part of only the pragmatic dimensions.[13] Rather than assuming that explanation is the aim of science and is involved directly in gaining scientific knowledge, he sees it as part of a human activity distributing already established knowledge. I agree with van Fraassen that the explainer passes her understanding to the audience unless, of course, the explanation is tentative. In order to explain something to somebody, one must already understand the connection between the information requested and the information offered in answer to the question. In this pragmatic sense explanations do not increase the explainers' knowledge but help them to apply their already established knowledge (or hypothesis) to answer a why-question.

Intuitively the underlying structure of why-questions looks something like this:

(2) Why (is it the case that) P?

where P is a statement that expresses the phenomenon to be explained. In this way "Why" turns statements into questions. A question arises only if P is true (or mistakenly believed to be). Let P be the statement "The conductor becomes warped during the short circuit"; the question Q "Why did the conductor become warped during the short circuit?" arises only in case the conductor actually does become warped. If P is false, and the conductor remains straight, we would dismiss Q by saying that the alleged phenomenon does not occur. So a why-question of the form (2) presupposes the truth of P, or at least that the interlocutor as well as the explainer believe P is true.

While van Fraassen thinks that explanations are answers and not propositions, he also maintains that a theory of questions (and answers) is closely connected with a theory of propositions.[14] A question is an abstract entity expressed by an *interrogative* sentence, just like a proposition is an abstract entity expressible by a declarative sentence. And just like the same proposition can be expressed by different declarative sentences, the same question can be expressed by different interrogative sentences. The converse also holds for both questions and propositions. The same declarative or interrogative sentence can express different propositions and questions respectively. No matter whether the form of the sentence is declarative or interrogative, the *context* determines which proposition or question is expressed.

Here van Fraassen draws on the insight of Hansson that the same interrogative sentence can pose different questions by asking why P rather than Q.[15] The general form of a why-question becomes therefore:

(3) Why (is it the case that) P in contrast to (other members of) X?

where X is the *contrast-class* consisting of the set of alternatives, which may or may not include P that forms the *topic* of the question. Moreover, (3) has different truth conditions depending on which contrast-class the questioner has in mind. In most cases contrast-classes are not explicitly described because the context makes clear what the contrast-class is. Hence, the context usually picks up which of the alternative sets of truth conditions to associate with the meaning of an interrogative.

Of course the information given as the answer to an explanatory request must be relevant to the question. Here van Fraassen adds to Hansson's treatment that the context also plays a major role in determining what is relevant to a given explanation. Recall the example with the car accident where various people have different interests in why a driver died in the car crash. The physician is interested in information about the body and the way it was thrown about causing a haemorrhage, the mechanics looks for information about how a defect in the brake caused the car to crash, and others, for example the highway engineer and the lawyer, have different interests. What counts as *the* salient feature, and therefore *the* cause, is always a result of a person's orientation, interests, and how he gets to know the problem. On this background van Fraassen, contrary to Hansson, suggests that the context determines what information counts as relevant, going well beyond the limited condition of what is statistically relevant.[16] Thus one can see how van Fraassen believes that context determines not only to which contrast-class one is referring but also what information is relevant, implying a number of interesting consequences.

The first is that whether or not an explanation is considered as "scientific" is solely a matter of the context in which it is offered. As van Fraassen correctly observes: an explanation is scientific if it relies on "scientific theories and experimentation, not on old wives' tales."[17] We don't need more. Even within science the contrast-class is not automatically determined because different scientists may have different interests in the same phenomenon. And since scientists evaluate scientific theories differently, they may come up with alternative explanations. Van Fraassen argues that pragmatic aspects provide reasons for theory choice independently of truth and empirical adequacy. The languages we use in scientific activity and in particular in theory choice demand pragmatic considerations, because (a) "the language of theory appraisal, and specifically the term 'explains' is radically context-dependent"; and (b) "the language of the use of theories to explain phenomena, is radically context-dependent."[18] In contrast to Searle's view, the language in which scientific theories are formulated is *not* context-dependent, but it becomes radically context-dependent only when we appraise a theory and we use it to explain phenomena. Therefore a theory taken in isolation can be studied at the semantic or syntactic level concerning its content and how it relates to the world. But, as we have seen, explanation is only partially a function of the content of theories; it is also partly determined by how we choose between alternative theories and how we use them in explanation. All these factors involve reference to the questioner and the explainer and to the linguistic, historical, and physical situation of the explanation.

Hence, contrast-classes and relevance relations constitute an essential part of scientific and non-scientific explanations. And since the contrast-class and the relevance relation are contextually fixed, only the context can determine whether an explanation is scientific or not. So if the context consists of scientific theories and methods determining the contrast-class and relevance relation, the answer to a why-question will be regarded as scientific.

The second important consequence of the contextual account of explanation is that no explanation is ever complete.[19] Since a discourse is an explanation only with respect to a certain relevance relation and contrast-class, an explanation will never be complete in covering every aspect of a phenomenon. This is in stark contrast to Hempel's view that an explanation is complete just in case we can subsume the explanandum under a general law. Van Fraassen illuminates his point with an Omniscient Being. It might seem that an Omniscient Being would possess complete explanations, whereas the contextual factors of relevance relations and contrast-classes merely reflect human limitations in grasping complete explanations, but van Fraassen argues this is a mistake. If the Omniscient Being has no particular interests (physical, chemical, medical, legal, economic, etc.), then no why-question ever arises for him in any way at all. He would never *request* explanations in answer to why-questions because being omniscient he has no need to ask any questions. If he does have specific interests, then his why-questions will be as context-dependent as ours. The difference is that

In either case, his advantage is that he always has all the information needed to answer any specific explanation request. But that information is, in and by itself, not an explanation; just as a person cannot be said to be older, or a neighbour, except in relation to others.[20]

Explanation is thus a *relational* concept; only in relation to a specific interest can a piece of information count as explanatory, and therefore explanation is essentially context-dependent. We have to remember, however, that Hempel never denied that scientific research had an inescapably pragmatic character; he abstracted the logic of explanation from it. Thus as long as omniscient incompleteness does not exclude an application of the covering law model, he would not need to oppose this view of omniscient incompleteness to defend explanatory completeness. The two views can both be accepted side by side because one's interests help pick out the relevant laws for explaining the phenomenon in question.

A moment ago we noted that an explanation-seeking question presupposes the truth of the *explanandum*, but according to van Fraassen, why-questions require other presuppositions as well. The particular why-question being asked by a given interrogative sentence is a function of three factors, the topic P_k, the contrast-class $X = \{P_1, P_2, \ldots, P_k, \ldots\}$, and the relevance relation R. The why-question Q: "Why P_k?" may then be identified with the ordered triple consisting of these three factors: $Q = \langle P_k, X, R \rangle$. In addition, a proposition, A, is "relevant" to Q exactly if A bears relation R to the ordered couple $\langle P_k, X \rangle$. Van Fraassen goes on to define a direct answer to Q as follows:[21]

(*) P_k *in contrast to* (the rest of) X *because A*.

A few things should be said about van Fraassen's answer. First, the answer expresses a proposition, asserted in the *same* context that picks out Q as the appropriate proposition expressed by the corresponding interrogative ("Why P_k?"). Second, the answer presupposes that P_k is true and the other members of the contrast-class are false (it makes no sense to say why Peter rather than Paul has paresis, if they both have it.) Here van Fraassen seems to overstate his case, because what is needed is only that the other members are *believed in fact to be false*, since I (the explanator) must believe that they *could have been true*, if A was not the case. My background knowledge must lead me to believe that some of the other members of X might have been true; otherwise the contrast would make no sense. If no other alternative state of affairs is believed to be possible, there would be no need to ask why P_k. Third, the answer says that A

is true. And fourth, the use of the word 'because' indicates that A is a reason, i.e. A is relevant, that it bears R to $\langle P_k, X \rangle$. There also seems to be a fifth, tacit, presupposition lurking in the background. Elsewhere, van Fraassen tells us that "as *in all explanations*, the correct answer consists in the exhibition of a single factor in the *causal* net,"[22] implying that the relevance of that factor changes, depending on the context, from one sort of efficient cause to another. Thus, a direct answer to Q also requires that A is relevant if, and only if, A refers to some salient features in a causal net or cites "*events leading up to*" the one mentioned by P_k. In other words, an answer demands reference to a causal relation between the events to which A and P_k refer.

Having the presuppositions in place, and by claiming that (*) counts as a direct answer, only if A is relevant, van Fraassen defines a direct answer to Q.

B is a direct answer to question $Q = \langle P_k, X, R \rangle$ exactly if there is some proposition A such that A bears relation R to $\langle P_k, X \rangle$ and B is the proposition which is true exactly if (P_k; and for all $i \neq k$, not P_i; and A) is true.[23]

The above definition of a direct answer to a why-question Q implies that

(i) the topic (P_k) is true,
(ii) in the contrast-class, only the topic is true (each P_i in X is false if $i \neq k$), and
(iii) at least one of the propositions that bear the relevance relation to its topic and the contrast-class is also true.

Taken together (i) and (ii) form the central presupposition of van Fraassen's account.

Notice that van Fraassen also allows an answer to be "complete." Any proposition, which implies a direct answer, is a complete answer. However, it is not complete in an absolute sense, but once the relevant context is specified, it is complete for this context.

A relatively complete answer to Q is any proposition which, together with the presupposition of Q, implies some direct answer to Q.[24] Since the presupposition contains a relevance relation and a contrast-class, and both are contextual factors, therefore an answer is complete only relative to a certain context. This characterization may then be generalized so that "a complete answer to Q, relative to theory T, is something which together with T, implies some direct answer to Q."[25] However, it seems that van Fraassen requires the reference of A really to bear a causal relation to the reference of P_k, and not merely that the interrogator and responder believe it bears this relation. But since my knowledge that the reference of A bears this causal relation can never be certain, the explanation can never be known to be 'complete' even in this relative sense. Others besides van Fraassen have made the contextual nature of explanatory completeness a key issue in every pragmatic theory of explanation. Michael Scriven, for one, makes this point in arguing that invariably an explanation is said to be "complete" relative to a context and to the level of knowledge already possessed: "[T]he notion of the proper context for giving or requesting an explanation, which presupposes the existence of a certain level of knowledge and understanding on the part of audience or inquirer, automatically entails the possibility of a complete explanation being given. And it indicates exactly what can be meant by the phrase 'the (complete) explanation.'"[26] Of course, Hempel would never have found acceptable a strongly relativized notion of a complete explanation like this.

Van Fraassen also treats the issue of how why-questions arise: the context in which a question is posed involves a certain body, K, of accepted background theory and factual information, depending on who the explainer and the questioner are, and determining whether the question can arise. Hence, a why-question may arise in one context, but not in another. This consequence apparently enables van Fraassen to solve two important problems in traditional theories of explanation; namely, their inability to account for (a) seemingly legitimate rejections of requests for explanations and (b) the asymmetries of explanation, such as the notorious flagpole and its shadow example.

Background knowledge determines whether a question arises or not by its relation to the central presupposition. If K implies that the topic P_k is true, and that every other member of the contrast-class is false, then the question genuinely arises. If K does not entail the central presupposition, the question never turns up. For example, the question "Why did Lee Harvey Oswald not act alone when killing President J. F. Kennedy?" does not occur in my epistemic situation because for all I know, it is not an established fact that he did not act alone. Moreover, it should be noted that if K entails that there is no answer, no question arises either. Some questions are inappropriate because they cannot be answered. To a question of why a radioactive nucleus decayed at a particular moment, it is impossible to offer an answer because current theory holds there is no reason why a nucleus decays at a particular moment.

Unfortunately, van Fraassen's solution to the inappropriateness of a question creates more difficulties for his theory than it solves. The solution succeeds only because he puts a much too strong demand on the topic P_k, namely that K must *imply* it. There is no reason to assume this because P_k could be a newly empirically discovered fact not implied by anything we already believed. Indeed this is *usually* the case when we ask why P_k, in a few situations it may be that K implies the topic, although we may not realize it before we ask why P_k. I come home finding my daughter at home, and by seeing her there I ask why she is not at school – although I already have all the information needed to know that she has a day off. But most often our beliefs are transparent enough that we are not really surprised, because if K implied P_k, then we would not find P_k surprising and so not be likely to ask why it occurs. Assume that I find my car where I did not park it. I know exactly where I parked it yesterday, and I know that cars do not move by themselves. So K cannot imply P_k in this case. Of course we would expect K could not rule out P_k, but even that may not always be the situation. How could we ever be forced to revise our hypotheses and background assumptions if all new empirical discoveries had to be consistent with them?

Sometimes it may be necessary to overthrow a large portion of K to permit a newly discovered P_k. In 1847 Ignaz Semmelweis observed that if doctors and students washed their hand in a chlorinated lime solution, the maternal mortality rate due to puerperal fever would drop significantly. His observation went against the current scientific opinion of the time, which blamed diseases on an imbalance of the basic four humours. In this context K did actually rule out P_k. Semmelweis's explanation of P_k in terms of transmission of 'cadaveric matter' was therefore neither a part of the general medical knowledge at that time, nor a part of his own background theories. Only when Louis Pasteur and Joseph Lister introduced the germ theory of diseases and antiseptic twenty years later did Semmelweis' explanation become consistent with accepted medical theories.

Van Fraassen is able to deal rather nicely with the infamous explanatory symmetries like the flagpole example, where the height of the flagpole explains the length of the shadow but apparently not vice versa. Since content determines relevance, if the asymmetries of explanation result from a contextually determined relevance, then it should be possible to imagine a context that reverses the asymmetry (e.g., explaining the height of flagpole from the length of the shadow). Indeed, he does just that with the story of "The Tower and the Shadow," too well-known to repeat here.[27] So the asymmetry is contextual, and both explanations of the length of the shadow and the height of tower are perfectly legitimate;

both are correct as answers to *different* why-questions with *different* contrast-classes and *different* relations of relevance. Van Fraassen is thus able to deal with one of the crucial problems of the formalist model.

Summing up, we may say that van Fraassen's theory defines explanations as answers to why-questions, which arise in a given context depending on the background knowledge and beliefs of the people involved in the specific situation. The context dependence of both the contrast-class and the relevance relation make explanation radically context-dependent, but even though explanations are so dependent, they can be as complete as possible in any given particular context. A piece of information is only deemed explanatory in relation in a specific request, arising only in case somebody genuinely requests it. Explanatory asymmetries therefore are also contextual. This advantage and the ability to incorporate contextual matters into a theory of explanation make van Fraassen's theory a substantial improvement in relation to the formal-logical approach. Nevertheless, it suffers from some serious shortcomings.

4 Critiques of van Fraassen

Van Fraassen's theory has received several criticisms, some more severe than others. Philip Kitcher and Wesley Salmon, for example, have criticized van Fraassen's treatment of the relevance relation. They argue that we must put some constraints on the relevance relation to avoid all sorts of foolish irrelevant 'explanations' that make astral forces or spells explanatorily relevant. Others have disagreed with his insistence that why-questions are essential to scientific explanations. As we have seen, many explanations are answers to why-questions, but there are many others that do not address any why-question at all. Some have argued that it is always possible to express any request for a scientific explanation as a why-question, but in the succeeding chapter we shall see that there are many cases of scientifically legitimate how- or what-questions that cannot be changed into why-questions without altering their meaning. It has also been argued that, despite van Fraassen's explicit claim to the contrary, his theory is not a genuine pragmatic approach. This, of course, depends on the definition of "pragmatic," but Peter Achinstein has accused van Fraassen's theory of failing to be 'pragmatic' in a sense that Achinstein considers essential to the meaning of "pragmatic."

First, I will examine the criticism of Kitcher and Salmon. Their objections are of a formal nature but have consequences going beyond the purely formal structure of van Fraassen's theory. They attack the notion of the relevance relation R between the topic and the contrast-class required for the explanation to work. They argue that if no constraints on the relevance relation are enforced, it "allows for just about anything to count as the answer to just about any questions."[28] A concept of explanatory relevance must be spelled out to elaborate on what it means for an answer to be relevant to a question. However, as soon we attempt to define "explanatory relevance," we are confronted with many of the same problems that haunted the traditional accounts of Hempel and Salmon; thus van Fraassen's account is not the improvement he believes it to be.

In detail they argue that if we let P_k be any true proposition, we can construct a contrast-class X as any set of propositions such that P_k belongs to X and every member of X, apart from P_k, is false. Let A be another true proposition. We can construct a relation R as the unit set of ordered pairs such that the only member of R is the ordered pair $\langle A, \langle P_k, X \rangle \rangle$. Given van Fraassen's definition of explanation as the direct answer "P_k in contrast to the rest of X because A" to the why-question "Why (is it the case that) P_k in contrast to (other members of) X?" the sentence "Because A" becomes an elliptical explanation of P_k. If the only restriction put on the choice of P_k and A is that they are true, it would seem that anything can explain anything as long as we are dealing with true propositions.

As an illustration of this quandary, Kitcher and Salmon offer an example in which somebody asks why John F. Kennedy died on 22 November 1963. Suppose that someone gives the elliptical answer "Because A" where A, the core of the direct answer, consists of a true description of the configuration of the planets, the Sun, and the Moon at the time of Kennedy's birth. Astrologers would then claim that they possess a theory that, given A, shows that Kennedy's death was highly probable or even deterministically certain. We have a why-question and an explanatory answer, but most scientists will reject it outright as an acceptable explanation.

Van Fraassen does actually provide three standards for judging the quality of explanations.[29] Kitcher and Salmon argue that on those criteria the astrologers' answer comes out as a good explanation. The first is whether A is more probable in light of our background knowledge K. The second criterion is the extent to which A favours the topic P_k against the other members of the contrast-class, X. The third standard concerns the comparison of "Because A" with other possible answers to the question. This implies three sub-questions: (i) Is any answer more probable? (ii) Does any other answer more strongly favour the topic? or (iii) Does any other answer render this one wholly or partially irrelevant?

It is fairly easy to see how Kitcher and Salmon think that the astrological explanation of Kennedy's death scores high marks on all these standards of evaluation. Because of our astronomical beliefs, it is reasonable to assume that as a description of the configuration of the planets, the Sun, and the Moon, A is very likely true. The notion of "favouring" means, roughly, that the answer must raise the probability of the topic and lower the probability of the other members of the contrast-class. This requires that the astrological theory, on which the answer "Because A" is based, has such an effect. Indeed, astrologers will claim all they have to show is how the date of Kennedy's death follows from the astrological theory (and their interpretation of Kennedy's horoscope). Since Kennedy died on 22 November 1963, we know already that the topic is true; and therefore all the other members of the contrast-class are false. So the astrologers score high marks on the second criterion as well. Furthermore, the astrologers score well on the third criterion. No answer is better than "Because A": since A is true; it has probability 1, so no other answer could be more probable. For an audience of astrologers, the configuration of the Sun, the Moon, and the planets (in the horoscope) at the time of a person's birth is a primary determinant of his or her fate. Therefore no other answer can favour the topic more strongly than "Because A." Within astrology A is relevant and nothing can render it partially or wholly irrelevant. Kitcher and Salmon conclude if no constraints are imposed on the relation R, then anything can explain anything.[30] But they also argue that if we try to confine R to any relation meaning something like 'objectively, scientifically testable', we face the problem of characterizing an objective notion of relevance, since there may be more than one. Relevance, as seen in the fate of the deductive nomological model, is not guaranteed by logical deduction. Inevitably we run into the familiar problem of demarcating lawful regularities from accidental ones. The upshot of Kitcher and Salmon's discussion is that van Fraassen has not evaded the traditional problems revolving around the notion of objective relevance.

I think that Kitcher and Salmon's points are essentially correct though I doubt they are as damaging as van Fraassen's critics seem to think. In fact, I believe that, contrary to their judgment, they actually point to a great strength in van Fraassen's theory of explanation. In my opinion it counts in his favour that he does not impose those restrictions on the relation R, which they so dearly want. I assume, he would not deny that the astrological explanation is anything but bad, even if it may score high marks in this evaluation. The reason why we reject astrological explanations is not to be found in our definition of explanation, but in the context in which the explanation is offered. If the explanation is presented to an astrological society, it meets with approval. In other contexts, however, and in scientific contexts in particular, it figures at the low end of the evaluation scale because astrological assumptions are inconsistent with scientists' conceptual and theoretical background. Kitcher and Salmon want van Fraassen to say more about what objectively constitutes a scientific explanation, other than that it is based on scientific theory and observation. Apart from a few informal remarks he does not elaborate, it would appear that he holds that any evaluation takes place within a conceptual framework and with respect to a certain belief system. Another obvious reason why van Fraassen does not pay more attention to the possible evaluation criteria is that he is really interested in empirical adequacy, and more generally anti-realism.

Undoubtedly, Kitcher and Salmon's reasoning rests on the questionable assumption that only one explanation can be considered as *the* explanation, since not everything can be relevant for the topic to be explained. But they ignore that every causal explanation is context-dependent and different causal explanations of the same event may reflect different purposes. Recall the example of the fatal car accident. The mechanic explains the death by a malfunctioning brake system, and the civic planner as a result of tall shrubbery at the turning. So, different people give different answers to why the person died in the accident. And, as van Fraassen notes, the answers cannot be combined into a single explanation. Each subject keeps a certain aspect of the car accident 'fixed' in order to pick out the feature that is salient to him and his interests. The civic planner keeps the mechanical constitution of the car fixed and gives his answer in the conviction that regardless of the faulty brakes, the accident could have been avoided if the shrubbery had not obscured the driver's vision. Conversely, the mechanic keeps the physical environment fixed: his conviction is that despite the presence of a tall shrubbery obscuring the driver's vision, the accident would not have happened if the brakes had been working properly.[31] No one of these explanations (or any of the other possible explanations) is *the* explanation of the driver's death, but this does not entail relativism, for once we determine what our interests are, the answer will be *objectively* relevant in the sense that anyone with the same interest would count the same information as relevant. What Kitcher and Salmon consider as a weakness, I consider as a matter of strength. In a somewhat ironic way Salmon's quote that one person's counter-example is another person's *modus ponens* applies to him as well.

Therefore I conclude Kitcher and Salmon haven't been able to pin down any serious problem in van Fraassen's concept of explanation, though they have emphasized that explanations often need criteria according to which they can be evaluated. Some of those criteria ultimately represent our epistemic situation in the world because judging an explanation is about examining how well it fits with the world as we know it. But, as we shall see, judging explanations is certainly also about how well they do as a communicative act of providing information and understanding to the questioner. This is the point on which Peter Achinstein has criticized van Fraassen's theory as not really being a pragmatic theory.

Achinstein's accusation is that van Fraassen fails to include reference to an explainer or an audience in the truth conditions for explanation sentences.[32] He points to the difference between

(1) Account A explains fact X.
(2) Explainer S explains X to person P by given account A.

Achinstein uses the term "explanation-sentence" to refer to any sentence using "explain" or "explanation." So an explanation-sentence of the form (1) is 'non-pragmatic,' because it lacks truth-conditions referring to an explainer or an audience, whereas (2) does have such truth-conditions and therefore is a 'pragmatic' explanation-sentence. It might appear that (1) represents Hempel's view, but that van Fraassen's was in agreement with (2) because the latter holds that explaining requires answering, in a way relevant for this context, why this event occurs rather than some other member of the contrast-class. But Achinstein argues that despite the contextual nature of the relevance relation and the contrast-class, what van Fraassen counts as a satisfactory explanation-sentence does not reflect (2) but (1).

Achinstein illustrates his objection with an example from his "hometown lore" about a flag atop Fort McHenry. By the early dawn Francis Scott Key sees the flag is up and asks:

Q: Why is our flag still there?

On van Fraassen's account, this interrogative sentence can be used to pose different questions, relative to which contrast-class is being emphasized:

Why is *our* flag (rather than some other flag) still there?
Why is our flag still *there* (rather than somewhere else?)
Why is our *flag* (rather than something else) still there?

And so forth. The most likely contrast-class, which we can tell from our knowledge of Scott Key and his whereabouts that morning, is between our flag and the British flag being there.

Once this contrast-class is fixed, the relevance relation determines the answer which is relevant to Q. Suppose the explanation is

(3) The hypothesis that the British failed to capture Fort McHenry during the night's battle explains the fact that our flag is still there.

But this explanation does not *directly* answer Q, since it does not specify a relevance relation and a contrast-class. Therefore it is incomplete in the relative sense that van Fraassen takes completeness to require. So Achinstein suggests we write (3) to get:

(4) The hypothesis that the British failed to capture Fort McHenry during the night's battle explains (by citing "events leading up to") why our flag is still there (rather than the British flag being there).

Achinstein states then that this explanation (or rather explanation-sentence) satisfies van Fraassen's concept of explanation.

Obviously (4) does not have the same form as (2); since it does not overtly refer to any explainer or an audience, it is not *explicitly* pragmatic. But Achinstein denies it could even be implicitly pragmatic, because that would require that the explication of its truth conditions contain terms referring to the explainer and the audience. The fact that we look at the intentions of the explainer, Francis Scott Key, and his beliefs to determine what the relevance relation and the contrast-class are, does make these intentions or beliefs reappear in the truth conditions of (4).

Achinstein wants to emphasize that even a non-pragmatist, like Hempel, could agree that we cannot find out which question someone wishes to answer, or which events someone wishes to explain, without invoking an essential reference to the explainer and questioner. But only pragmatists hold that once the question has been identified, it is impossible to determine whether an explanation-sentence explains without paying attention to the explainer and questioner. Van Fraassen differs from Hempel only in that they supply different truth conditions for explanation-sentences. The former holds that an answer of the form "P in contrast to X because A" is true if (and only if) the topic P is true, the other members of contrast-class X are false, A is true, A does bear relevance relation R to P, and A refers to the events "leading up to" the event in P. The latter, however, urges us to accept as truth-conditions for an explanation sentence those given by the D-N or I-S model. *Neither* Hempel *nor* van Fraassen's truth conditions contain or require terms for an explainer or an audience.

I conclude that by and large Achinstein is correct about van Fraassen's theory of explanation. It may not be possible to obtain direct (complete) answers of the form (4) to a question like Q without involving the explainer's intention and interests. But once the explanation-sentences are completed, there is no need for further reference to any (particular or type of) explainer or audience in order "to understand what they mean, or to determine whether or not they are true." Van Fraassen's theory is "pragmatic" only insofar as explanation-sentences have context-depending truth conditions, but this does not make it a pragmatic theory. The only possibility I see for rescuing van Fraassen is to point out that he is not talking about hypotheses as such but about *answers* to explanation-seeking questions, and answers are speech-acts which the explainer produces with the intention of informing the questioner. There is no reason, however, for van Fraassen to be happy with such a defence on his behalf, for though explanations are not the same as propositions, but answers to why-questions, they do express propositions, and that depends on a context. The main task of a context is to determine for each sentence the proposition it expresses "in that context," and such a proposition has a truth-value in each possible world.

The disagreement between van Fraassen and Achinstein is very much about how questions should be understood in general. Van Fraassen's main interest is in the general logic of why-questions, which he believes can be abstracted from the particular contexts; Achinstein is more occupied with the communicative situation in which an explanation is proposed in answer to a particular why-question. In the beginning of this Chapter, I mentioned two very different concepts of what a question is. We can see now that van Fraassen follows the effective view concerning questions and answers and that Achinstein advocates a real pragmatics in line with the intentional view. Later we shall look into Achinstein's pragmatic approach and attempt to develop a pragmatic theory ourselves. But first we shall consider other forms of question than why-questions.

9
Not Just Why-questions

We have isolated various types of response to a why-question. But other questions not initiated by "why" may also invite an explanatory reaction, for example, what-, which-, or how-questions. Take a sentence like "What makes the Sun shine?", "Which one of the two chemicals caused her death?", and "How did the Universe begin?" Such information-seeking questions certainly enter into scientific discourse. If we characterize them as a different form of explanation-seeking question, then we must assume that answering how-, which-, and what-questions cannot always be reduced to answering why-questions; that is, such questions are genuine irreducible explanation-seeking questions. Some philosophers have argued for the reductionist view that all explanation-seeking questions can be "reduced" to why-questions, and some for the anti-reductionist view, holding scientific explanations respond to other forms of questions that are irreplaceable by any why-questions. As a pragmatist I reject the pro-reductionist arguments.

Philosophers like Hempel, Ernst Nagel, and to some extent the earlier Wesley Salmon, who all focus almost exclusively on why-questions, favoured the reductionist view. And van Fraassen is no exception: "An explanation is an answer to a why-question. So, a theory of explanation must be a theory of why-questions."[1] But what is so attractive about why-question if, as we have already seen, they fail to guarantee a uniform kind of responses? Does not science also explain *how* things work and *how* different features in nature are related? Surely, there is more to scientific explanation than asking and attempting to answer why a certain thing or state of affairs happens to exist.

There are indeed anti-reductionist philosophers, such as William Dray[2] and Michael Scriven,[3] who have noticed that science poses other explanation-seeking questions than why-questions. Also Sylvain

Bromberger,[4] Peter Achinstein,[5] and the later Salmon[6] denied that all explanations are answers to why-questions. I agree with them that what distinguishes description from explanation concerns only pragmatics, not logic or semantics, but I believe that at least some of my arguments are original.

I totally agree with van Fraassen's contention that an explanation is not identical to a proposition, or an argument, or a list of propositions, but is an *answer*. Nevertheless, such an answer must yield relevant information as a response to several forms of questions. In my opinion if we consider explanation as part of a communicative practice, the traditional restriction of explanation to answers to why-questions is revealed as an unappealing leftover from positivism. Of course not every *information-seeking* question requests an explanation; there are informative answers that do not come close to giving an explanation. To be explanatory the information conveyed by the answer in a particular context must address an explanation-seeking problem in the mind of questioner.

Aristotle postulated four "explanatory reasons": *causa materialis, causa formalis, causa efficiensis,* and *causa finalis*. It has been said that when modern science arose in the Renaissance scientists gave up on the other forms of explanation and appealed only to efficient causes. As long as scientists mixed them and thought that all were necessary to understand natural phenomena, science remained immature. Certainly, it had an overwhelming heuristic value for scientists to abandon the standard that required an understanding of inanimate things to include a *causa finalis*. But was Aristotle altogether wrong in assuming that some scientific explanations are not responses to why-questions, and so not reducible to causal explanations? I propose to retrieve Aristotle's four causes as possible forms of explanations in modern science.[7] We may draw the rough parallel that *causa efficiensis* and *causa finalis* enter our responses to explanation-seeking why-questions, whereas *causa materialis* or *causa formalis* typically enter explanatory responses to what- or how-questions. *Causa materialis* is found in accounts in which we explain what kind of stuff or material the explanandum is made of, and *causa formalis* is applied when we explain in terms of a structure, a mechanism, a style, a figure, or a genre.

In modern cosmology there has been a heated dispute between those cosmologists who claim to explain why we find so many fine-tuned constants in nature by introducing the strong anthropic principle and those who believe that this alleged explanation is circular and abandons the standard of explaining only in terms of *causa efficiensis, making* possible the achievements of post-Renaissance science.[8] The strong anthropic principle, formulated by Brandon Carter, attempts to invoke human observers in the explanation of fine-tuning by saying "the Universe (and hence the fundamental parameters on which it depends) must be such as to admit the creation of observers within it at some stage."[9] However, explanations of the past by invoking the development of future observers invoke *causa finalis* or design arguments. Both sides here see the aim of science as producing answers to why-questions. But in this dispute other cosmologists argue that traditional science makes no attempt to answer why-question, but pursues answers to how-questions. Responding to the anthropic answer to why we have these fine-tuned parameters, astrophysicist David Schramm objected: "Physics tries to answer the 'how' questions, and in some sense it is a philosophical question rather than physical undertaking to have a go at these 'why' questions, since they are unanswerable by the techniques of physics."[10] Hence, according to Schramm, the equation of all scientific explanations with answers to why-questions is a philosophical illusion. Perhaps it is time that philosophers should come to terms with the fact that advanced sciences are more interested in addressing how-questions than why-questions?

Nevertheless, whether it is how- or why-questions characterizing scientific knowledge in terms of the form of its questions is foolish. I can say "How does X happen?" or "Why is it the case that X happens?" Most scientific problems can be put in a variety of interrogative forms. In any case, to generalize about science from cosmology would be a mistaken inference because it is in a unique position. In other sciences if I explain why A happens in terms of B, you can reasonably ask, then why B? The so-called "initial conditions" are "initial" only relative to the occurrence of the effect for which we seek an explanation. And so on, always to more "initial" conditions. But in cosmology one is concerned with retrodicting the really initial conditions. And, if you ask "Why those conditions?" (or "How did these conditions come to be?") by definition you have stepped outside of science.

Those who advocate that all explanation-seeking questions can be expressed as why-questions hold some more or less explicit arguments in favour of their position. Van Fraassen seems to argue that explanations are responses to why-questions because why-questions are contrastive. This is one of six such arguments which I shall attempt to show fail to hold water. I do not claim that they have no bearing on each other; merely that each one runs into trouble in trying to establish that all requests for explanation can be posed as why-questions.

1 Why-questions in science

Why do the planets move in elliptical orbits around the Sun at one focus? Why does the Moon look much larger when it is near the horizon than when it is high in the sky? Why are children of blue-eyed parents always blue-eyed? Why did Hitler go to war against Russia? All these questions seek explanation, but not all why-questions call for *scientific* explanations, nor are all scientific explanations responses to why-questions. If someone asks, "Why should I *believe* that the Earth is spherical?" the proper answer would not be to explain its shape as caused by gravitational forces and the general evolution of the solar system. The proper answer would state the reasons for believing this in terms of evidential support of pictures from space, the disappearing of ships on the horizon, etc. Although it has been claimed that questions of this sort are requests for evidence rather than explanation,[11] this kind of question is really a request for an *epistemic* explanation. Other why-questions call for moral or legal explanation, as when a judge asks the offender why he has killed another man. Such questions do not work differently from other why-questions although they request for intentions and motives rather than physical causes.

There are also rhetorical why-questions which are really requests for consolation or comfort as when a bereaved wife asks her sister in despair why her husband died. A proper response would clearly not be a medical explanation. Instead the sister should offer comfort and consolation and not explain anything at all. Thus a why-question can be properly answered without aiming to explain anything. This shows, at the very least, that the proper response to a why-question depends very much on the context. If it had been the doctor to whom the wife raised the same question, the proper answer would be for him to say, if true, that her husband's vital organs failed because a tumour had metastasized. This answer would be most relevant in the context of her question, and it would function as an explanation on the death certificate.

Thus, the context-dependence of why-questions is evident from rhetorical questions in this form, which are not interrogative sentences but emotive expressions. They cannot be distinguished from other why-questions merely by their syntactic or semantic nature because they are the same as in genuine why-questions. Only the *context* of the question allows us to tell the difference. Moreover, the example of the judge's question also indicates that whether or not a why-question is asking for a *scientific* explanation depends on the context in which it is posed. What interests us, however, is what makes a response to a scientific why-question a *scientific* explanation, and whether or not such a response is context-dependent.

In everyday life reference to singular events often suffices in response to a why-question. The mechanic explains why my car will not start by saying that *this* particular car has a dead battery, and the doctor says that the boy is sick because *this* particular boy caught mononucleosis. In supplying an explanation one doesn't necessarily appeal to any empirical generalization, maintaining that under such and such circumstances every event of such and such a kind happens because an event of this or that sort happens. When *needed*, and when the circumstances are not too particular, such generalizations can be reached by simple induction from the observations of singular cases.

Sometimes we do turn to empirical generalizations seeking an account of a particular fact *qua* its membership of a certain *types* of fact. For instance, all swallows in England disappear each year before the winter comes. Why? The explanation is that in the autumn many species of birds on the northern hemisphere migrate to the southern hemisphere because they cannot find the appropriate food up north during the winter. The same answer can be used if we want to explain why a particular swallow flies south in the fall. The appeal to a generalization will in many cases enhance our beliefs in an explanation: it is not only this car with a dead battery that will not start; experience tells us that no cars with a dead battery will start.

In science, however, laws are commonly taken to have explanatory force. By understanding these laws we supposedly gain knowledge of the world in which we live, especially that part of it outside our normal sensory range. Using laws and initial conditions scientists can explain a huge variety of phenomena including the ocean tides, the electrical conductivity of copper, plant photosynthesis, and bird migration, and this explanatory virtue is taken to be the primary goal of scientific activity. By explaining all kinds of phenomena, covering laws provide an understanding of the underlying mechanisms and structures of the world, visible or hidden from our unaided eyes.

More often than not sciences propose explanations relying on either *statistical generalization* or *statistical correlations*. Sometimes we use statistical correlations because we are ignorant of the precise causal relationships in a highly complex system or organism; thus there is a statistical correlation between smoking cigarettes and developing lung cancer, but given an individual who smokes, say, twenty cigarettes a day, it is difficult to predict successfully whether that individual will get lung cancer. Because the causal mechanisms between smoking and lung cancer are very complex, this particular individual may not develop lung cancer. Instead we have a large amount of statistical data supporting the connection between smoking and lung cancer. In the social sciences statistical correlations are frequently used to explain relationships between different groups in society. There are, for example, statistical correlations between low-income groups and unhealthy life styles, people in rural areas tend to favour the right-wing political parties whereas in urban areas the votes are for more left-wing parties, the well-educated tend to live in the big cities and the less educated settle more frequently in rural areas, and so on.

Such statistical correlations often serve as the basis of statistical generalizations; in their simplest form they can be phrased as something like "Most Fs are Gs," and if we know the ratio r of Fs that are Gs, we can write it as $P(G,F) = r$ (i.e., the probability of an F also being a G is r.) Such generalizations may reflect our ignorance of what makes something that is an F also a G, but some scientists and philosophers have claimed that in some cases phenomena are inherently probabilistic. For example, the laws governing radioactive decay make the chance of a specific radioactive atomic nucleus decaying within a finite time span calculable only with a certain probability. No matter how much information we adduce, it will never be possible to determine the exact moment at which the nucleus will decay. In a similar manner, Heisenberg's Uncertainty Principle states that it is impossible to simultaneously measure with complete accuracy the position and the momentum of a particle. This is not because we lack knowledge or the measuring equipment is too crude – at least according to some interpretations – it is not an epistemic uncertainty, but an ontological indeterminacy. The more precise the location of a particle the less precise is the momentum and vice versa. This means that there is an ineliminable element of chance in nature and we can understand this only by probabilistic concepts. But the hope to reduce all statistical explanation to non-statistical deterministic explanation takes on less importance if we abandon the hope of explicating scientific explanation solely in terms of the logical form of a group of propositions.[12]

One epistemic goal prominent in the physical sciences is understanding entities at one level in terms of their smaller constituents at a lower level, and to discover the general laws holding among these constituents. For example, suppose one asks why water evaporates when heated. In answer, the physicist explains that water is made of constantly moving tiny molecules. At room temperature the intermolecular forces are sufficient to keep the molecules close together in the liquid state, but when the water is heated and the molecules increase their kinetic energy, the mutual attraction of the molecules is not strong enough to overcome their momentum. Thus they fly apart and escape into the atmosphere. Such an explanation is common within the physical sciences and is called a "reduction." Because of the prestige of physics, and the prevalence of a materialistic ontology it was believed to support, reductive answers of this structural sort were often assumed to be the correct response to *all* scientific why-questions. Since efficient forces on the micro level are the real things possessing explanatory power on the macro level, ultimately an explanation must cite facts involving atoms and molecules. All responses to why-questions in science should therefore reduce, in principle, to nomological accounts where the laws cited govern entities on a more fundamental level and, especially, those explanations which chemistry and biology proposes should always appeal to causes on the physical level. Nevertheless, partially because of its questionable metaphysical assumptions, this sort of structural reduction remains a controversial issue.

2 The paradigm argument

The paradigm argument assumes the common distinction between description and explanation, and claims there should be a noticeable difference between mere descriptions and explanations; if not, the term "explanation" would not add anything new to our descriptive practice. It is one thing to know what is the case, quite another why it is the case. The assumption is that other questions than why-questions are not explanation-seeking, but merely requests for descriptions of various phenomena. Those advocating this view do not deny that science also aims at descriptions of the world; but insofar as explanations are concerned, all are answers to why-questions. All other questions are requests for either description or something else; for example consolation or even epistemic justification.

But this very sacrosanct distinction can be doubted. The distinction may be based on either *internal*, i.e. linguistic, differences between explanation and description or on *external* differences related to different types of questions. Scriven has made some enlightening remarks on the former, contending that explanations are not something 'more than' descriptions but simply complex descriptions.[13] He considers Hempel and Oppenheim's example with the mercury thermometer immersed in hot water where the mercury level first drops a little before it rises quickly again. Hempel and Oppenheim give the following account of the explanation:

> The increase in temperature affects at first only the glass tube of the thermometer; it expands and thus provides a larger space for the mercury inside, whose surface therefore drops. As soon as by heat conduction the rise in temperature reaches the mercury, however, the latter expands, and as its coefficient of expansion is considerably larger than that of glass, a rise of the mercury level results.[14]

This is a narrative description of what happens, and it also counts as an explanation. Scriven understands why Hempel and Oppenheim take this narrative as indicating a hypothetico-deductive covering-law structure; nevertheless, such words as "thus," "however," and "results" are only reminiscent of an argument or demonstration. Scriven argues that these words are not part of an argument or demonstration but of an explanation, and they occur in simple descriptions as well.

To drive his message home, he considers the sentence "The curtains knocked over the vase." This is arguably a mere description of an event. But it is also explanatory because it includes a causal claim that could be rephrased, perhaps not so elegantly, as "The curtains brushed against the vase, *thus* knocking it over." Or even more obviously: "The movement of the curtain *caused* the vase to tip over." It is not at all clear that this explanatory account is anything more than a description of the curtain knocking over the vase. Thus the question is how and when certain descriptions also count as explanations. An adherent of the ontic view may reply something like this: "Yes, I agree that explanations are descriptions, but they are always causal descriptions. This is what makes them explanatory." He may then add: "Causal descriptions count as explanations, only if they are responses to why-questions." But in Chapter 5 we have already seen reasons for rejecting the causal option. Another example may therefore help us to proceed: Explaining *how* fusion processes enable the Sun to maintain a regular heat output consists exactly in describing these processes and their various interactions. An astrophysicist simply offers us a complex description when asked to explain *how* the Sun is able to sustain its heat production.

So sometimes explanation just requires giving an appropriate description. Hempel would object that this example is not given specifically as an answer to a why-question. But the same description can also be used to respond to a why-question: "Why does the Sun produce heat?" Thus, says Scriven, what makes our answer explanatory is that it gives the right sort of description, which is "the one that fills in a particular gap in the understanding of the person or people to whom the explanation is being directed."[15] On this view, the difference between explanation and description is that explaining is giving appropriate pieces of information, where "appropriateness being a matter of its *relations to a particular context*. Thus, what would in one context be 'a mere description' can in another be 'a full explanation.'"[16] I fully agree with Scriven on this; a particular context may contain a how- or a why-question, but this fact does not by itself suffice for assigning one answer to a description and the other to an explanation. The distinction depends on whether the answer conveys new and relevant information to the one who needs such information.

But is relevant or appropriate information identical with explanatory information? In ordinary life we pose questions like: 'What time is it?', 'What kind of dress did she wear?', and 'How did he manage to break into the house?' These questions can be answered by describing a fact. It is often said that we do not explain anything by just stating a fact. For instance, answering that today is Monday, or that she was wearing a red coat and skirt, intuitively would not count as an explanation. But again it depends on the interlocutor's background knowledge, beliefs, and assumptions whether a simple description or a fact-stating answer such as 'The time is five o'clock' or 'Today is Monday' work as an explanation. To see answers like these as giving explanation rather than merely information, we need to know more about the context in which they were expressed. For instance, we can imagine a context where a tourist, standing in the front of a museum, realizes to her surprise that the main entrance is closed, then a response like 'Today is Monday' functions as an explanation if she already knows that most museums in town are closed on Mondays. The tourist has forgotten all about what day it is; but as soon as she realizes the day, she can connect this fact with the fact that the door is locked, and then the fact-stating answer becomes explanatory.

Scientists raise various types of explanation-seeking questions using different interrogative forms: "*When* did life begin?" "*Where* did the embryo form of life develop?" are serious scientific inquiries. Some genuine explanation-seeking inquiries are in "*What is...*" form, like "*What* is the habitat of reindeer?" "*What* is the chemical composition of water?" "*What* is spin if it is not a classical angular momentum?" "*What* is the significance of Planck's constant?" and "*What* is the difference between a W- boson and a neutral Z meson?" are also. Others are expressed using "How": "*How* did the universe come into existence?"

"*How* far away are the quasars?" and "*How* rapidly is AIDS spreading in the United States?" All these non-why questions should also be considered as serious requests for explanations in particular contexts. For any of these questions what makes an answer *scientific* is how it is justified (or not) by the methods by which it is reached, rather than the interrogative form of the request. Let us say that the geneticist and the tealeaf reader both predict that Mary will die young. At Mary's funeral we would count one as a scientific answer to "Why did Mary die young?" and one as not, precisely because of how the answer was arrived at. Indeed someone might say that this is of course because of what you believe; you believe in genetic science and not in tealeaf reading, but if it were the reverse, then the tealeaf reader's response would be in a position of providing a scientific explanation. To this I would rejoin that anyone who believes in tealeaf reading doesn't have enough of a conception of scientific explanation to request a scientific explanation of Mary's early demise.

Nonetheless, someone might still hold that the different forms of non-why-questions are information-seeking, but not explanation-seeking. But, again, the pragmatist would always reply, it is the *use* of language, not its syntax or semantics, which determines whether or not a particular question is explanation-seeking. First of all you can simply put the word 'Explain' in front of a long series of descriptive sentences, thereby making the sentence explanation-seeking, even though it is not even in interrogative form. It is clear that an explanation is not only a response to a question if one realizes that all kinds of *wh*-terms can follow 'explain.' For instance, "Explain why, how, who, what, when and where..." Even more generally any request for explanation could have the simple command (imperative) form "Explain X", where X can be an event, a state of affairs, a process, or a universal relation. The word "why" is totally unnecessary, and all scientific discourse could very well be rephrased in a way in which the word "why" never appears as such. According to the speech act theories we may produce indirect speech acts in linguistic communication. Many sentences can function as explanation-seeking questions in the right context in spite of the fact that they are not phrased as a question. In a certain context a declarative sentence such as "You are too late!" can act as a request for an explanation, whereas a question like "Can you reach the salt?" is a request to pass the salt and not a genuine question at all.

Thus, there is good evidence that we associate explanations with informative answers to a wide variety of kinds of questions (or sentences beginning with "Explain...") of which why-questions represent but one kind. Scientific practice shows that spoken sentences like "Could you explain how U-238 decays?" or "Explain to me what is the significance of Planck's constant in quantum gravity?" are answered by giving a complex description that contains information relevant to enlighten the topic of a how- or a what-question. This observation indicates what can count as explanations in the physical sciences can be a complex description as a response to the request for further information that is posed within an appropriate scientific context and which therefore is proposed as providing an appropriate scientific answer.

In sum, the paradigm argument depends on a clear distinction (and non-overlap) between explanations and descriptions, but as we have seen, such a distinction cannot be drawn, neither on linguistic features nor on the kind of question addressed; so the argument fails to work.

3 The relation argument

In light of the above discussion, I submit that we ask questions whenever we seek information about something of which we have insufficient knowledge, or when we do not know what to believe. We may be interested to know *where* something takes place, *when* it takes place, *what* the case is, *how* something is as it is, and *why* something happened as it did. Answers respond to these requests for various forms of explanation, supplying us with the proper information to fill in gaps and lacunae in our field of beliefs. Thus, the answer to a when-, where-, what-, how-, or why-question becomes equivalent with the ability to give an appropriate description of when, where, what, how or why.

Of course, even taking account of context, not every kind of information transferred by a speech act is explanatory. Explaining a fact is not the same as stating a fact; a fact is explained by other facts. Quite often a response to questions like "What time is it?" "When did you arrive at work?" and "Where are you?" only states the single fact that is requested. Answers like "The time is 2 o'clock," or "I arrived at 9:00 a.m.," or "I am at Heathrow" is not explanatory in most contexts. They merely state facts answering the question the interlocutor has asked. However, explanations inform the interlocutor about one fact in virtue of its relation to at least one other fact. Some have held only responses to why-questions exhibit this feature. Answering a why-question appropriately is explanatory because the answer requires relating the topic of the question to something else. As noted earlier, explanations often provide a narrative context around the fact being questioned. Does this indicate that what-, when-, and where-questions cannot also act as explanation-seeking questions?

An answer offers information as an *acceptable* explanation whenever it gives us a story making the fact stated in the *explanandum* consistent with specific parts of our background knowledge, and therefore this fact becomes more likely, significant, or less surprising in the light of some other facts stated as part of the information expressed by the story told by the *explanans*. Imagine a question like:

(1) When did the Big Bang take place?

Raising this question certainly presupposes that the interlocutor knows something about the origin of the Universe. Assume the response prompted by this request is:

(2) Approximately 13.7 billion years ago.

In most contexts this answer may not be seen as an explanation. The questioner may indeed be completely satisfied by just being told this fact, if it is a fact, but he may also expect to hear an explanation of that fact. The fact to be explained is the time of the origin of the Universe.

The explanation we are looking for must have the following form:

(3) The Big Bang took place 13.7 billion years (rather than X billions years) ago *because*...

Now, the respondent may assume (1) is requesting the prior necessary and sufficient factors that caused the Big Bang to take place at that time or requesting an explanation of how we know when in time this event took place. Perhaps both questioner and respondent already know that physics cannot state any such prior necessary and sufficient factors. So if the answerer states (3) and connects the claimed fact to other facts about the expansion of the Universe, the red shift of the distant galaxies, the Hubble constant, the deceleration parameter, the temperature of the background radiation, etc., and mentions the uncertainty of some of these numbers, the questioner seems to receive an explanation to his inquiry. (3) might be the typical response which a student of cosmology would give on a question like "Explain when the Big Bang took place" given the data mentioned. Thereby the respondent excludes alternative times. The details of the explanation given depend on how much knowledge the listener already has about cosmology, that is, whether the interlocutor is a cosmologist himself, a physics student, or a bus driver. The story being told is presumably the same: the one to a colleague is not more correct than the one to the bus driver – it only contains information of a different order.

In principle, the way we address such when-questions in science is no different from how we address similar everyday when-questions like "When shall we meet tomorrow?" Again, the respondent could just say 6 o'clock, but most people would not give a time with no reason. Rather they prefer to offer an explanation in a form like "Let's meet at 6 o'clock *because*..." The respondent would thereafter give her reasons of the choice by saying that 6 o'clock suits her best since she is going to have tea with a friend at 5 o'clock, and it will take only five minutes to walk from the cafe to the restaurant. She connects the selected time with other facts to explain her choice.

This kind of explanation is not limited to responses to when-questions; it also holds for who-, where-, how-, and what-questions. If a student raises the question to his teacher "What number does gold have in the periodic system?" the teacher may give an answer in the form of an explanation "Gold's atomic number is 79 *because*..." Here the teacher explains to her student gold's position in the periodic system by referring to its physical structure. The student knows that gold has a number in the periodic system, but does not which one it is.

Schematically we may define an explanation-giving answer to all interrogative statements, including why-questions, as

(E) P (rather than X) *because* A.

Van Fraassen's suggestion concerning what an incomplete answer to a why-question looks like is strikingly similar to (E), which is assumed to constitute an answer to all *interrogative* questions. But there seems to be an important difference between how much the questioner may know about P. With respect to why-question the topic P expresses what the questioner already knows, and A states the information she is asking for. The information contained in A is that P is related to X, Y, and Z in a specific way. But in the case of who-, what-, when-, where-, and which-questions, the questioner knows less than everything said by P, but she knows something, say that the Big Bang happened, but not when it happened. So P also expresses the intended information, and when we give either *reasons* or *evidence* for P (including the hoped-for information), we gain an explanation.

So the difference between explanations answering why-questions and other forms of explanations seems to be that in the first case the questioner, already knowing P, asks for what explains P and the respondent explains P by relating it to A, whereas in the second case, where the questioner know less than P, she asks to be informed about P, and the respondent informs about P by explaining P in virtue of its relation to A. Nonetheless, there seems to be a problem, for when the questioner is first told P (about the time of a meeting, the time of the Big Bang, or gold's place in the periodic system), then the respondent explains P as if she was addressing a why-question.

It would be natural to object that an explanation-giving answer fulfils (E) only because it does not respond to the original question (1), but to an implicit why-question. It is only when P refers to the particular date of the Big Bang that the answer fits the form of (E), but then it is not really addressing the why-question:

(4) Why did the Big Bang take place 13.7 billion years ago?

For all we know (4) has no answer, or if it does, it may beyond human capacity to know. A more plausible candidate would be:

(5) Why do physicists believe that the Big Bang took place 13.7 billion years ago?

It now seems to ask for *adequate grounds* for believing that Big Bang took 13.7 billion years ago. And the argument continues: it is exactly those grounds, which come to light with the use of the term "because." There is still, I think, an important difference between an answer such as (3), which seems to be an appropriate response to question (1), and an answer like

(6) The physicist believes that the Big Bang took place 13.7 billion years ago (rather than X billion years ago) *because*...

The latter seems to address the interrogative sentence (5). The truth conditions of (3), in contrast to the truth conditions of (6), do not refer to the physicists' beliefs. Rather the truth conditions of (3) contain only terms of processes, interactions, and structures, which allow cosmologists to calculate the age of the Universe.

Nevertheless the whole analysis is flawed by assuming that only an appropriate response to a why-question guarantees the reference to a relation which is enlightening in an explanation. It has not been shown, nor can it be proven, that we can request such an explanatory relation only by means of a why-question. First and foremost we ask for an explanation to be enlightened about how a particular fact, described in the *explanandum*, is significantly connected with other facts stated by the *explanans*. But the term "because" does not tell us why these facts are connected, how they are connected, or what constitutes the connection. Assuming that explanatory understanding is identical with having beliefs about the relevant connections between facts, it is not obvious that connecting facts with other facts is always identical with identifying the cause for the existence of some fact. Moreover, an explanatory relation cannot be reduced to a deductive relation. Facts are related in many different ways. They can be temporally, spatially, causally, semantically, structurally, mereologically, functionally, or intentionally connected, and they may be connected in even further types of ways. Knowing any of these connections is a possible candidate for the explanatory understanding we desire. Yet not all kinds of connections count. If facts are merely connected by all happening on a Tuesday, this doesn't qualify as an explanation in almost all contexts. In general, the nature of the missing knowledge, as well as the context of the question, determines the type of connections we want to include in an explanation and the type we want to rule out.

Beliefs concerning these facts and relations are not only inferentially but also explanatorily related. Thus logical inference is not the most relevant form of epistemic connection when discussing explanation. First-order beliefs are usually attached to a system of beliefs by second-order beliefs about how the first-order beliefs are related to the system. Therefore it makes much more sense to say that what is going on in an explanation is that we connect our beliefs about facts by means of telling different stories about the facts. Explanation in the above example is the same as telling a temporal story about how far back in time the Big Bang took place. A story is not an argument. It is a narrative, which need not specify the epistemic grounds for believing a certain fact. It describes how things are connected and how things may change by interacting with other things. However, it does more than merely explain by relating things, it also gives the explanation structure and character.

If someone asks "How did Socrates die?" an appropriate explanation may take the form "Socrates died in peace, faithful to his principles, after having been condemned to drink hemlock." The questioner knows that Socrates died; she may even know that the great philosopher drank hemlock, but she does not know the way he died before being told. The respondent answers both by relating Socrates' death to another fact (drinking hemlock) and by telling us, as part of the explanation, about the manner of his death.

Summing up, stating a single fact provides no explanation in all those contexts in which the interlocutor lacks the epistemic background to connect this fact with other facts, but an explanation can be given by a single description of a fact, about which the questioner seeks information, with descriptions of other facts which the questioner may or may not know in details. It is not always the little word "because," but a story that does the work of establishing a connection between one description and some other descriptions. The kind of connection we need to gain understanding depends on the particular question being asked, the type of question it is, and the context of the response.

4 The reason argument

Another argument begins by assuming any explanation gives *reasons*, i.e. we have an explanation only if we can find a reason why something is as it is, or occurs as it occurs. Therefore all scientific explanations are responses to why-questions.

The *Oxford English Dictionary* gives the verb "to explain" two seemingly different meanings. One is to make something plain or clear; the other is to give or be a reason for something. This reflects the distinction between description and explanation; thus there are description-giving explanations and reason-giving explanations. These different kinds of explanation seem to correspond to different kinds of questions. The description-giving explanation responds to a how- or what-question, whereas reason-giving explanations respond to why-questions. Unfortunately, things are not so obvious and straightforward as they seem.[17]

No doubt the appropriate answer to every why-question gives reasons. This is the essential feature of the logic of discourse of a why-question. Consider a question like:

(1) Why do some birds *migrate* to Africa in the autumn?

An appropriate reaction would be:

(2) The reason that some birds migrate to Africa in the autumn is that in Europe they would not be able to find food during the winter.

Of course, the question and the answer only make sense within the broader context of the geography of the Earth and the annual climate changes on the Northern Hemisphere. In fact the answer that cites the lack of food is no more relevant than one that refers to the lack of daylight, cold temperature, or snow cover. All these facts are parts of the same overall causal story where the lack of food is the perspicuous result. Another appropriate answer is:

(3) The reason that some birds migrate to Africa in the autumn is that they have an instinct to do so.

This answer (and question) makes sense only in an even broader context, including biological evolution and natural selection.

I think that it is because the stated reason in either (2) or (3) is meant to *justify* believing in the migration that explanations are often only associated with responses to why-questions. This element of justification, connected with stating a reason for believing-that, intuitively gets people to think of explanation as a reason-giving answer and, therefore, being an answer to a why-question.

Indeed *many* requests of knowledge in terms of a how- and a what-question can be re-phrased as a why-question but not all. A what-causal question as

(4) What causes some birds to migrate to Africa in the autumn?

can be replaced by (1). Hence (2) and (3) are both possible answers to (4). However, a what-question such as

(5) What do you mean?

may be translated in a particular context as

(6) Why are you saying so?

Indeed, (5) is simply ambiguous as a question out of context. It may be requesting what it literally asks for: a clarification of meaning for a problematic term or proposition, which explains only how words are being used. Or the questioner may fully understand what the interlocutor said, but finds it astonishing, and therefore desires a *justification* of the statement, which caused him to ask (5). Thus, it seems, not every how- or what-question can be translated into a why-question without a loss of meaning.

However, one may argue that translation between how- or what-questions and why-questions is not really the issue. The important issue is whether or not a response to a how- or a what-question can *always* be construed as if it also were a response to a why-question. A positive claim does not require showing that non-why-questions cannot be constructed for explanatory answers, nor that these explanation-seeking non-why-questions are somehow disguised why-questions. The suggestion is only that if a proper answer to, say, a how-question should count as an explanation, you are *always* able to replace it with a why-question requesting the same answer. I shall call this requirement the *replacement condition*, holding that a response to a how- or what-question is sufficient as an explanation, only if the same response would address a proper why-question that could replace the how- or the what-question. In other words, the replacement condition holds that an explanation is necessarily a *potential* response to a why-question; however, it is not necessarily an *actual* response to a why-question.

The idea behind this suggestion is that an explanation can be an answer to all kinds of questions as long as you can pose a corresponding why-question requesting that same answer. Consider the following question:

(7) How do birds from the Northern Europe actually *migrate* to Africa?

This how-question cannot just be replaced with

(8) Why do birds from the Northern Europe actually *migrate* to Africa?

Replacing "How" by "Why" changes the meanings, i.e., they are requests for different sorts of responses. The how-question asks for the *actual manner* in which birds migrate, whereas the why-question asks for the *actual reason* why birds migrate. Both are requests for information about North European bird migration, but they are seeking different pieces of information. A straightforward, but highly relevant, response to this how-question would be that birds fly (instead of walking, swimming, etc.), whereas the proper response to the why-question cites the lack of food in Northern Europe during the winter (instead of the lack of daylight, cold temperature, snow coverage, etc.) Thus, since the answer to (7) will not include a reference to a 'reason of how,' it cannot be an explanation.

Several authors have recognized that *how-possibly* questions are genuine explanation-seeking questions; consider for example:

(9) How is it possible for birds to migrate to Africa?

One possibility, Hempel's interpretation, is that (9) is posed by a person under the mistaken impression that this occurrence is physically impossible or highly improbable. But there are several other adequate interpretations. The correct understanding depends on the explanatory situation; for instance, (9) may just as well express not disbelief, but a lack of knowledge about birds' abilities that enable them to be heading in the right direction. In this context the intended meaning would be something like "How do birds find their way to Africa while migrating?" or "How are birds able to navigate their way to Africa?" Here the relevant answer depends on whether the birds are only nocturnal migrators, only daylight migrators, or both. Therefore a response may refer to the birds' internal star maps, or their magnetic sense, and/or their ability to correct the course by the sun as well as by landmarks. One answer is

(10) The reason that it is possible for nocturnal migrators to migrate to Africa is that they can navigate with the help of the stars.

Here we can easily find a why-possibly question matching this response to (9):

(11) Why is it possible for birds to fly straight to Africa?

One may then argue that since (9) and (11) are answered by the same reason-giving explanation – to that extent they are equivalent.

Alternatively, (9) could be interpreted as posed in an explanatory situation in which the questioner was thinking of the *distance* between Northern Europe and Africa:

(12) How are small birds able to fly such a long journey?

In this case (10) is no longer appropriate. Instead, an answer like,

(13) The reason it is possible for birds to migrate the long distance to Africa is that they can find enough food while resting,

seems to reflect the intentions behind the question. Again (13) is an appropriate response to the following why-question:

(14) Why is it possible for birds to travel the long distance to Africa?

Thus, any satisfactory answer, also those to how-possibly questions, gives us a reason for birds migrating to Africa. Consequently, it seems possible to stipulate an appropriate why-possibly question corresponding to any how-possibly question.

Let us now turn to the how actually-question, taking another example taken from Salmon. Instead of (7) we could say:

(15) How did mammals (other than bats) come to be in New Zealand?

The answer is that human beings came in boats and later imported other mammals. Salmon regarded this answer as a genuine scientific explanation. It is not an explanation of why they came there, but an explanation of how they got there. Thus (17) cannot be reformulated as

(16) Why did mammals (other than bats) come to be in New Zealand?

Here an appropriate response is

(17) The reason that mammals except for mice and rats came to be in New Zealand is that people wanted to use them.

In contrast, the proper response to (15) cannot be expressed as a reason-giving answer. Does this means, then, that telling how mammals came to be in New Zealand does not count as an explanation? Salmon said no, but gave no reason.

Some may hold that answering a how actually-question merely gives us a description because such a response provides no reason as answering of a why-question normally requires. In my opinion such a reply is wrong. Only a response to a why-question (or a matching what-causal or how-possibly question) expresses a reason. If one restricts explanation to giving reasons, then answers to how actually-questions cannot act as explanations, case closed. But if one allows an explanation to be an answer selected from a huge repertoire of possible responses, then the case is still open.

I maintain that many answers to how- and what-questions, which cannot be replaced by a why-question, function as genuine scientific explanations, such as "How did the Universe begin?" "How did the Egyptians build the Pyramids?" and "What kind of chemical bond connects Na-atoms and Cl-atoms?" You may object that a response to such questions simply describes a fact, but those facts are not merely "described"; the descriptions also are claimed to function as the 'reason' for the fact expressed in the explanandum. So, for instance, the fact that the Universe began as a Big Bang is the reason we find space-time expansion and background radiation in the Universe today. As always, the appropriate response depends on the context in which the question is posed. Before we get to this explanation of the origin of expansion and background radiation, we need to know that the Universe began in a state of overwhelmingly high energy, as a tiny seed of expanding space-time (called the Big Bang). In every explanation we explain one fact, say, the beginning of the Universe, by describing it in relation to another fact, the Big Bang, in contrast to a whole class of possible descriptions, the contrast-class, in which the beginning of Universe is associated with very different physical states. This answer counts as an explanation in this situation because it answers the epistemic problem of how the Universe began that initially gave rise to the question.

In some explanatory situations, we regard an answer to a how- or a what-question as a description, but in others as an explanation. This difference arises from whether or not the question poses a lack of knowledge to be overcome. Explanation-seeking does not start with the question but with what leads up to posing the question. It begins when a lack of knowledge is felt as an urgent problem that the questioner does not know how to overcome in the particular situation. Lloyd Bitzer has named the problem that gives rise to a question an "exigency."[18] Thus the exigency reflects the kind of deficit the questioner experiences, and an "explanation" is any description that helps overcoming this exigency. The answer to (7), that birds fly, is seen more as a description than an explanation because it is common knowledge that nearly all birds fly, and this is how they move around over longer distances. This is not something discovered by science; it is part of our common background knowledge. The way birds migrate would not normally be an exigency for anybody, except children. We have no expectation of discovering that migrating birds don't fly, but, again, the question could signal the presence of a real epistemic problem in some unusual contexts. Other answers to (7) are possible, responses we may take to be explanations. For example, situations where the questioner wonders about whether or not birds migrate in all kinds of weather, flying day and/or night, making stops or flying non-stop, etc. If (7) expresses an exigency, an appropriate explanation relies on scientific investigation.

Accordingly, the answer to (15), namely that human beings came in boats and later imported other mammals, functions as a scientific explanation. Here the answer provides information that is not common knowledge.

How mammals came to New Zealand represents an exigency not merely to an individual but to the scientific community as a whole. The information solving the problem requires scientific research. Therefore it is always possible that biologists and historians one day might discover that mammals already lived in New Zealand before humans arrived.

Our discussion shows that responses to how-actual and what-questions also function as explanations although the logic of these kinds of questions does not always request reason-giving answers. As already emphasized, the distinction between description and explanation is a pragmatic one. If a response addresses an exigency raised in a question in a way relevant and informative in that context, the answer yields an explanation. If the answer does not approach any exigency, because the question fails to express one, it merely functions as a description.

5 The translation argument

Philosophers of science, who maintained that requests for scientific explanation must be put as why-questions, have long claimed that requests for explanation formulated in other ways can always be translated into why-questions without distortion of meaning.[19] This is a bit stronger than the *replacement condition* introduced in the previous section, which only requires that *if a response to a how- or a what-question should be considered to be an explanation*, there must be a why-question, which is answered by the same response. The *translation condition* demands that such a response is an explanation, only if the content of the original question can be recaptured by the content of the why-question. In other words, any request for genuinely scientific explanation in form of a how- or a what-question is believed to be reducible to posing a why-question. The idea is that most information-seeking questions can be addressed satisfactorily by a descriptive response, and mere descriptions cannot operate as explanations. An appropriate response to a what-question, for instance, often requires only a descriptive answer, but if it requires more, then it can always be restated in terms of a why-question.

Indeed, whether or not all scientific explanation-seeking questions can be translated into why-questions can easily become tautological: a question is a request for scientific explanation if, and only if, it can be translated into a why-question, and if not, it is not. Nobody wants to say this, but we can avoid this trap by looking into explanatory practices in science to see whether scientists do pose non-why-questions and, if they do, whether they can always be translated into why-questions. As already indicated, I challenge the view that they can.

Evidently, some tokens of what-questions are not translatable into tokens of why-questions because this type of question does not communicate a request which is communicable by a why-question. Assume van Fraassen's position that explanation-seeking questions are essentially equivalent to a set of answers. This implies that what-questions are not all translatable into why-questions if it is possible for the truth-conditions of a direct answer to a what-question to be different from the truth conditions of a direct answer to any possible why-questions. Suppose a cosmologist asks "Of what stuff is the universe predominately made?" based on a nucleosynthesis calculation suggesting that dark matter in the universe does not consist of ordinary matter. No why-question can convey the topic of this question. Several proposals to explain the nature of dark matter have been given, but none have gained consensus. Whatever the answer, obviously the explanation will consist of a very complex story requiring calculations based on observations, hypotheses, conjectures, and theories. Such complex answers act as genuine scientific explanations. How- and what-questions apparently respond to exigencies where we want to explain the nature of things, concepts, theories, relations, logic, mathematics, structures, fundamental laws, rules, etc.

By the same token a question like "How did the Universe begin?" cannot be translated into a question like "Why did the Universe begin?" even though they refer to the same topic. The questioner realizes that the former question cannot be reformulated in terms of the latter since they request different answers. Many people have believed that the universe has no beginning or end and is truly infinite. The accumulation of evidence in favour of the Big Bang supports the view that the *actual* universe cannot be considered infinite, thus it takes on the properties of a finite entity, possessing a history and a beginning. The Big Bang began to be taken seriously as more than just a mathematical possibility because it explains why distant galaxies travelling away from us at great speeds, and subsequently it explained the cosmic background radiation (the glow left over from the explosion itself). In the first question the questioner asks about the state in which Universe started out. According to the classical Big Bang model, the answer is that the Universe began in a state with an extremely high energy density and with a space-time volume close to zero. However, the second question inquires into what caused the Big Bang. Here the answer may be that the Big Bang is the reversal of an "earlier" Big Crunch, that it occurred as a huge excitation of the quantum vacuum state, or what have you.

Admittedly, many what-question and how-question tokens are unproblematically intertranslatable with why-questions tokens. Therefore, just as what- and how-questions can be rephrased in terms of why-questions, we can also reformulate why-questions in terms of what- and how-questions. But no direct translation from a what- or how-question to a why-question is possible by just changing "what" or "how" into "why." Whether or not such a possibility is open depends on the meaning of the question, which is partly determined by the questioner's intention. A little argument shows that the translation of a what- or a how-question into a why-question depends on the intention behind the question, and not on the kinds of response possible. Sometimes there exists a corresponding why-question to which one possible answer has the same truth conditions as an answer to a how- or a what-question, but sometimes when we bring in the context of the question (including the questioner's intention,) the meaning of the question *cannot* be translated into a why-question.

To illustrate, recall Socrates' death and let us pose three, apparently, different token-questions:

(1) How did Socrates die?
(2) From what did Socrates die?
(3) Why did Socrates die?

How should a historian answer these questions? For the sake of argument, assume all three questions have the same topic, so they have a common set of truth conditions. A Hempel-style historian will answer, "Socrates drank hemlock, and everybody drinking hemlock dies; therefore Socrates died." This answer covers all three questions if they are regarded as scientific questions. A historian inspired by van Fraassen will reply, "Socrates died because he drank hemlock." He can argue that the interrogative of (3) contains the interrogative of (1) and (2). Therefore the direct answer to (3) is also the direct answer to (1) and (2). The direct answer of (3) is also the scientific answer. However, the insensitivity of these historians to the nuances of meaning would produce a caricature of the real-life explanations made by historians.

Assuming each question arises from different intentions, the requested answer to each must reflect this diversity in order to be relevant. (1) can still be posed in a context in which the questioner already knows Socrates was executed and that the execution was poisoning by hemlock. What then could be the intention behind (1)? Possibly the questioner may want to know something about Socrates' state of mind: was he upset or calm, did he or did he not regret what he had been accused of, etc. Alternatively, the questioner could wish to know something about the physical situation. Was he physically forced to take the hemlock or did he drink it as an act of free will; did he die alone or surrounded by friends; or/and did it happen in a jail or at home? Turning to (2), obviously this question cannot be posed as genuinely seeking an explanation in a context where the questioner already knows that Socrates drank hemlock and that hemlock is poisonous for humans. The questioner must know, of course, that Socrates has died. But apart from this information she does not need to know more about Socrates. Thus, the answer to (2) could be "Socrates died because he drank hemlock and hemlock is poisonous to people." Finally (3) can be posed in a third context in which this answer to (2) is not satisfactory, such as a situation where the questioner knows that Socrates died prematurely, and perhaps is even aware of his drinking hemlock. What she does not know, but wants to know, is whether it was suicide, an accident, or something else. So the answer "Socrates died because he drank hemlock and hemlock is poisonous to people," which Hempel and van Fraassen would consider a scientific answer, is in the given context not an answer to a why-question but a what-question. And in the same context the same answer would not of necessity be the kind of answer sought by the corresponding why-question. Thus it is impossible to grasp the practice of scientific explanation if we hold that what- and how-questions can always be translated into why-questions owing to stipulating that an explanation-seeking question reduces to a set of answers, which are assumed to be the same for these different types of questions. We have seen contexts where this assumption fails and where the intention behind the question determines its content.

6   The contrast-class argument

The question "Why P?" always appears to be elliptical for "Why P rather than P^*, P^{**},...?" We simply explain P by excluding the possibility of a class of alternatives that might have been the case, and we do so by relating P to other facts that determine P. Perhaps this particular feature endows answers to why-questions with their explanatory force and leads people to believe scientific explanations are always answers to why-questions? Of course it is possible to imagine contexts where the questioner desires an explanation to exclude alternatives. But in other contexts, for example if the questioner has made a new and unexpected discovery, the contrast-class may not be known or is non-specifiable. In this case a contrast-class is uninformative, as in a question like "Why P rather than nothing?" The elliptical nature of why-questions really conceals the fact that posing a question requires possessing *some* background knowledge, beliefs, or assumptions about possible alternatives. The contrast to these alternatives is the reason answers to why-questions are considered to be informatively rich and why we regard such explanations of great value in science.

When specifying the contrast-class there are two possibilities. We can think of it as a class of *propositions* confined by the entire logical space of alternative propositions, i.e., a set of propositions expressing all logically possible alternatives. Or we can regard it as a set of *descriptions* confined by the context, i.e., by the explanatory situation giving rise to the actual exigency. If we consider a contrast-class as all logically possible propositions that cannot be simultaneously true, every contrast-class will contain many propositions not included in our background beliefs. However, when viewed pragmatically, the contrast-class narrows down to only those descriptions that could be possible answers to a particular question, given one's background knowledge, beliefs, and assumptions. Not surprisingly, the content of the contrast-class depends on the context of the explanation-seeking question, including the questioner's background knowledge, beliefs, and assumptions. If we think of the contrast-class in the logical sense as what gives us a satisfactory complete explanation, an explanation would not only have to rule out all known alternatives, but also all possible alternatives. It seems unlikely that this could ever be done, even assuming "fixed" background knowledge.

Interestingly, van Fraassen makes both of the following two claims:

(a) Explanations are responses to why-questions.
(b) Why-questions can be given a contrastive analysis.

Are these two claims somehow interdependent? Although van Fraassen subscribes to both, he does not argue explicitly for either. He explicitly claims (a) but does not argue in its favour. He seems to think that (a) and (b) are connected in the following manner: (1) explanations are merely answers to why-questions because (2) only why-questions can be given a contrastive analysis, and they can be given such an analysis, only because (3) they appeal implicitly to a causal element or at least a reason.[20] Therefore the contrastive nature of all explanations implies they must offer a cause or a reason. But I doubt there is a necessary connection between (a) and (b), because nothing in logic prevents us from holding that explanations are responses to many types of questions, but that why-questions distinguish themselves from others by their contrastive form. My point is that if the contrastive class view of explanations is correct, its correctness is not determined by the logic of the question and its response. It would have to be established as correct some other way, presumably by establishing that in actual practice all explanations involve ruling out other alternatives. However, such a claim seems preposterously at variance with the facts about scientific behaviour when actually explaining things.

Nonetheless, I can see no reason why questions other than why-questions cannot also be given a contrastive formulation. How-questions can be analysed in terms of contrast-classes, and this applies to both *how-possible* questions and *how-actual* questions. Salmon mentions an example of an accident involving a DC-9, which went down shortly after take-off and landed upside down.[21] We might ask "How is it possible for a DC-9 to turn upside down?" As Salmon remarks, this question cannot be translated into a why-question because we are not asking for an explanation of the *actual* situation but an explanation of how it is *possible* for an airplane to turn upside down. However, this type of how-question can be given a contrastive form because (i) there is a topic (namely DC-9 turned upside down); (ii) we can establish a contrast-class {the DC-9 did not turn over, the DC-9 turned vertical, and the DC-9 turned upside down}; (iii) it can produce an explanation appealing to a reason favouring the explanandum at the expense of the contrast-class. In the actual case the explanation was that both wings had been covered by ice but only one was de-iced before take-off. Thus, "How is it possible for a DC-9 to turn upside down?" is answered by "Because only one wing was de-iced, or because the plane experienced a strong turbulent crosswind when it took off, etc." Normally there is not just one correct answer to a how-possible answer, whereas we often (falsely) think a why-question has one and only one correct answer.

Can we give a similar contrastive account of how-actual questions? Take the question: "How did this DC-9 actually turn upside down?" that is raised after the disaster. Just as we saw in connection with the how-possible question, this particular how-actual question entails the same topic and exactly the same contrast-class. But it does not seem to require a reason excluding other members of the contrast-class from occurring. Another particular how-actual question such as "How did this accident happen?" might give rise to a causal answer, but not this one. The how-actual question does not only ask for what brought about the incident, but for the manner in which it happened: the airplane found no stability in the air (because the lift on one wing was different from the lift on the other), so the heavier, icy wing tilted towards the ground, causing the plane to flip over. We get an explanation that contains a causal part but which also includes more than the causal element, namely one element from the contrast-class that consists of manners of crashing.

In principle all questions can be given a contrast-class, since they express a central topic. Take the question "Of what stuff is the universe predominately made?" The topic is: "The universe is predominately made of a certain kind of stuff." The contrast-class is then {The universe is predominately made of matter; The universe is predominately made of radiation {The universe is predominately made of visible stuff; The universe is predominately made of invisible stuff {The universe is predominately made of baryonic stuff; the universe is predominately made of non-baryonic stuff {The universe is predominately made of..., etc.}}}} This contrast-class for an explanation-seeking what-question consists of sets that are embedded in each other, distinguishing it from the contrast-class of a why-question, and seeming to indicate that a direct answer to such a question must deviate from a why-question.

As with why-questions, the rejection of the topic means the rejection of the corresponding what-question. The contrast-class may help specify what *we* take to be explanatorily relevant in a given situation. However, a reply to non-why-question does not invariably explain by providing causes or reasons. These types of questions do not ask for a contrastive account because their possible answers do not state that we should favour P at the expense of the other members of the contrast-class. The desired information does not alone give us a reason (or a cause) to select this element rather than other members of the contrast-class. Nonetheless, all types of questions can function as explanation-seeking questions as long as the requested information may provide understanding to the questioner by eliminating a particular exigency. Contrastive accounts are indeed significant because they are rich in information in virtue of their reference to facts which determine that P while excluding $P^*, P^{**}, \dots$.

7 The relevance argument

Two main issues in the theory of explanation have been (i) the *asymmetry* between *explanans* and *explanandum* and (ii) the *relevance* of the particular explanation to the particular inquiry. The only type of question, the argument goes, which guarantees the relevance of the answer stands in an asymmetric relation to the topic of the question is a why-question. Scientific explanations are nomic or causal explanations, and nomic or causal explanations respond to why-questions, hence scientific explanations are responses to why-question. Thus scientific explanations are nomic, if not causal, explanations; therefore, all answer why-questions, the context contributes only to which of the alternative causal features that are elevated to be *the* cause. This perspective regards an answer as explanatorily relevant because it is tied to the appropriate question either by an inferential link between sentences or by of a nomic or causal link in nature to which the appropriate answer refers. If explanation is nothing but a tool for understanding, neither logic nor causation can determine whether an answer is relevant or not. I suggest that the link is epistemic and intentional, permitting other types of questions than why-questions to request an explanation.

Explanatory relevance must be relevant to somebody; no answer or a fact is intrinsically relevant as such. For instance, power failure is not *intrinsically* relevant for explaining why the light went off. It is relevant only if power failures can be put into a relation to other facts, which the information-seeking person knows as part of her background beliefs or assumptions. In other words, the appeal to an unexpected power failure becomes explanatorily relevant only if the person requesting an explanation believes in a causal connection between electric power and artificially created light. So the relevance on an answer is determined by the kind of contrast-class the questioner has in mind. Basically to answer the explanation-seeking question the respondent needs to identify the questioner's contrast-class. But sometimes the respondent may have to persuade the questioner that the selected contrast-class is 'ill-selected,' by showing that some of his or her beliefs and background assumptions are false and attempting to provide some alternative set. When the questioner is also the respondent, he has a direct access to his own beliefs about the contrast-class, and may eventually revise them to make his explanation accord with observation and experiments.

Asymmetry very much depends on the question of relevance. If some piece of information is relevant to answering a certain question, by the rules of discourse, its relation to the question is asymmetric; in other words, the relevance relation is primary in understanding explanation. So the real question is: if explanations are information-providing responses, how do we make sure that this information is relevant?

I maintain that the relevance of the available information given by an explanation depends on the type of question being posed and the context in which *this* particular question is raised. How-questions and what-questions determine different contrast-classes, which in turn determine relevance relations different from those that would be appropriate for answering why-questions. Different types of questions presuppose different contrast-classes and relevance relations. However, given the content of a particular question, the context determines the particular contrast-class, and therefore the particular relevance relation. Thus, the types of answers that can function as genuine explanations in everyday contexts depend partly on the kind of knowledge we are seeking, and partly on the accepted background assumptions and the factual information forming the context of the question.

The same holds in scientific contexts. When scientists want to explain something about the more permanent states of affairs of the world, the attributes of various entities, the constituents of materials, physical constants, etc., they may request an answer to a what-question. The appropriateness of a particular answer to that particular question *is* only a question of whether it provides them with some information, which is relevant given their theories and cognitive interests. Since background knowledge, beliefs, and interests change, so will the answers regarded as relevant to a certain question. Even answers to the same causal request may differ from person to person, according to what that individual regards as the salient feature among the causally relevant circumstances. This contextual element in explanation does not entail explanatory relativism as long as the various explanations are not logically exclusive; however, it does highlight the rhetorical nature of explanation.

But before presenting my rhetorical account of explanation, we need to bring the discussion of this Chapter to an end. The most direct and uncontroversial way of requesting an explanation is simply to request "Please explain X" or simply "What is the explanation of X?" But since one cannot use "explain" to explain 'explanation' this analysis is unhelpful. So philosophers have to rephrase requests for explanation as why-questions. The word "cause" also offers a natural way to request explanation: "What causes X?" But many philosophers had problems with 'cause' and therefore wanted a theory of explanation that avoided saying that explanations were responses to requests for 'causes.' Of course the word "cause" can be circumvented simply by saying "What makes X happen?" or "What conditions are necessary and sufficient for X to occur?" or something similar. Because of their nervousness about 'causes,' empiricists like Hempel and Nagel fell back on the strategy of beginning by *assuming* that all requests for explanation are responses to why-questions, such as the presentation Nagel gives in the opening chapters of *Structure of Science*, a popular textbook in the pre-Kuhnian era. So expressing requests for scientific explanation as why-questions was really just a strategy employed by philosophers of science in the past, but it was only one possible strategy that got erected into the dogma that *by definition* "explanations" are answers to why-questions. Surely all natural languages have a wide variety of linguistic forms through which a desire for explanation can be expressed; these vary in mode (interrogative or imperative) as well as degree of politeness or formality, very strongly depending, as I have emphasized, on context. A student asking a professor for an explanation is in a very different situation from a critic of science challenging the scientist by demanding an explanation for some X, which scientists widely admit to be unexplained (so far). There is no magic land of explanation. Explanation-seeking questions require no particular linguistic form of question, but a particular context in which the question expresses a certain exigency. All sorts of descriptions may work as explanations in the right circumstances as long as they answer a question that addresses a specific exigency or a series of exigencies, which may come in an unlimited variety of forms.

10
A Rhetorical Approach to Explanation

The pragmatic view of explanation defended in the preceding chapters may be more precisely called the "pragmatic-rhetorical theory of explanation."[1] The underlying idea of this approach is that explanation should be understood in terms of its function of providing understanding rather than its content or logical structure, as traditionally assumed. I hold that explanation is a part of a communicative practice of answering explanation-seeking questions, which is intentionally directed, context-bound and persuasive; therefore, it should be understood accordingly. The standards of scientific explanation depend on different epistemic contexts, including technical, legal, political, religious, ethical or everyday explanations. What is an acceptable explanation in one context may not be acceptable in another because the particular evaluative standards we uphold in a given situation depend on the topic to be explained, our assumptions, background knowledge, interests in the answer, and level of cognitive sophistication. Sometimes interests are personal, but often they are communal as well, at least with respect to scientific explanations. Such interests may include truth, empirical adequacy, unification, derivability, predictive power, beauty, etc. However, none of these particular interests dictates any particular criterion as necessary for an acceptable explanation.

1 The problem context

The pragmatic approach is rightly or wrongly identified with different theories. One is the model-theoretical theory associated with van Fraassen. Another is the *epistemic* theory of scientific explanation, which attempts to account for explanations solely by non-pragmatic means, in other words, without appealing to such practical concerns as specific human interests. I take Peter Gärdenfors and Matti Sintonen to be among its proponents. As Gärdenfors explains it:

The central idea is that the explanans should increase the belief value, i.e., the probability, of the explanandum in a non-trivial way. The belief value of a sentence is defined in terms of a given epistemic state. This state is not the one where the explanation is desired, but instead the contraction of that state with respect to the explanandum statement.[2]

However, I believe this suggestion fails to convey what is essential about explanation. A person who asks why something is so already knows *what* the case is, and the explanation will not alter the belief value concerning this fact, which is already *one*. If I observe one morning that my dahlias have wilted overnight, my belief in this fact does not increase when I am told that this is so because there was a severe frost during the night. I am quite sure that my dahlias are dead no matter what explanation is given. The explanans does nothing to alter my belief in the explanandum, what it does do, is to fill in some information previously missing in my belief system.

Somebody may protest that this is a rather uncharitable rendering of Gärdenfors' idea. He would obviously say that the direct evidence for the wilted dahlias must be ignored when carrying out the analysis (thus the demand for raising belief value "in a non-trivial way"). The truth of the explanandum (for the sake of argument) is a stock issue and well known to everyone, so presenting his theory as though he is unaware of this point is unfair. Gärdenfors is after something much more sophisticated, namely, the sense in which belief is at stake when *seeking* explanation. His point of departure is an epistemic situation in which I could have exclaimed: "I can't believe my dahlias have wilted! How could this happen?" This might be in a context, for example, where I didn't see the wilted dahlias, but while I am away from home, my wife tells me on the phone that the dahlias have wilted. Gärdenfors' analysis rolls the analysis back to the point just before my wife informs me – when I know neither that they have wilted nor that there was frost (or, perhaps, the connection between frost and the health of dahlias). The explanation then makes the event "more believable" even though I don't, of course, really doubt my own eyes (or ears).

Although this may very well defend Gärdenfors' intention, it misses my point. The actual epistemic situation when we ask for an explanation rarely is one in which our belief is at stake; i.e., we do not doubt the fact we want explained. In most contexts we might just want additional information to help us understand, for example, the causes of, or ways to avert, or ways to live with, an event such as global warming, a fact which in this context is not doubted and is generally accepted. I believe Gärdenfors errs in thinking that he could treat all epistemic situations as occasions when we ask for an explanation to increase the credibility of uncertain or improbable phenomena. Certainly, obtaining relevant information about a problem is not the only reason we are interested in explanation.

Apart from delivering information, explanation can sometimes make the claim of an unlikely fact, or a belief in an uncertain phenomenon, more likely to be accepted into the explainee's belief system. In particular, both nomic and causal explanations provide reasons why we should believe in a given phenomenon or state of affairs. So something may be said in favour of Gärdenfors' claim: If a phenomenon, which is otherwise unbelievable or improbable, can be shown to fit into a causal or theoretical pattern, it definitely increases the epistemic credibility of that phenomenon. Nonetheless, it is also possible to imagine occasions where scientists might say, "*A* was caused by *B*," even though knowing *B* would not necessarily enhance the credibility that *A* would occur. For example, I might say that the reduction of birth rates in Europe was caused by the advent of the birth control pill, even though at the time of the pill's first appearance no one foresaw that it would lead to this steep decline.

The epistemic approach has one important thing in common with the formal-logical approach. They both conceive of the *explanandum* as a proposition, but this narrows the usefulness of their analysis. While the epistemic approach does seek a contextual analysis of explanation (i.e., one that goes beyond syntax and semantics), it leaves such pragmatic issues as human interest on the side.

Matti Sintonen, who calls his theory "erotetic" or "interrogative," includes the extra-logical contingencies of explanatory discourse in a five-placed analysandum. His formula can be reworded as

S explained to H why q by uttering u in a problem context P.[3]

The rationale behind the utterance u can then be stated as follows:

> The role of u as intended by S, is to cause in H an epistemic change vis-à-vis H's question.

This analysandum rules out that u has to meet certain formal requirements to cause an epistemic change. But u could cause such a change in H without being relevant. If S responds to H's question "Why q?" by saying "Because *trolls* cause q," it would hardly be considered an explanation, even though it could cause the desired change in H's belief system. Not every possible answer will do. Rather one must imagine that if u successfully changes H's belief states, it must somehow be relevant to the question "Why q?" Thus, a formal logic of explanation *à la* Hempel could possibly guarantee u's relevance, unless elements of the problem context P somehow excludes that any inferential characteristic of relevance can be abstracted from the pragmatic context.

Of course in a particular context formal relevance might be *a* factor for a subject's experiencing a change of epistemic state *vis-à-vis* a particular belief. (Someone might show me it can be logically deduced like a mathematical theorem from other beliefs I hold strongly.) But it is hardly the only way in which an explanation can change a subject's epistemic state with respect to another belief. There are many other factors potentially relevant to change of beliefs in a variety of different contexts of explanatory communication. No single description covers them all.

In determining the relevance of response u to the question "Why q?" the basic notion is the problem context P, which has both a material and an epistemic side. A problem arises in the tension between what is known and what is not known. Thus, the problem context, it seems, can be characterized in terms of a set of propositions stating a series of known facts belonging to H's background assumptions. Furthermore, the problem context includes metaphysical commitments, such as the principle that regularities will continue to remain regular. Finally, it also contains a number of propositions the truth of which H does not *know*, but which she suspects to be relevant to solving the problem. Her reason for this suspicion is an open question, which the content of the problem context does not help us answer. Therefore neither Sintonen nor Gärdenfors have shown that the epistemic account is superior to the formal-logical account. Their analysis confines itself by considering only epistemic and aletheic factors. But to win the day, self-declared pragmatists must argue convincingly that some specifically pragmatic elements of scientific practice are relevant for understanding scientific explanation, i.e., elements that cannot be explicated non-pragmatically.

Missing from their analysis are S's and H's own cognitive and personal interests in the problem, which initially give rise to the explanatory question. The speaker's fundamental interest in presenting an explanation is not only changing the cognitive state of the explainee but also changing it in a certain way. Furthermore, what the explainee considers as relevant is partly determined by her background knowledge and partly by what she already knows and what she seeks to find out. The explainee accepts an explanation by picking up that which best suits his or her interests and background knowledge.

Since different auditors might have different background knowledge, obviously it will appear to some explainees that the speaker is indeed explaining the phenomenon and to others that the alleged 'explanation' is an impostor (in a particular context), which is really no explanation at all. Sometimes this may be an accurate description, but in other situations it seems more in keeping with actual practice to say that the speaker is "explaining" the phenomenon to all his auditors, some of whom find it acceptable while others do not; i.e., those who don't accept it would say "S is explaining P, but I don't accept his explanation." Some might think S is speaking truthfully while others may think he is mistaken.

Thus, cultural as well as personal interests must be added into the problem context. This means that the problem context contains both beliefs concerning how the world is and beliefs about how one wishes the world to be. Consequently, no explanation is entirely objective; they are always viewed from a certain cultural and personal perspective. Since explainer and explainee might have different interests, the speaker might say that he is explaining q by u while the auditor might deny this or find the explanation unacceptable. The subjective factors, however, are more relevant to the "acceptance" of the explanation (also for the speaker) than to the definition of the explanation *per se*.

2 Individual cognitive interests

Philosophers of science are so focused on the public reports of science, which present new discoveries and explanations in the form of data and theoretical analyses, that they often overlook one of the mainsprings of scientific explanations. Behind the answer, which the individual scientist offers as an explanation, stands a set of presuppositions that motivates the *kind* of answer he or she accepts as an explanation of which scientists may not be explicitly conscious, or even suppress, when they make their explanation public.

Gerald Holton was one of the first to acknowledge such individual presuppositions in the process of theory formation; he called them themata and set up a theoretical framework to analyse them.[4] While distinguishing between 'public' science and 'private' science, he points out another source of errors when the historian tries to understand the scientists' motivation for pursuing their research problems, their choice of conceptual tools, or their treatment of data from their books, printed papers, and other published sources. He explains why in these words:

> In all these cases, one may discover that during the nascent, "private" period of work, some scientists, consciously or not, use highly motivating, very general thematic presuppositions. But when the work is then proposed for entry into the "public" phase of science, these motivating aids tend to be suppressed, and even disappear from view. Even though thematic notions arise from a deep conviction about nature, on which the initial proposal and eventual reception or rejection of one's best work may be based, they are not explicitly taught, and they are not listed in the research journals or textbooks. That has certain advantages, insofar as silence about personal motivations and thematic preferences avoids any deep, unresolvable disputes in the public phase. Consensus is more easily reached if thematic elements are kept out of sight.[5]

Holton's observation here tells us why so many philosophers of science believe that the pragmatic features of explanation can be safely ignored. Unlike historians, philosophers of science do not look into 'private' science. They address only 'public' science and therefore think this is all that matters for a theory of explanation.

But so-called "private" science cannot be ignored because it constantly influences 'public' science. Thematic assumptions form the cognitive interests of the individual scientist, and partly shape the problem contexts in which explanation-seeking questions arise. Whether tacit or explicit parts of the problem context, these presuppositions guide scientists in finding answers that accord with them. Most of the time when scientists present an explanation to the public, they leave out their motivations, because the themata, if recognized at all, are thought to be irrelevant for the objective explanatory purpose.

Holton's distinction between 'private' science and 'public' science reflects two distinct explanatory cases: (i) the explainer and the explainee are the same person, and (ii) the explainer and the explainee are different persons. When the explainer and the explainee are the same, the interests of both are naturally identical. The interests of the individual scientist both motivate him in selecting a certain research problem and lead him to accept a specific hypothesis to explain the problem in accord with his understanding of the world. The scientist wants to find an explanation in harmony with what he believes are the facts and, at the same time, satisfies the interests which initially motivated the problem selection. The explanation is a result of both the scientist's objective knowledge of facts, and the more subjective presumptions constituting his themata. But when the scientist turns to the public sphere to offer his view, he is no longer the explainee, so his interests change. Now, by playing only the role of an explainer, he needs to change the cognitive state of his audience in a certain direction. In this situation it could be counterproductive to include *his* private motivations in the account. *Either* the thematic assumptions are shared with the explainee, in which case it would be superfluous to bring them in, *or* they are not shared, in which case they might obstruct the intentions behind the account. Therefore, consciously or unconsciously, the explainer leaves out his thematic presuppositions. Presumably he does so because *he* (unless dishonest) always finds the explanation acceptable.

A scientist's thematic presuppositions do not *arise from* data or theory, but are *imposed on* them, as they constitute the mode of thought through which he understands the world. They are neither verifiable nor falsifiable. For a particular individual they will often remain fixed for long periods and can be replaced by other thematic assumptions only with great difficulty. Giving up on them might engender the same sort of feelings as rejecting one's faith in a religious authority. The general indispensability of themata explains what Holton describes as "the willingness of the scientists to adopt what can only be called a suspension of disbelief" in case their hypothesis is confronted with possibly falsifying data.[6]

Themata concern metaphysical, epistemological, methodological, ethical, or aesthetic issues. In his study of the themata guiding Einstein's theory construction, Holton isolated a number of motivating issues on which the investigator makes some presuppositions or takes a stand: formal rather than materialistic explanations; unity and unification, logical parsimony and necessity; symmetry; simplicity; causality and determinism; completeness, continuity; and constancy and invariance. In addition, one could mention: value-definiteness, locality, separability, and the objectivity of theoretical descriptions. Other scientists may hold opposing themata. It is no secret that Einstein was not satisfied with quantum mechanics because it did not fit his criteria for acceptable physics. In contrast, Niels Bohr, his strongest opponent and the major figure in the development of quantum mechanics, did not hesitate to accept explanations motivated by discontinuity, indeterminacy, value-indefiniteness, superposition, non-separability, entanglement, and intrinsic probabilities. Moreover, Bohr preferred physicalistic rather than formal explanations; he regarded classical concepts as essential for any unambiguous communication of experimental results in physics, and he denied that wave and particle visualizations objectively represent the system as it really is.[7]

Because of each scientist's allegiance to his themata, in each specific problem context he is strongly interested in producing a hypothesis that can answer the epistemic problem in a way meaningful for him. A scientist selects an explanation only insofar as it makes sense by respecting his explicitly or implicitly held interest of preserving his personal modes of thought. This does not exclude a few cases in which individuals go through a Kuhnian like conversion without explicitly deciding to accept an explanation (or not). As Laudan once pointed out, "If I can't have what I want, I learn to accept what I can have." This may well be true with respect to explanation.

3 Explanation as speech act

Scientific explanations cannot be grasped in terms of formal logic or semantics alone. Explanation includes a real pragmatic dimension, which cannot be ignored, since it forms an essential part of a complex understanding of explanation. This dimension is important because explanation is an appropriate answer to an explanation-seeking equation, and pragmatic elements like intention and the context determine what counts as an appropriate answer.

This kind of insight motivates Peter Achinstein's theory of explanation as a speech act. He argues that explanation can be understood as either a process or a product. The *product* is the content of the linguistic performance that the person makes when *producing* an explanation. But the process is primary because characterizing the product must consider the *intention* behind the explanation. Hence, he calls his account *the illocutionary theory*: "Explaining is what Austin calls an illocutionary act. Like warning and promising, it is typically performed by uttering words in certain contexts with appropriate intentions."[8] While this approach escapes some problems besetting Gärdenfors' and Sintonen's epistemic theories, it fails to flesh out what notions like "explanatory context" and "intentions" really mean. Can we say something more precise about the context and the intentions?

In developing his illocutionary view Achinstein stipulates two criteria for explanation:

> If S explains q by uttering u, then S utters u with the intention that his utterance of u renders q understandable.
>
> If S explains q by uttering u, then S believes that u expresses a proposition that is a correct answer to Q.[9]

Oddly, this preliminary formulation does not explicitly say that an illocutionary act is always directed towards somebody. S explains q to an audience. Thus, the kind of intention the speaker S has is to make a certain fact q *understandable* to a particular audience.

But it is problematic whether this notion of understanding adds anything new to our concept of explanation. If 'understandable' means raising the probability of the belief concerning the fact to be explained, then we are no better off with Achinstein than with Gärdenfors. But this is not Achinstein's intention. He gives this definition of "understanding":

> A understands q, [if, and] only if there exists a proposition p such that A knows of p that it is a correct answer to Q, and p is a complete content-giving proposition with respect to Q. (Here p is a proposition expressed by a sentence u uttered by A.)[10]

As an example, let us consider it a fact that Nero played his fiddle after he set Rome on fire. The question would then be "Why did Nero play the fiddle?" and a straightforward complete content-giving proposition with respect to this question is "The reason Nero fiddled is that he was happy."

How shall we understand the phrase "to be a correct answer"? According to Achinstein, a correct answer has to be *true* as well as *relevant* to the question. However, I believe Achinstein makes a serious mistake by thinking of explanations as '*correct* answers.' There are several objections to his view: First, a person A may understand q whether or not she knows that p (or any other proposition) is the correct answer that explains q. Second, A may *believe* that p is a correct answer to Q, *but* in many cases, we cannot say that she *knows* that p is true, and therefore that p is a correct answer. Third, that A knows that p is a correct answer to Q cannot be that which establishes that p is a relevant answer to Q. As Salmon pointed out, this would be highly question begging.[11] Anyway, as just mentioned, A need not know whether p is true. Fourth, there may be no one complete content-giving proposition with respect to Q. If there are many, which one should we choose, and on what criteria? And fifth, why should one correct answer to a question about a particular q also be a correct response to all instances of Q? Achinstein's analysis ignores that explanation always takes place in a context, and the same answer u (expressing p) may be a correct response in one context but a wrong one in another. Take a question such as "Why is the global mean temperature currently rising?" Suppose it has a definite correct answer in the context of the present. But at another time or in another place the question would have referred to a numerically different global mean temperature and the answer might have been different.

Salmon suggested that there must be a causal mechanism providing an objective relevance relation. But, as we have seen, not all explanations are causal explanations – and even causal explanations only partly consist of descriptions of objective relationships. Hence, this relationship can only be one out of several constraining factors that could determine the relevance of a given answer. Since explanation is both a persuasive act of communication and is goal-oriented and context-bound, we can understand the relevance of the explanatory content only by knowing the goal and the context involved. I suggest that Achinstein is mistaken in approaching explanation as an illocutionary act; it is instead what John Austin called a *perlocutionary* act. Explanation is not only a matter of saying something and meaning it, it is also a matter of intending to *change the mind* of the explainee, or to enlighten him on a subject about which he is ignorant.[12]

The distinction between *illocutionary* and *perlocutionary* speech acts can be used to illustrate the distinction between giving a description and this being an explanation.[13] A description considered as an illocutionary act is successfully executed if the answer states a relevant state of affairs as a response to the explainee's question and if the answer has the power to dispel the ignorance of the explainee. Whether this description also functions as a perlocutionary speech act, and therefore acts as a genuine explanation, depends on whether the speaker *intends* to inform the explainee in order to reach a new or better understanding. But being a *successful* explanation requires even more. The questioner or audience may very well understand the intention behind the answer, but that is not sufficient to call an answer to an explanation-seeking question a *successful* explanation.

Imagine a situation in which the explainer provides the questioner with a response, which is stated in the right circumstances and a direct response to an explanation-seeking question; moreover, the questioner or the audience understands the speaker's intention, namely as the speaker's wish to provide an answer to the question. But if the answer does not improve the questioner's understanding, it seems not to be a *successful* explanation. For such an answer to be successful as an explanation the perlocution must succeed, that is, an answer to an explanation-seeking question is 'successful' as an explanation if, and only if, this linguistic act *changes the cognitive situation* of the questioner.

So a *successful* explanation can be characterized by its perlocutionary *effect*. Whether or not a response to the question functions as a perlocutionary speech act, and therefore acts as a successful explanation, depends on whether the speaker *intends* to inform the explainee so that she can reach a new or better understanding, giving the speech act its *perlocutionary purpose*. Moreover, the response succeeds in its perlocutionary effect if it actually has an informative effect on the explainee by changing her cognitive state. And finally the explanatory speech act is successfully executed if the explainee begins to believe more or less as the speaker wants her to believe, which depends not only on the explainee's understanding of the linguistic meaning of the answer, but also on her background assumptions, knowledge, and beliefs.

Austin himself believed that in order for it to become an illocutionary speech act, the circumstances in which an expression with illocutionary force is stated impose some requirements on its expression. He called these 'felicity conditions': There has to be a certain setting before a speaker is allowed to pose an utterance with a definite illocutionary force. The speaker must have the right and the authority to conduct the speech act, and some external circumstances must be appropriate for invoking the particular action that is invoked. The speaker must be sincere regarding her utterance. For instance, in case we are talking about performing a propositional act, she must believe in the truth of her proposition and feel obliged to argue for its truth. However, if it is a perlocutionary speech act, the speaker must give her response with the purpose of informing and convincing the audience. But this leaves us with the predicament that sometimes the explainer and the explainee may be the same person, and therefore in these cases explanation cannot to be identifiable with a perlocutionary speech act. I suggest that in such cases when a scientist arrives at the explanation he has been seeking, his cognitive situation has changed.

4 Explanation as a rhetorical means of communication

Here I intend to argue that as a perlocutionary speech act, explanation becomes a rhetorical practice because it is an intentional act of communication. "Rhetoric," in my usage, concerns expedient communication that is context-bound, directed and intentional, potentially persuasive, etc. An explanation is a fitting response to a question of the explainee, intended to inform him of what he does not understand by providing some missing information, by making the *explanandum* probable, or by making abstract issues concrete. The respondent's answer brings insight to the questioner by putting the information into the broader context of what he both already knows and is willing to accept.

Philosophers of science often focus on the individual scientist who himself may seek to explain a phenomenon, and then proceeds to do so by performing experiments and, based on the results, constructing a story of what causes the phenomenon. But this traditional focus ignores the fact that the scientist can raise such questions and answer them only as a member of a linguistic and social community. This membership allows him to understand what he is doing, as well as what it means to raise a question and give an answer, and which standards the answer must meet for the community to regard it as relevant and appropriate. As an appropriate answer to a question, explanation is constrained by the public rules of speech acts between people, and thus part of more general communicative practices.

Thus, a fuller rational account of explanation must address the rhetorical features of this explanatory practice. As a recognizable speech act, explanation is successful when it follows any recognized practices for raising information-seeking questions and giving appropriate answers. In other words explanation involves far more diverse communicative practices and cognitive processes than merely logical ones. Although this explanatory practice is not rule-governed, certain underlying features of explanation seem independent of the particular problem-context and characterize the explanatory practice as a whole. These features are necessary because for a linguistic action to count as an explanation, it must be able to fulfil general commitments that anyone who wants to explain anything must accept; however, they are nothing but linguistic norms for responding to an explanation-seeking question. Explanatory commitments are embedded in explanatory practice; they make it intelligible. But I agree with Wittgenstein that an explanatory practice cannot be justified in virtue of these commitments, since they are part of this practice itself and therefore cannot count as necessarily true presumptions. They are semantically embedded in our explanatory practices, so their apparent universal normativity evaporates as soon as we learn that their actual manifestations are polyvalent, depending on context. The following universal characteristics of explanation share only a "descriptive" universality and not a "prescriptive" one.

First, explanation provides *understanding*. We intend an explanation to make sense to the listeners to give them a psychological feeling of knowing – just as it very often puts us in a state of actually knowing something. As we have seen, philosophers' tendency to ignore understanding led to the attempts to see all explanation as having the logical structure of a formal argument. But facts of the world are not arranged in a structure of premises and conclusions, so are there good reasons to claim that understanding, and thereby explanation, always comes from logical arguments? None! Rather, explanation provides information that somehow increases our grasp of the issue. If we know why something is the case, we don't need explanation; the response adds nothing new to what we already know. If the interlocutor does not experience something new, in his eyes the respondent does not provide an explanation. The respondent must, indeed, recognize what she thinks is an appropriate response before she can offer an explanation. (Of course in everyday contexts we respond with the right sort of explanation more or less reflexively without stopping to think.) In other words, what counts as an explanation for her is perhaps not an explanation for him because her response must fit into his background beliefs to provide him with understanding. For both to understand in the same way, they must share a common epistemic background and the same explanatory standards – something that would commonly be the case if they belong to the same research field, professional group or linguistic community. A mathematical explanation relies on mathematical terms, a physical explanation on physical terms, etc., and it yields an acceptable kind of understanding only to those who speak the same language, and to that extent share the same world-view.

Second, explanation is *fact-oriented,* referring to facts, or what are at least taken to be facts. The information offered in an explanatory account is concerned with what is rightly or wrongly assumed to be the case. But not all information conveyed by the facts counts as explanatory. Factual information is necessary but not sufficient. An explanation does not consist of merely a citation of a fact; rather, it tells us something about one fact by informing about other facts. An explanation takes the form of a story putting the requested information into a wider context.

Third, explanation is *truth-appealing*: We presuppose that the epistemic value of explanation is not merely that it provides alleged information conveyed by facts, but hopefully true information. Understanding the *true* explanation is the cognitive goal of the questioner, so the explanation offered in response to his question must rely on true information. This does not indicate, however, that explanatory force requires truth. I distinguish between *force* and *value*. Many explanations are false though they still serve as explanations to certain questions in certain contexts. Aristotle's accounts of the movement of an arrow and his account of the fall of a stone are, regardless of being false, nevertheless explanations. If truth were essential for explanatory force, then much information provided by modern science as explanations would probably not be explanations after all – in spite of the fact that we are currently justified in believing this information; the seemingly meaningful phrase "false explanation" would become self-contradictory. Truth is too strong a demand for something to function as explanation. A *correct* explanation is relevant and true, whereas an *incorrect* explanation is explanatory but false.

Fourth, explanation is *context-dependent*. Thus the same answer may on one occasion explain a certain phenomenon, but not explain the same phenomena on another occasion. Since explanation is relative to context, the requirements for explanation change with context. The standards of explanation are much higher in scientific contexts than in everyday contexts. This is true with respect to what counts as an explanation, i.e., the kind of answer wanted, what makes it relevant, how detailed and complex it must be, and how good a fit is demanded between the explanans and accepted background theories. Not only do the standards of explanation change between scientific and everyday contexts, but they may also change between the natural sciences, the social sciences, and the humanities. The word "explain" functions very much like indexical words, the meaning of which depends on when and where they are used, and also who is using them. In a similar way the meaning of "explain" varies with the context in which it is uttered, determined by the 'conversational context,' defined as what has been said, asked for, or implied, in that particular conversation. Thus, the rhetorical approach to explanation well suits a general contextualist view of knowledge. One could even argue that a pragmatic-rhetorical approach to explanation is an integral part of the contextualist view that epistemic standards are context-dependent.

Fifth, the *explanans* must be *explanatorily relevant* to the explanandum: we must have good reasons to believe that the story being told is somehow connected to the fact being explained. Thus, referring to the decrease of storks in Denmark after the Second World War is not an appropriate response to the question of why there is a strong decline in the birth rate during the same period – these facts are simply assumed not to be relevant to each other in most, if not all, scientific contexts.

Sixth, explanations are *asymmetrical* because the information explaining a fact is not also explained by this very fact, on pain of circularity. The height of the flagpole together with the sun's position in the sky explains the length of the shadow, whereas the length of the shadow does not (in most contexts) explain the sun's position in the sky or the height of the flagpole. The direction of this particular account is not merely established by convention in order to avoid circularity. *Here* the explanatory asymmetry rests on the underlying causal asymmetry.

Nevertheless, such ontic asymmetries do not rule out that the particular choice of the direction of explanation may in the end prove to be context-dependent.

The ultimate test for any account of explanation is incorporating these requirements into a satisfactory theory of the explanatory practice. I claim that my pragmatic-rhetorical view, which also attempts to focus on the practical interests of the interlocutor and the respondent, stands up to such a test.[14] It sees an explanation as a reaction to a question concerning an issue where the interlocutor lacks information. Explanations are determined by the rhetorical practice of raising questions and providing answers, so explanations are intentional communicative actions, they are concrete answers to definite questions, answers that have to fulfil certain rhetorical demands of purposiveness, relevance, asymmetry, etc. seen in the context of the background knowledge of the explanation seeker, not necessarily "our" background. From a rhetorical point of view, the persuasive force an explanation gains from its ability to induce understanding successfully in the inquirer depends on the relevance and the plausibility of the response as perceived by the explanation seeker.

5 The rhetorical situation

I hold an explanation is an answer to an explanation-seeking question that the explainer puts forward in a problem context with the intention of solving the inquirer's problem by an information-giving answer. To avoid the accusation that this definition is question-begging, I must define an "explanation-seeking" question independently of what I mean by "explanation." *I hold an explanation-seeking question expresses an epistemic problem or entails lack of information by the explainee.* However, this does not suffice. Not knowing the time is a cognitive problem for the person who has an appointment at a certain hour. But asking for the time is asking for a fact, not for an explanation of a fact. To avoid cognitive problems of this sort, we may add a further requirement. The epistemic problem must terminate when the question is answered with reference to other facts, and when this connection, by being brought to the questioner's attention, improves her understanding of the *explanandum*. Thus, an explanation solves a cognitive problem by putting the puzzling fact in a story where it becomes related to other facts.

We noted that the context of the problem in which an explanation is given is insufficient to determine explanatory relevance. Besides the problem context, explanatory relevance also relies on interests and perspective. As a communicative act, an explanation depends on the intention of the explainer, the problem of the inquirer, the background beliefs and interests of both, and, not least, the facts of the matter provoking the problem. Taken together these elements create a particular rhetorical situation that I call the *explanatory situation*.

The notion of *the rhetorical situation* was made famous by Lloyd F. Bitzer, who in 1968 characterized situations that invited a discursive response. Bitzer argued that rhetorical situations are governed by 'exigencies,' that is, urgencies that call upon a speaker capable of modifying the urgency for a particular audience, if persuaded to do so. In the case of scientific explanation, these exigencies can be identified with the epistemic problem. According to Bitzer:

> A work of rhetoric is pragmatic; it comes into existence for the sake of something beyond itself; it functions ultimately to produce action or change in the world; it performs some tasks. In short, rhetoric is a mode of altering reality, not by the direct application of energy to objects, but by the creation of discourse which changes reality through the mediation of thought and action. The rhetor alters reality by bringing into existence a discourse of such a character that the audience, in thought and action, is so engaged that it becomes mediator of change. In this sense rhetoric is always persuasive.[15]

This is exactly why I take explanation to be a rhetorical act. The aim of an explanation is to *induce new beliefs* in anyone who seeks an explanation; to understand the notion of explanation we must take this intended function into account.

Bitzer characterizes the rhetorical situation by three elements existing prior to any discourse: (1) the *exigence*; (2) the *audience* to be constrained in decision and action; and (3) the *constraints* that influence the rhetor and can be brought to bear upon the audience.[16] Any *exigence* "is an imperfection marked by urgency; it is a defect, an obstacle, something waiting to be done." There are numerous exigencies, but only those that can be modified or changed are rhetorical ones. Furthermore, a rhetorical exigence must be modifiable only by a discourse; other means of change are not rhetorical. Bitzer claims, moreover, that in a rhetorical situation there will be a least one exigence that controls and organizes the situation: "it specifies the audience to be addressed and the change to be effected." A rhetorical discourse requires an audience because it must influence people to change their epistemic state, and thereby to make decisions and actions. Finally, in every rhetorical situation there exists a set of *constraints* created by persons, events, objects, and relations that are able to confine the decision and the action necessary in order to change the exigence. Bitzer mentions beliefs, attitudes, documents, facts, traditions, images, interest, and motives as some of the main sources of constraints. Each particular respondent adds further constraints, apart from the manner in which his discourse follows the constraints already given by the situation, such as his personal character, his logical arguments, and his style as well as the particular respondent's background knowledge and entire worldview.

How does the rhetorical situation help us to understand explanation, and in particular scientific explanation? Bitzer actually denies that scientific discourse requires the same kind of audience as a rhetorical discourse, arguing that science needs no audience to produce its ends, since scientific discourse expresses and generates knowledge without engaging another mind.[17] However, this view conveys a superficial understanding of scientific discourse, most obviously because Bitzer ignores the highly social nature of the scientific enterprise. Surely a single scientist can establish some empirical knowledge without engaging an audience. She may report low-level facts such as that a mercury column, at the same temperature, is higher at the sea level than at the top of Mount Blanc. But a scientific explanation hardly consists merely of observation reports. It expresses some hypothetical beliefs, because, in general, scientific explanations appeal to invisible entities or states of affairs that the explainer claims explain the phenomenon in question. The scientific community as a whole must accept any such theoretical assumptions to elevate them to scientific knowledge.

Of course an explanation always has at least one proper audience: the person who originally raised the explanation-seeking question. Bitzer defines his rhetorical audience as persons "who are capable of being influenced by discourse and of being mediators of change." This description fits the inquirer even when she is the same person as the explainer. The answer she eventually produces changes her beliefs or modifies her state of mind from ignorance to understanding. But scientists present their explanations to a larger forum. Through journals or conferences she will express her response to her fellow scientists, who may have asked the same question and struggled to find a proper explanation. In the end, if she successfully convinces them of her suggestion, it is not only her mind that has undergone changes but also the entire scientific community's.

What further can we say about the rhetorical exigence? Bitzer sees it as an imperfection that specifies both the audience to be addressed and the changes to be made by the discourse. In explanations the rhetorical exigency is a lack of understanding, signalled openly, say, by asking 'Why P?' This both controls and organizes the situation. A person's lack of understanding is an imperfection remedied by an explanatory response.

Of course giving a reason why something is the case is constrained in many ways. Every constraint that confines ordinary explanation can also influence scientific explanations proper. First, the explanatory situation is constrained by the fact of the matter. A successful scientific explanation cannot deny or ignore obvious facts. Indeed, a scientist can deny "obvious facts," *i.e.*, those believed by the audience, but then he would not be successful. Second, the contrast-class constrains the explanatory situation. The explanation given should be more probable than any other from the contrast-class. The explainee will not be convinced by a response that is far-fetched with respect to her other beliefs, that is, if the response appeals to assumptions that are highly improbable given the common scientific background into which she has been trained and socialized. There are exceptions, of course. She may accept an improbable response if new theoretical considerations or new empirical evidence support it. But her personal beliefs, interests, and perspective also play a role in accepting a response as the explanation. Simply the fact that no explanation can cite all the appropriate facts at once implies that an explanation involves selection. But empirical underdetermination implies that it may be impossible to select rationally the best explanation among equally well empirically supported theoretical proposals. Similarly a constraining factor can be the specific kind of *actions* the inquirer may want to take on the basis of the explanatory information. In general, we may say that the explanatory situation demands that the response is relevant, but not only objective facts count as 'relevant features.' Social and personal facts count as well, so 'relevance' is highly context-dependent.

Approaching explanation as a kind of rhetorical discourse helps us to grasp the essential notion. The explanation is solicited in a situation: the situation that an explainer understands as an invitation to present an explanation. We have called this *the explanatory situation*, and not every response will fit it. A fitting response establishes its relevance by providing the required information. Seeing a situation as inviting a fitting response makes sense only if the situation itself somehow determines either implicitly or explicitly the kind of response that fits. The requirements for fitting the situation are partly objective and partly subjective.[18] The exigency that prompts an explanation is the epistemic problem of not knowing why something is the case, and the explanation is intended to solve this problem. The questioner may also address and solve the problem, but the explanatory situation requires that this solution is always formulated in terms understandable and communicable to other scientists struggling with the same problem. This is also the case with non-scientific explanations. Most often someone in a particular community raises a question and somebody else answers it.

6 Explanatory relevance

Any response to an explanation-seeking question must be relevant to be explanatory. An irrelevant answer does not function as an explanation. What establishes explanatory relevance, and how much does it depend on the problem context? No single feature is essential, but formal, semantic, methodological, and pragmatic elements of the explanatory situation all play a role in determining its relevance. These features are descriptive as well as normative, but relevance is always measured against the explanatory situation including (1) the cognitive background of the inquirer, (2) the epistemic problem expressed by the interlocutor, and (3) the objective state of affairs that generated the epistemic problem.

The cognitive background of a scientific inquirer includes metaphysical beliefs, theoretical assumptions, empirical knowledge, practical skills, and social training, as well as epistemic and methodological values. To a large extent different inquirers will share the same beliefs, practices, and values, depending on their cultural background, etc. But some features may vary from one field of science to another (or from science to humanistic studies), and even within one field they may diverge from scientist to scientist. Communication serves different functions and purposes, and different idioms or languages are used for different purposes: we use one when speaking of physical entities and another when concerned with mental items. Each researcher is familiar with and uses terms and predicates of a particular idiom or language: the physicist is familiar with the language of physics, the chemist with that of chemistry, and the economist and literary scholar with those of their respective fields. Like all functional systems in modern society, each language becomes more and more specialized and works with its own specific form of rationality. Science works with truth/false, economy with profit/non-profit, politics with influence/not-influence, and so on. Thus different languages specify different frames of rationality. Often we see classes between those rationalities, even within science broadly understood: ecologists have a frame of rationality different from economists. Furthermore, different scientists may have different metaphysical views, and this influences what they consider a relevant response. As mentioned, Einstein never accepted quantum mechanics as an adequate theory of atomic processes because he believed that the world was deterministic, whereas Bohr did not share the same predilection for determinism. He considered quantum mechanics as the only proper account of atomic phenomena.

A similar difference may exist between *methodological* values, including simplicity, accuracy, consistency, inter-theoretical unity and coherence, and fruitfulness. Kuhn correctly observed that methodological values are vague, and different scientists may apply them differently, but even if there were no ambiguity, these values would sometimes conflict with one another.[19] Some scientists prefer a more accurate explanation, while others look for explanations having a broader perspective and better explanatory resources. Kuhn's example *par excellence* was what happened when astronomers had to choose between the geocentric and the heliocentric explanation of the planetary movements before Kepler's and Newton's contributions. The geocentric explanation had its adherents because it was *consistent* with the physics of that time. But others supported the heliocentric explanation, since it was overall *simpler* than the geocentric explanation.

Two scientists may share the same methodological values; they may apply them in the same way and put them into the same hierarchy of importance in case of a conflict. Nevertheless, disagreement is possible with respect to the relative weight these values have in cases where both are relevant. Scientists may agree on everything concerning these methodological values and yet prefer different explanations if the responses are empirically underdetermined. So apart from common criteria of relevance shared due to common scientific training, we also find individual criteria. These rely on the scientist's previous experience, the type of work she has done previously, whether she has been successful within her earlier work, the kind of concepts and techniques she masters, and so on. Some scientists may prefer mathematically developed explanations; others seek more visualizable accounts. Among the individual criteria figure non-scientific values too. For instance, Kuhn argued that the young Kepler accepted the heliocentric cosmology because he was occupied with hermetic and Neo-Platonist thoughts at that time.

The second feature of the explanatory situation shaping the response considered relevant as an answer is *the problem* that creates the question. The scientist accepts a lot of facts, but something necessary for understanding is missing, and an explanation is relevant only if it can provide information about what she is missing. The adequacy of a response is a function of not only the information provided but also how it is told. She may not understand why a certain phenomenon exists, why a certain anomaly appears, and she will ask for an explanation reflecting the kind of epistemic problems she has. The nature of this problem points to *the genre of explanation*, that is, to the question of the proper format of a relevant response. Explanatory genres include nomic, causal, functional, functionalist, structural, intentional, and interpretative explanations. All respond to why-questions, but the particular problem in question prescribes the relevant genre.

A scientist may ask why a particular event occurs; hence a causal explanation will be the relevant kind of account. Another scientist might hope to grasp why a certain property helps an organism or an artefact to be successful. Here an appeal to its actual effect, rather than its cause, may be considered relevant to gain the appropriate understanding. The effect is not intended to explain why this particular feature exists, but to explain the particular *function* of that feature, and therefore why possessing this feature contributes to the object's (or its species') continued existence. The same is true with respect to understanding people's actions. The epistemic problem is to get to know why they did what they did. Usually we believe people have motives for fulfilling certain goals, and their actions are regarded as means for realizing those aims. Hence a social scientist may regard a relevant response as one explaining the action in relation to its intended effect and not the actual cause that motivates it.

Similarly, the *real world* constrains the explanatory situation and thereby affects the relevance of an explanatory discourse. All serious requests for knowledge are formed as information-seeking questions, and if the answer consists of more than just stating a fact, it must give us explanation. Thus, explanations are answers that provide information about a fact by relating it to other facts.

If a response reflects a fact that is regarded as not possibly having any *real* relation to, or any influence on, the fact stated by the *explanandum*, the response will be irrelevant and therefore not an explanation. For instance, as I have previously argued, it is a legitimate explanation to claim that certain patterns in the English cornfields, which have been reported now and then, are due to aliens from outer space. Such an answer is relevant, although highly improbable, as an explanation, because it refers to something we think could make such patterns if aliens were real and had visited the earth. Facts like the height of the Eiffel tower, the date of my birth, that some mammals lay eggs, or that supernovae are exploding stars, cannot figure legitimately in a response, if the answer counts as an explanation. These do not belong, as far as we know, to the right ontological categories that can stand in the appropriate connection to the cornfield patterns. In other words, an answer will be considered as an explanation only if it does not commit a category mistake within a given categorical scheme. There is here no *a priori* "ontological" distinction to be drawn here. The kind of fact deemed relevant depends on our beliefs and general imagination regarding the kinds of facts we will accept as possibly having a causal influence on the *explanandum* fact. Nobody can exclude *a priori* any explanatory connection; the ontological categories we regard as being explanatorily relevant depend on our prior experiences and background assumptions.

Among possible explanatory relations, causal connections seem to be by far the most effective. It is not spurious, it is real, and it seems to be observable. The explanatory virtue of causes is that causes exist in the world, they connect facts or events together, and that we think of the cause as what brings about the effect. Because of these features any causal answer is regarded as highly relevant, assuming the cause is of the right ontological sort, and therefore able to explain the effect. But we have to remember that causation takes place in a network of circumstances, and no single fact or event in this network is objectively *the* cause. The entire causal network as such constrains the explanatory situation, while we select the particular causal factors we find most interesting. Furthermore, there are real relations other than causal relations that can constrain the explanatory situation and condition any explanatory answer. Not every relation in nature is causal. And in a world of structures, functions, meanings, rules, and interpretations, there are other kinds of relations playing the same kind of constraints. In these cases the exigence of the explanatory situation is the lack of knowledge. Beneath some of these relations there are causal ones, but this is irrelevant as long as our cognitive goal is getting to know things in terms of their structures, functions, meanings, rules, and interpretations.

7 Explanatory force

How can an explanation explain a certain fact? It describes the event to be explained and the events invoked to account for it. The explanans itself is only an explanans *relative to* the explanandum: its explanatory force is directed towards one goal. We can ask: must every explanans stand in the same kind of relation to its explanandum in order to be explanatory? The rhetorical-pragmatic theory offers some interesting insights on this issue.

An important distinction must be made before we proceed. Remember the distinction between *explanatory relevance* and *explanatory force*. The former notion implies that the explanatory answer fits the explanation-seeking question because there is an appropriate thematic connection between the two. Whether the accessible information is seen as relevant or not is determined, as we have seen, by our background knowledge, our interests, the nature of the epistemic problem, and the world. However, the latter notion implies that an answer is successful in getting the interlocutor to believe it answers her question, and therefore to believe the facts are as stated in the explanation.

The pragmatic-rhetorical view holds that logic alone cannot account for the explanatory force. The formal-logical approach assumes explanatory force is solely a function of the inferential link between explanans and explanandum. If the explanans logically entails the explanandum, then, and only then, does the explanans have the power to explain the explanandum.[20] It assumes that a logical fact of the matter gives an explanation its explanatory force, and whenever the interlocutor grasps this objective state of affairs, she understands how and why the explanation explains. But we have already argued that there need be no such *deductive* link between a theory, a set of propositions, and those stating the facts of the *explanandum*. Even if there were such an inferential connection, fundamental laws alone would never explain anything because they do not describe this actual world.

The pragmatic-rhetorical theory insists that truth has little, if anything, to do with explanatory force. Neither does it have anything to do with explanatory relevance. A theory of explanation should be able to specify what an explanation is, regardless of whether or not it is true. No doubt truth is not sufficient because the explanans is never unconditionally true, if it is true at all, relative to the explanandum. Nor does truth seem necessary.

Looking into the history of science, more often than not we see that frequently a hypothesis is promoted as yielding an appropriate explanation of observed facts, and generally accepted as such, but later turns out to be false. Among the assumptions that today's scientists uncontroversially accept, some are probably false too. In spite of their falsity, nevertheless they are thought to offer genuine explanations. False hypotheses are the *rule*, true ones the *exception*. Hence, we can hardly require that truth be a necessary condition. When the scientific community thinks an assumption is true, nobody doubts that it can explain what it claims to explain. But a false explanation gives us wrong and useless information, which will mislead us if we want to act on this assumption. When the hypothesis is eventually discovered to be false, scientists then reject it as the desired explanation.

So truth, in itself, is not essential for explanation, but to make an explanation 'acceptable' we require the *belief* that the explanation is true. Undoubtedly we often acknowledge explanations we know to be false because we see the answer is relevant to the explanation-seeking question. For instance, Lamarck's suggestion that acquired attributes could be inherited in a new generation explained the development of the biological species. After Darwin, biologists recognized his hypothesis was false; however, it is still an explanation, but as we now believe, a *wrong* explanation. This and many similar cases indicate that we can accept a response to a why-question as a *possible* explanation without believing it is true. Indeed, the explanatory request is not for any possible explanation; we want an explanation true of the actual world, since only such an explanation will benefit us. But the history of science seems to show that unlike knowledge, being true is not a part of the meaning of explanation.

If neither truth nor correctness matters in making an answer to an explanation-seeking question count as an explanation, one may wonder how to distinguish between possible and actual explanations. I suggest the following: A *possible explanation*, for the explainer, is a perlocutionary response plausible in the light of his beliefs concerning the evidence, his background knowledge, assumptions, and cognitive interests. An *actual explanation* for the explainer is a perlocutionary response, which he thinks to be true in the light of his assumptions, background knowledge, cognitive interests, and beliefs concerning the evidence. Indeed, on these definitions, both a possible and an actual explanation may be false, and therefore explainees may reasonably assume they are false.

But can we show that truth doesn't matter? Normally we run out these examples to extremes depending on our intuitions, so we might construct this case: Suppose a man is having an affair and gets lipstick on his collar as a result. At the time he doesn't notice, and as he comes home on the bus, a woman with heavy makeup bumps into him. He still hasn't seen anything on his collar, and when he gets home, his wife asks him how it got there; he remembers the incident on the bus and tells her about it. There are two possible explanations: he chooses the least controversial. Both accounts could be true; neither is "known" to express the *actual* cause (indeed, in one version of this story, the other or *both* women may have left lipstick on his collar). Thus, truth is irrelevant to which of these two accounts are explanatory. If his wife doesn't believe him that is because she thinks the most probable explanation is that he has had an affair. She might have background beliefs making this account more likely than the bus story (even that may be perfectly true and explanatory in the basic sense). She seems justified in inferring that he is having an affair. But this cannot be because of this single piece of evidence. Rather, whether she finds his version of story improbable depends on the larger pattern of his behaviour; it is a matter of his personality and culture.

We may distinguish between *look-alike* explanations and *proper* explanations to say that potential explanations are only look-like explanations. In real life, lying to his wife about the lipstick on his collar, a man may tell her that the bus had stopped very abruptly, and a lady's face bumped into his shoulder. He knows that this is not true, and she does not believe him. She thinks it is a lie, concocted for explaining away illicit activities. It appears to be an explanation, but is it a *proper* explanation? Can a response to a why-question be a 'proper' explanation without being true? Intuitions seem to be divided. If one holds that believing the truth of the response is essential to having a 'proper' explanation, then the man's lie provides no explanation. It is just a matter of "if true, it would explain;" *i.e.*, it could explain the lipstick, but it doesn't, because it is false.

Assuming that 'look-alike' explanations are syntactically and semantically indistinguishable from 'proper' explanations, we must specify in which respects these 'look-alike' explanation are similar or dissimilar to 'proper' explanations. 'Look-alike' explanations and 'proper' explanations must have some essential features in common because both provide reasons why something is the fact. In this case it must be the format of a causal story. The man appeals to a possible, causal connection between the lipstick on his collar and an imagined episode on the bus, hoping his wife would believe him. In other words, his response is relevant because it refers to this possible connection, although it is false, and his wife does not believe him. Hence, it is justifiable to say that any relevant response to an explanation-seeking question may act as an explanation.

This analysis implies that an explanation is a response considered to be relevant to an explanation-seeking question. An explanation may lack explanatory force even if the so-called "facts" claimed in the explanans are as mentioned. The man could speak the truth, or he may tell a lie; in either case his "explanation" would explain the fact. But if his wife doesn't believe him, the "explanation" has not fulfilled its purpose. She refuses to embrace the "explanation" not because she doesn't see it as relevant, but because she has not been convinced that things are as she is told. It does not matter that she has no problem imagining things happening as stated in her husband's explanation. But she possesses no evidential support and may indeed have counter-evidence, perhaps love letters to her husband from an unknown woman, disrupted phone calls, etc. What is missing and what would make the "explanation" work as an explanation is trust in her husband. The man has lost his trustworthiness as an explainer. Therefore the wife thinks his "explanation" is very implausible; i.e., while a possible explanation (if true it would explain the lipstick), it is not accepted as explanatory in this actual real world.

Thus, as an act of rhetorical discourse, an explanation has explanatory force if, and only if, it can persuade the audience to regard it as true. Its explanatory power over an audience rests on both its explanatory relevance and the explainer's ethos. It is relevant in relation to the problem context addressed, but this is not sufficient to make the explainee believe that things are as the explanation claims. The function of an explanation in explaining the facts in question is partially due to the rhetorical situation, including the explainer's ethos. In science people can accept explanations simply because of the explainer's ethos. Historians of science have shown that an experiment or an assumption is given more credit than it deserves if a famous scientist supports it. Even a wrong formula can be long accepted in the scientific community for no other reason than the reputation of the person(s) who claimed it. A couple of years after the advent of the relativity theory, several physicists, among them Max Planck and Einstein himself, separately formulated the laws of thermodynamics in accord with special relativity. Their treatment was adopted by many textbooks over the years as late as 1963, until H. Ott and independently H. Arzeliès, discovered in 1965 that the old formula was unsatisfactory. This was so because Planck and Einstein used generalized forces instead of true mechanical forces in describing thermodynamic processes.[21]

With respect to how the explainer's ethos plays a role in the audience's belief in an explanation, Møller describes the laws of relativistic thermodynamics in terms fitting neither the formal-logical nor the ontic approach. Let me quote at length:

> The papers by Ott and Arzeliès gave rise to many controversial discussions in the literature and at the present there is no generally accepted description of relativistic thermodynamics. This is because many different formulations of the thermodynamical laws are possible, since the principle of relativity alone does not determine them uniquely. In fact, from this principle we may conclude only that the

classical laws of thermodynamics are valid in the momentary rest system S^0 of the matter, independently of the motion of this system with respect to the fixed stars. However, there is a wide spectrum of possible ways of describing relativistic thermodynamics in any other system S since the basic laws may be assumed in a rather arbitrary way to depend explicitly on the velocity of the matter relative to S. In this situation we must have recourse to arguments of simplicity and convenience.[22]

Here we see Møller maintains that beyond a particular problem context, selecting a particular relativistic thermodynamics is both empirically and theoretically underdetermined. Therefore the particular one of the many possible formulations one prefers depends on methodological and pragmatic criteria such as simplicity and convenience. Again we see there is not *one* correct covering law, and therefore not only *one* explanation.

To close this Chapter let me anticipate some criticism that might be raised against the pragmatic-rhetorical view. I claim explanations gain their explanatory force from a successful ability to induce understanding in the inquirer. As a protest one might argue that if explanatory force comes from the persuasive *success* of a statement, it seems as if a statement counts as an explanation only if it provides understanding to the explainee. Are we prepared to ascribe explanatory force to a statement if the explainee actually fails to reach a state of understanding? Say, a professor explains a problem of thermodynamics to a sophomore. The student knows that she has received an explanation of the problem, although the answer fails to provide understanding. She may respond something like: "Yes, I understand that you have tried to explain it to me, but I still don't understand the problem." It is reasonable to say this, not because the professor's explanation lacks explanatory force, but because his answer possesses no explanatory relevance in her mind given her state of knowledge. The student attributes a strong ethos to her professor because she believes he actually knows the answer; therefore, she accepts his answer as true before she even understands its meaning. But her physical knowledge is not adequate for grasping this point. In contrast to the distrusting wife who understands her husband's story but doesn't believe it, the student has confidence in her professor, but doesn't yet understand the relevance of his answer. Thus, an answer is *successful* as an explanation only if it is seen as relevant *as well as* creating understanding believed to be true.

Because explanation yields understanding, it helps us to account for what is sometimes called the completeness of explanation. Every theory of explanation as a communicative act, rather than as its abstract result, must address the following quandary: How can an explanation be an act of discourse intended to change or improve the epistemic state of the inquirer, unless the answer is no longer an explanation the moment the goal is achieved? Nobody would deny that an explanatory act continues to be an explanation even after the explainee is fully informed. The Newtonian model of the solar system, say, explain the planets' motion around the Sun regardless of whether anyone understands the problem. Thus, if we can successfully produce an explanation, A, of X to a person who already knows the relevance of A to X, and therefore does not have any epistemic problem, then A is not able to explain by supplying the information solving an exigency in the first place. A does not merely constitute an explanation in virtue of addressing an epistemic problem. A model continues to explain, or keeps its explanatory force, even though the original epistemic problem prompting it has been solved.

I regard this to be the solution: Explanation always reflects the explainer's own understanding of the cognitive situation, although he does not necessarily share the explainee's feeling of exigency. When some philosophers say Newton's laws explain the tides, it is because Newton's theory provides a general conceptual framework of how to describe the tides by means of the Moon's gravitational attraction. A scientific theory codifies a certain descriptive insight about nature reached by the scientific community. Thus, Newtonian dynamics can be explanatory only if it is the basis of somebody's understanding of mechanical phenomena. Aristotelian dynamics, for instance, no longer possesses this explanatory capacity, because today no one understands the world through an Aristotelian framework.[23] So the explainer selects an explanatory answer from a huge repertoire of possible answers, in the light of his cognitive situation in which he holds Newtonian mechanics to be the appropriate theory to explain mechanical phenomena, and is certain a Newtonian answer addresses the inquirer's exigency. After an explanation is produced, the propositional content of the explanation preserves its explanatory relevance without the former epistemic problem. The same content will convey understanding that we can carry over to similar cognitive situations, where another explainee faces the same kind of exigency. Hence the understanding expressed by a theoretical description keeps its explanatory relevance in other situations, where this theoretical description is applicable, whereas its force is always determined by the explainer in relation to the explainee.

An explanation is considered as a communicative response if we see it as an utterance, but the meaning of any utterance is normally not fixed solely by linguistic conventions. As always, the context in which understanding arises is relevant. The content of an answer depends on the reference intended, the sense of the words, and the speech act performed on the occasion, as well as the explainee's cognitive and emotional interests in the problem. In general, perhaps more than any others, scientists seek explanations, fulfilling standards holding in many contexts, and which therefore are likely to cover epistemic situations similar to the one that raised the original epistemic problem. Thus a particular response to an explanatory question may still keep its explanatory relevance because of its propositional content in other contexts similar to the one prompting the explanatory request. We stick to Newton's theory as long as we think it gives us an adequate description of mechanical phenomena, and we use mechanical models to actually convey this understanding to people who do not yet understand.

Thus explanations answer explanation-seeking questions. An acceptable answer must reflect the intention of providing information to the explainee, and be an answer with the informative content that solves the exigency of the situation. The answer must be relevant, where 'relevance' is determined by the explanatory situation including the exigency, the real world, and the cognitive interests and cognitive background of the explainer or inquirer (if the explainer is the same person as the inquirer). To be successful the explainer must have the ability to persuade the explainee that the answer is both relevant and correct. This happens only when the answer is trusted as *true* by the explainee and yields an epistemic solution to her problem. So what counts as an explanation to the same explanation-seeking question may differ from one context to another. Nevertheless, some answers are certainly more 'robust,' i.e., more context-independent, and therefore tend to vary less throughout many contexts.

11
Pluralism and the Unity of Science

Human nature takes a significant part in developing a scientific practice because the way people gain knowledge and understanding is basically innate. This is the naturalistic side of the sciences. But the coin has another side: the pragmatic side of the sciences that rests on human perspectives, cognitive limitations, and epistemic interests. Thus naturalism has been my underlying philosophical orientation in approaching scientific understanding. This attitude may appear to point towards reductionism, scientific monism, uniformity, and the unity of science. But my contextual and pragmatic approach to explanation, interpretation, and representation – with its strong emphasis on individual as well as shared interests – signals disunity and scientific plurality. Now and then I have emphasized the limitations of models and theories, limitations delineated partly by nature and partly by our cognitive interests. Hence some people might think that these two ways of grasping scientific practice are entirely incompatible with each other. Here I intend to show that this is not really the case.

Epistemological naturalism is an obvious perspective from which to defend common epistemic grounds for understanding all sciences. It understands science from the perspective of the natural evolution of human beings. When you talk about applying 'naturalism' to science, the result of that process of 'naturalizing' is a 'naturalization' of science. This naturalization of science has a methodological as well as an ontological dimension. In the ontological sense, naturalizing may be taken to imply the possibility of reducing the social and human sciences to the natural sciences.[1] And perhaps more generally – as Quine suggested – implying that epistemological questions should be handled by the natural sciences. 'Naturalism' in this context holds that traditionally normative disciplines (especially but not exclusively epistemology and philosophy of science) ought to be handed over to the empirical, descriptive sciences *en bloc*. New trends such as neuro-aesthetics, neuro-linguistics, and neuro-economics can be seen as parts of this development. In contrast, naturalization of epistemology and philosophy of science refers to all explanations given by any empirical science regardless of whether or not it belongs to the natural sciences. In that sense, the naturalization of epistemology is often used to refer specifically to passing on of what-used-to-be a normative discipline, once perceived as the prerogative of philosophers, to *the social sciences*, notably to sociology of science and sociology of knowledge (in modern science studies). This seems to be the real motive behind social constructivism's attempts to supplant traditional philosophy of science.[2]

The reductionist understanding of naturalization often goes hand in hand with the traditionally rationalistic belief that the whole of science aims toward a single comprehensive theoretical system. This was the epistemic ideal envisioned by those Enlightenment thinkers present at the birth of modern science. From a few basic axioms we should be able to deduce and predict everything that happens in nature, including ourselves. This effort has been the ultimate goal of science in the past, and still is in much current science. Hempel's deductive-nomological model of explanation was symptomatic of the trust in this ideal.[3] Nevertheless, even within physics we use many different theories built upon very different concepts between which there is little connection. Even a grand 'theory of everything' seems to lack any form of empirical evidence.[4] Across disciplines the situation becomes even worse. Science appears to be a conglomerate of loosely connected disciplines and subdisciplines each giving us different kinds of knowledge about specific concrete issues.

Epistemic disunity is real both now and over time. But one sign of at least a kind of unity is that in almost all of the disciplines, except physics, there is a healthy borrowing of the substantive conclusions of other disciplines. Chemists rely on findings of physics; biologists on chemists, etc. And many of the hybrid sorts of disciplines make heavy use of the substantive claims of not only physics and chemistry but also other hybrid disciplines like meteorology, geology, astrophysics, etc. The fact that these different disciplines can co-operate to solve a puzzle (why did the dinosaurs disappear; why is the planet warming, etc.) suggests that while they might not all be on the same page, at least most are writing in the same book. So this historical fact suggests that although on an epistemic level there is great disunity, we find on the substantive level they really do all seem to be working on the same big picture of the natural world, but of course at various levels and in various domains of human experience. This kind of unity is in good keeping with the arguments of this book.

Thus I have prepared the epistemic grounds for grasping science through a *non-reductive* understanding of the naturalization of human cognition. Human intentions, norms, and values – considered as naturalistic categories on a par with selection, adaptation, evolution, and physical forces – play an independent and non-reductive role in explaining how science developed as a part of human cognition, communication, and culture. In particular in this book I have articulated a line of thinking supporting *pragmatic pluralism* due to both the general cognitive limitations of human beings and the divergent cognitive interests among those who do scientific research.

Epistemic pluralism is committed to some versions of instrumentalism or ontological pluralism, in establishing common grounds for multiple theories and models. For some philosophers of science epistemic pluralism is taken to imply a strictly instrumentalist stance, attempting to purge science of any conceit deemed "metaphysical." For others it may imply a non-reductive plurality of ontologies. These positions have different implications for the question of realism versus relativism. I think it is possible to be a contextual realist and a domain pluralist.

1 The epistemological issue

The logical empiricists were no friends of metaphysics, but their physicalistic view had obvious metaphysical implications only in connection with a reductionist stand which the positivists themselves did not necessarily embrace. Nonetheless, they strongly advocated the unity of science in the sense that all scientific claims should be intersubjectively evaluable. Hence, if physicalism is assumed to be the fundamental descriptive basis of reality for the disciplines of philosophy, history, psychology, and sociology, some statements within any of these fields must be capable of being expressed purely in terms of the physical sciences, including anatomy, physiology, and neurology. However, the logical empiricists did not hold that every statement of these disciplines is translatable into a purely physicalistic language. Had they thought so, the implication then would be that in the discourse of science every assertion about social and human phenomena, like "communication" or "intentions," is a meaningful expression only if a physicalistic translation is possible. The logical empiricists abstained from arguing that each statement of science could be reconstructed in terms of the same set of basic concepts.[5] Indeed, the logical positivists were deeply aware that the unity of science program had not been carried through in any of these fields, but they were confident that the human sciences could be developed successfully in this direction.

The logical positivists were motivated to defend the unity of science primarily for epistemological reasons: some scientific statements should be expressed by empirically verifiable sentences such that their truth value could be determined by direct experience. Hence, there must be 'correspondence rules' which allowed somehow translating theoretical terms into observational terms, and according to the physicalist ontology observational terms were sufficient for describing all reality. The positivists thought that only by using this language was it possible to express what was intersubjectively accessible and publicly testable by experience. This language, and only this language, has cognitively meaningful expressions. So, requiring the language of physicalism to be the ultimate language of science implied that every theory of a particular domain of phenomena had to be constructed in terms linguistically connected to observational, physicalistic terms. Although theories could be formulated in many different terms, for the positivists the terms of any scientific theory invariably could be construed in physicalistic terms, in order for the theory to be verifiable and so genuine science.

Given their commitment to the unity of science – as they saw it – logical empiricists naturally opposed *semantic* and *epistemic* pluralism: one scientific language has a privileged status and this is the physicalistic language, the statements of which we can determine directly to be true or not. We use the methods of inductive inference to reason from this empirical foundation. But the positivists missed two important facts. First, as *Homo sapiens* human cognitive capacities are restricted in space and time and adapted to occupy standpoints in a very limited region of space and time. Therefore human beings tend to describe things according to different standpoints. Second, as social and intentional beings, every description we express always serves some purpose; it is produced to fulfil certain cognitive and practical interests.

Today most philosophers of science have abandoned the positivists' unity of science program and commonly argue for some sort of scientific pluralism. Paul Feyerabend, whose views were more radical than anybody else, was among the first. He emphasized the necessity of using alternative theories to discover theoretical problems and new phenomena that could not be revealed otherwise. Today John Dupré, Ronald Giere, Nancy Cartwright, Helen Longino, and many others embrace forms of disunity. Increasingly more and more philosophers ignore the goals of the unity of science movement.

Most defenders of scientific pluralism also subscribe to epistemic pluralism, by arguing for what I call *account plurality* that holds some natural phenomena cannot be fully explained by a single model or fully investigated by a single approach. Some entities can be represented differently, within a single domain, by incompatible models and theories, presumably because they do not attribute the same properties to these entities. The classic example of account plurality is the competition between the wave theory and the corpuscular theory of light. The basic linguistic rules of wave theory define the terms "wavelength," "phase velocity," "period of oscillation," "frequency," "angular frequency," and "wave number" in terms of each other. Using these definitions more complex expressions like the wave equation of motion can be formulated. The linguistic rules of the corpuscle theory are Newton's 'laws of motion.' Since the seventeenth century these two theories have competed to give the most accurate description of light, when applied to models of transmission, absorption, reflection, refraction, diffraction, interference, dispersion, and polarization phenomena. Each theoretical model can be used to describe and explain some observed phenomena, and neither theory can be said to be "more accurate" than the other. What one chooses depends on the problem one wants to solve.

As is well known, quantum mechanics incorporates both pictures in its account of wave-particle dualism, crystalized in Einstein's conception of a light quantum defined by terms from both corpuscular (momentum) and wave (frequency) models. Nevertheless, even though both models for explaining physical phenomena are formally united in quantum mechanics, they constitute a 'duality' insofar as they exist side by side as different aspects of the observed systems. Schrödinger's wave function represents only a probability amplitude; it is a misunderstanding to believe it resolves the dualism by representing a physical wave in spacetime. Rather we seem to be limited to a perspectival grasp of wave-particle duality.

There are many reasons why so little of the unified science program has ever been realized even within the natural sciences; I shall mention only a few:

First, theories and models are made by human beings. Unless one embraces essentialism there is never exactly one 'correct' way to represent what we can, and do, observe. How we represent our experience of the world is always conditioned by our interests. How scientists represent scientific phenomena is determined partly by nature and partly by how they intend to use the representation. Their interests are a function of what they wish to understand. They choose a representation which they think will provide the desired information. For instance, they may be interested in understanding the dynamical features of a particular system, its structural features, or its functional features, and these different features might not be comprehensible in terms of one single model. Indeed, one of these accounts often presupposes some knowledge of the other accounts, but scientists choose a representation just as much on pragmatic grounds as on the phenomena to be represented.

Second, translating 'theoretical' terms into 'observational' terms always relies on interpretation. The translation is not only a question of derivability or explainability. In case we are confronted with two domains, one of which is claimed to be reducible to the other, we always have to show that all significant terms of the reduced vocabulary can be translated into the vocabulary of the reductive level. But deciding whether such a translation is successful or not rests on the collective agreement among scientists about the meaning of both classes of terms. H_2O is not just identical with water. We cannot even claim that "H_2O" or "nH_2O" and "water" have the same extension, albeit different intension, since the terms "H_2O molecule" does not refer to water but to the H_2O molecule. Translation requires an interpretation, and an interpretation always depends on background knowledge and cognitive interests. When scientists study H_2O to acquire information about chemical bonds and the detailed structure and dynamics of the molecule, the observational evidence they seek depends on their research question. The chosen theoretical model of observation connects observational data to what these data are taken to justify. In these contexts it is completely irrelevant that H_2O is the chemical basis of water. Also, as a technical term of chemistry, H_2O is quite precisely defined, whereas the definition of "water" would normally be quite imprecise.

Third, no explanation can prove its own truth. Different theoretical descriptions of the same phenomenon may equally be true. As long as scientists agree that a particular way of describing certain phenomena is meaningful, they may find that nature supports the truth of other descriptions that also have been found conceptually and semantically adequate. Therefore scientists may use alternative but empirically underdetermined descriptions the meaning of which cannot be translated into sentences about their common empirical support.

Fourth, we know that 'law' is a particularly ambiguous notion. Most causal laws are *ceteris paribus* laws, raising the question of how many qualifying clauses can be added to a universal statement before the expression becomes too specific to be considered a law. Can we set up as many strictures as we want and thereby narrow down the extension of the law statement to the point where it includes but one instance? It seems that the scope of causal laws is as much determined by the investigator as determined by nature. Therefore, there is no single and unambiguous translation of causal laws into a certain set of observational statements.

2 The methodological issue

Does the possible failure of physicalist reductionism mark the end of the positivists' dream of the unity of science? I don't think so. But the dream will not be one single overarching approach, as some scientists and philosophers may have once believed. We should abandon hope for a reduction-of-theory physicalism, but as I have argued elsewhere, what makes physics, biology, history, and literature alike can be found at the level of methodology.[6] I agree with the logical empiricists that both the social sciences and the human sciences should be considered part of the empirical sciences by adopting some of the same empirical methods as the natural sciences. Some methods are very specialized to particular objects of study, but all disciplines exhibit some basic methods for getting empirical knowledge, which we acquired through the adaptive selection of human beings. Evolution adapted human beings to basic procedures for testing beliefs and assumptions enabling us to judge whether they are in accord with our experience of the world. Modern science merely extends and refines ordinary human practices aimed at building representations of one's world and selecting between various hypotheses. Methods include both physical and mental procedures that help scientists produce data, set up hypotheses, and justify these hypotheses.

First, science aims to give us a theoretical and systematic understanding of nature, society, and human beings. Each and every discipline has developed its particular language for expressing such a theoretical and systematic approach. When we talk about scientific theories, usually we have in mind a set of linguistic rules explicitly defining certain basic properties in terms of other basic properties. In the natural sciences such linguistic rules are often called 'fundamental laws,' but as linguistic expressions, they have no descriptive content. Only when applied to scientific models are scientists able to match the model to something in reality to produce scientific explanations. So three things are really going on in arriving at an explanation: using the language and its linguistic rules on data or observation; then constructing a model based on these "interpreted" data; and then using the model to "represent" reality, i.e.,

Pluralism and the Unity of Science 277

some aspects of the world. I believe this holds in the natural sciences, the social sciences, and the human sciences.

Second, in all fields of research theoretical understanding is founded upon the systematic interpretation of data, evidence, or other observational phenomena. When scientists face a representational problem such as how to represent a newly discovered phenomenon or how a given representation or a set of data should be understood, they confront a representational problem of some kind, and this requires interpretation before the scientist can say anything further about it. However, interpretation is not just guesswork, as Popper would have us believe; the process of interpretation is guided by analogies and comparisons with already accepted beliefs. In the sense of constructing meaning, an interpretation creates a conceptual grasp of the phenomenon by placing it in the taxonomy of the proper field (or even by changing the entire taxonomical system). Conceptual reconstruction is an important element in any abductive inference where the extension and/or intension of selected terms of the hypothesis language are modified with respect to the data language. But sometimes modification is not enough, and completely new concepts and definitions have to be introduced.

After the solution of representational problems, scientific practices enter the stage of hypotheses formation. Here the unified understanding of science makes sense. Due to human beings' biological evolution there are common ways of processing sensory information. The particular beliefs based on sensory experience are generalized to similar phenomena by ampliative inferences, like simple induction and abduction, and then justified by further observation and actions (experiments).[7] Human evolution selects in favour of checking our beliefs against nature, because false beliefs can be very dangerous when they affect our survival. Hence, human cognitive capacity has developed by trial and error, i.e., the conceptual correlate to biological selection and adaptation. Contemporary scientific methodology is originally an abstraction and idealization of our ancient cognitive practices.

My brief analysis implies that the natural sciences, the social sciences, and the human sciences, as well as everyday "common knowledge," have something in common, namely naturally inherited methods for formulating hypotheses and checking their truth empirically. One of the first philosophers to abstracted modern methodological rules from biologically acquired practices of inquiry was Ibn al-Haytham (called 'Alhazen,' 965–1039/40,) whose treatise on optics formulated some methodological procedures for controlling and examining theoretical hypotheses and inductive inferences. They were (1) observation,

(2) formulation of a problem, (3) formulation of a hypothesis to solve the problem, (4) testing the hypothesis by experiments, (5) analysis of experimental data, (6) interpreting the results and reaching a conclusion, and finally (7) making these results public to secure criticism and discussion. Today these prescriptions have become some of the basic norms for any empirical science. In sciences where direct experimentation is not feasible, other methods replace instrumental manipulation. In medicine placebos are used to study the effects of drugs, and in history scholars have developed over centuries methods of source criticism.

Some may argue that this proves that there is no universal method. Certainly it does at one level. But these diverse methods are less general than those common to all scientific research, and I regard them as more specific, practico-instrumental methods, guiding action as much as reasoning. The general methods, common to every science, are ordinary rules of reasoning such as the methods of induction, abduction, inference to the best explanation, as well as deduction and falsification, all grounded in our cognitive capacity of reasoning in a way advantageous for our survival.[8] Hence methods of *data collection* are distinct from methods of *data treatment*. The methods for collecting data vary from one science to another, each designed according to the aims of the particular science; while the methods of data treatment are abstracted from general forms of reasoning independently of the scientific topics.

How each discipline characterizes its research objects determines what counts as a topic. The same object may yield different topics for different disciplines, and the way a topic is conceived may enforce certain methodological actions or make us refrain from some methodological actions. How scientists approach a particular topic is determined by their beliefs about which particular actions give the best and most relevant information about the topic. For example, under certain research circumstances a chimpanzee might be considered an animal with a particular metabolism but in others be considered a social being with certain group behaviour. These distinct ways of conceptualizing a chimpanzee show that the same object gives rise to different research topics, each requiring distinct methods of collecting data. The first case requires physiological data reached experimentally; the second case interprets data gained by field observation.

The general methods of data *treatment* are 'topic independent'; they can be used to reason about any topic to acquire justified true beliefs. Data *collection* methods are more specific to the topic being studied since they are tailored to our investigative actions. These practico-instrumental methods like collider experiments, chemical analyses, double blind tests, etc. are designed according to the topic, so unlike the general methods, they are strongly topic-dependent.

3 The ontological issue

In 1958 Putnam and Oppenheim advanced a very captivating argument for ontological monism.[9] This 'cosmological argument' builds on scientific cosmology and two metaphysical principles. Contemporary cosmology tells us that the present visible Universe began in a very dense, hot state 13.7 billion years ago with a Big Bang and has expanded ever since. The expansion caused the temperature to drop, and symmetry breaking processes led to the known forces necessary for creating elementary particles and their interactions. Over time these elementary entities enter into more and more complex structures leading to stars, galaxies, and us. Putnam and Oppenheim knew few details of this ontogenetic story, but proposed two metaphysical principles: *the principle of evolution* and *the principle of ontogenesis*. The first states that for every level of organization there is a certain time in the evolutionary process when only levels up to this particular one, and no higher, had evolved. The second one says that for each level of organization there is a time when this level did not exist but when all its components existed. From those two principles they inferred that everything at any given level is ontologically dependent on the level below. This claim we may call *ontological physicalism*: every physical object and every property of this object consist of the object's parts and the properties of these parts, including the relations between these parts.

These metaphysical principles seem unproblematic to me. There is only one world, and according to contemporary cosmology it has developed as they say. This metaphysical scenario includes room to see human culture as part of nature; it is just one of nature's many forms of manifestations. Human evolution and the brain-mind relationship have developed as a reaction and adaptation, possibly to climatic changes in our forebears' environment. The evolution of consciousness with cognitive elements like beliefs, feelings, perceptions, and intentions made possible representing, planning, and acting towards a goal. The evolution of human language was a step towards forming institutional intentions that are necessary for complex societies and civilized cultures. Therefore human societies and cultures are ontologically dependent on human brains and minds. But this ontological *dependency* does not imply ontological *reduction*. Putnam and Oppenheim's metaphysical principles together imply neither ontological reduction nor non-reduction, but to make the case for reduction they tacitly assumed a third principle for which they never argued. This third principle might be called *the principle of reduction*: Each level of organization, except the bottommost, is *completely determined* by the organization of the level below. The principle comes in two variants: (i) every level of organization is identical to a simple composition or aggregation of its parts (identity reduction); or (ii) every level of organization is completely caused by (or correlated with) its components and the relations between these components (bridge law reduction).[10] Only by adding this principle of reduction to the two others can one derive the conclusion of reductive physicalism.

Although there is strong evidence for the principles of evolution and ontogenesis, the principle of reduction seems unjustified.[11] The substantive content of today's various sciences yields little evidence confirming such a general principle. On the contrary, we conceptualize different levels quite differently; even the same level may be described from different perspectives. If we take a look at the way different scientists comprehend their objects, they seem to isolate a phenomenon that can be considered a relatively self-contained system, in the sense that the causal influences on that system are minimized by screening it off from its surroundings, by limited range of activity, semi-walls, etc. Thus we can argue that at every level of organization physical components and relations between them are "encapsulated" in a bigger system such that many causal processes of the subsystems are confined to those subsystems. Therefore they do not affect the larger system (of which they are parts) and its causal relations to other systems in its environment. Quarks and gluons are "encapsulated" in hadrons, and the strong nuclear force "confines" protons and neutrons together to the atomic nucleus, etc. In both cases the confinement is due to the short range of the strong force. So neither quarks and gluons, nor protons and neutrons, have any role in how the macroscopic world behaves, other than those determined by the surface properties of hadrons and the nucleus. Similar self-containment is found in biology where cells, organs, and organisms are systems enclosed by membranes, pellicles, or skins that separate the interior of the system from the outer environment. Sometimes a subsystem may cause significant changes in the larger system as in cases like radioactive radiation or atomic explosion. But more often a system interacts with other systems in its environment. Thus, from some important perspectives a system is – to a reasonable approximation – causally confined within its own boundaries, and may be treated as 'closed,' but from other perspectives that system is seen as interacting with its environment, and in these circumstances is considered an 'open' system.

Distinguishing between a system and its subsystems is an analytical tool based on certain objective delineations. In both physics and biology we regard spatially separated phenomena as exhibiting the behaviour of different subsystems of a single self-contained system or of constituting many different systems. For example, the Milky Way and the Andromeda galaxy, just by the fact that they have different proper names, show that they are considered different physical systems. This way of thinking is quite reasonable for many astronomical purposes because all the stars in the Milky Way are bound together in the gravitational field of Sagittarius A*, a supermassive black hole at the centre of the galaxy, and the stars of the Andromeda Galaxy are similarly bound by another supermassive black hole. Although most of their physical features are separated spatially from one another, we may also consider them as one binary system, because they are so close that they orbit around a common centre of mass. Thus the Milky Way and the Andromeda Galaxy are approaching one another and will become one single elliptical galaxy 3 to 4 billion years from now. Hence it depends on the research context in which we see them, whether we take them as two separated systems, or as together interacting with their further environment, and describe them as one big system. The same relative distinction can be made with respect to clusters of galaxies as well as super-clusters.

In general, similar distinctions hold in biology. The complexity of a system, which biology studies, is much greater than in astrophysics because it has many more degrees of freedom than a non-organic system. But this does not change the claim that the same biological entities can be considered as a system or as a subsystem depending on what scientists want to know. A cell, for example, is often considered as the fundamental building block of life, but it consists of many different organelles, and each of them can be studied as a single system. However, understanding their functions demands considering their relations to their environment, which is the entire cell. Similarly, understanding the function of the cell requires understanding its relation to the organ of which it is part.

Therefore, I suggest, a system is a structure consisting of material that participates in constant processes that actively keep the system running independently of what happens outside the system. These internal processes are mostly "indifferent" to the external processes in the environment because little interaction happens between the internal processes and the external world. For instance the Milky Way is 'isolated' from the rest of the Universe since we can describe most of its dynamical processes without considering the gravitational pull or radiation input from Andromeda or any other galaxy.

A parallel distinction can also be made concerning the relation between the brain and the mind. For some purposes the brain can be considered as a closed system; for other purposes it can be understood only as an open system. The capacity of having conscious cognitive intentions developed through biological selection and adaptation of organisms' brains to their environment. Intentions are presumed to correspond to certain structures of brain states, but intentions are directed towards something and are distinguished from each other by what they are directed towards. So they are ontologically dependent on something *outside* the brain. No structural pattern in the brain can be identified with an intention until we include the causal interaction of the physical, social, and cultural environment. Also their function cannot be completely specified merely as a neurological structure. One needs to know, even if each type of intentions has its own particular structure, the kind of object that is the object for a certain type of intentions. One needs to know their content; i.e., whether this or that brain pattern corresponds to the intention of drinking or eating or mating, etc. in order for neuroscience to recognize the meaning of different neural patterns and define their functional role. Therefore, the reductionist theory faces both an explanatory and a predictive gap.[12]

In earlier chapters we saw how representation and explanation depend on intentions and purposes. Many intentions have a functional role that is understandable only in terms of social norms and rules. My intentions of finding something to drink, to eat, or to mate with have a purely biological origin; but many, if not most, intentions are shaped by *institutional* goals, and thus are culturally selected. Selected intentions are socially constructed in the sense that they exhibit formative influences from the culture and the society in which one lives. Like individual intentions, institutional intentions are identical with certain brain patterns, but they are learned only by experience. These intentions are as much ontologically dependent on culture and society as on the individual brain. Consider when I stop at a red light. This is a shared, institutional intention. When I see the red light, I intend to stop, even if unconsciously; this intention is determined by my perception of the red light, and so is the corresponding brain state. My response to the red light of wanting to stop is a socially and culturally induced type of intention that society reinforces in me as the right intention to have in this situation. In this case, it simply depends on whether I understand intentions as particulars or as types, since I am not able to tell that a particular intention is the right type of intention, unless I put it into the right social and cultural context. If this is true, we are unable to say whether this particular intention is correctly determined by this type of brain pattern because it does not tell us that it is the right type of intention.

Therefore these considerations may invoke metaphysical notions like 'emergence' and 'supervenience' to replace the notion of reduction, hence supporting non-reductive physicalism. This can be put in terms of the *principle of emergence*: At each level of organization there are genuinely novel entities whose behaviour and properties either are not *identical to* the properties of their constituents and the relations between them (the strong version); or are not entirely *causally determined by* entities and properties of their subsystems (the weaker version). I hold emergent properties appear whenever a self-contained entity causally interacts with an environment consisting of similar entities and thereby forms a bigger system. So each level of organization becomes a different domain of research. A non-reductivist view of this kind is usually associated with the idea that the various sciences are autonomous, each having its own descriptive categories and distinctive methodologies. However, the latter need not be the case.

The principle of emergence is consistent with Putnam and Oppenheim's two metaphysical principles, and the view that at a higher level of organization new entities or new properties may come into existence that cannot be explained by what exists at a lower level. Thus we get *domain plurality* implying that one and the same system may belong to different ontological domains because we can represent it as being ontologically closed (i.e., we ignore "external" influences as "negligible") or being ontologically open. An entity categorized in different contexts as in different ontological domains demands distinct concepts and models to be understood and explained with respect to these different domains.

In conclusion we already know many different perspectives from seeing and representing in everyday life. We see things from different angles and different distances, and therefore the same thing can be described or depicted in alternative ways. Although we can imagine things only from a certain standpoint, this does not deprive these descriptions of their objectivity. Naturally human thinking is unable to comprehend all aspects of a phenomenon in terms of a single description or picture. In similar ways, since it is created by humans for humans, science uses a plurality of representations. Yet, it is amazing that this animal, *Homo sapiens*, has been able to grasp the world beyond its own habitat by invoking abstract concepts. But what is perhaps not so surprising is that concepts, formed long before we got to this level of abstract thinking called "science," are applicable only in specific scientific contexts. Therefore as intentional beings we select descriptions both according to the objects we want to understand and the purposes for which we use these descriptions. As we have just seen, there are also ontological reasons for pragmatic pluralism other than the cognitive capacities of *Homo sapiens*. We may treat a system either in isolation or in relation to its environment, due to the fact that every system is to some extent self-contained and consists of self-confined processes but at the same time has the capacity to interact with its environment in different ways. Moreover, every system may be studied diachronically or synchronically because everything has both a history and is present at each moment of its history. These different ontological aspects give rise to different epistemic foci, but these are still different perspectives on either an 'isolated' system or the same system in relation to its environment.

4 Conclusion

First and foremost pragmatic pluralism is an explanatory strategy. It designates an attitude manifested in actual scientific practice where we see that in any specific context scientists select their explanations on both epistemic and pragmatic grounds. They choose that model or hypothesis yielding the most accurate and adequate understanding of the phenomenon they want to explain, and what they consider a research problem depends very much on their scientific training and professional outlook.

Classical debates between realists and instrumentalists over whether or not scientific models explain experiments and observations are primarily epistemic discussions, which have been disconnected from scientific practice. The orthodox realist considers explanatory power a high virtue of scientific theories, which does not appear merely in connection with justifying theories, but is something regarded as valuable in supporting the general realist interpretation of science. The traditional instrumentalists deny that theories can explain phenomena; indeed they say explanation forms no part of the scientific enterprise. Rather, theories and models are but useful predicative tools for dealing with observable entities. Therefore we need to see what it means to say that *science explains* and how far such explanations go. This effort has motivated this book. In contrast to traditional instrumentalists, I claim science does provide us with explanations, but they are not necessarily explanations in the sense realists have desired, i.e., to explain they need not provide a true explanans from which one can logically infer the explanandum.

A theory of explanation and understanding must be able to characterize an explanation regardless of whether it is true or false. We have to face the problem that laws used to explain phenomena are not necessarily true. The Aristotelian model of motion explained why heavy bodies fall to the ground; Newton's model of gravitation also explained the falling body as well as the tides and the planetary movements; the phlogiston theory explained combustion and oxidation; Lamarck's model of evolution explained the creation of new species; and Bohr's model of the atom explained the hydrogen spectrum, just to mention a few cases. These models have later been proven false in the sense that they could not account for all of the phenomena they were expected to explain. A model either represents reality (for a given purpose) successfully or not. Obviously assertions we make based on the assumption that the model represents reality (at least in certain respects) may be true or false. So truth played no role in the ability of these models to provide understanding of the *explanandum*. It is perhaps yet more revealing that even our best scientific models may one day be regarded as false. But this does not imply that our best models don't provide what we accept as explanations today.

The simplest conclusion is to say that these models were once *used* to explain various facts until people realized that the assumption that the model adequately represented reality was wrong. As long as a model is considered as representing reality adequately, it is the basis of assertions that have explanatory relevance. Thus models once thought of as having explanatory virtues because they were believed to represent reality are now regarded as explanatorily impotent. This view seems to make explanation a very anthropocentric practice, since no proposed explanation will have explanatory relevance unless it is believed to be true. But obviously explanation entails more than mere credibility; otherwise, the mere fact that we believed A explained B would imply that A does explain B. Indeed, what we regard as a *true* explanation has credibility; however, a credible false explanation is still an explanation, as bad art is still art. Old models may simply have lost their explanatory power because nobody still believes they represent reality adequately. But when people did trust them, these models were not only believed to represent the world but regarded as giving a credible acceptable answer to the request for an explanation of certain phenomena. Consequently, a person, who believes that H is true, will consider this hypothesis explanatory, while another one, who does not believe that H is true, will think it fails to explain the phenomenon. On my account explanation is heavily dependent on persons, times, and many other contextual factors. So an adequate philosophical analysis of scientific explanation must ask what it is about these models, besides truth (since false models as well as true can provide understanding), that supplies a putatively explanatory connection between them and the phenomena we have been asked to explain.

The issue here might best be expressed in terms of 'credibility,' treated as a property of all assertions relative to an audience measured on a variable scale (some assertions are more credible than others). Bad art may be art, of course, but there is art so bad that we are tempted to say, "That's not art at all!" When does that point come with explanation? It seems if a purported explanation has zero credibility for me (obviously this is relative to an audience) then I will say "That's no explanation at all!" Also I will regard it as providing no understanding. So, as I have emphasized, it is not the truth or falsity of the statements that give the explanation its explanatory power, but whether they have any credibility to an audience. Explanations in terms of the actions of supernatural beings have no explanatory force for me, though presumably there once were people for whom they did. But explanations in terms of Newtonian mechanics or the Bohr atom are credible because they are the sorts of explanations that I deem possible, even though I now deem them to be based on false assumptions and mistaken models.

Throughout this book I have exhibited my disagreements with both the formal-logical approach and the ontological approach regarding their characterizations of the aim of scientific explanation. However, it is also clear that these non-pragmatic approaches, if successful, would provide robustly objective accounts of explanation. Perhaps this is where the rub really lies. No explanation can be completely objective in the sense of detached from the epistemic situation in which it is proposed. All explanations represent a certain perspective on the *explanandum* phenomenon. Of course this perspective is partly determined by objective matters like the issue itself, and partly by the scientific tradition and personal interests. Hempel did not deny that explanation is primarily an integrated part of scientific practice, but he, and those who followed him, believed that what is of explanatory value in this practice can be explicated by abstracting from the pragmatic context. But no one has ever succeeded in demonstrating that it makes sense to abstract and isolate explanation from the context of scientific communication. Both approaches take for granted that behind the explanatory practice of science there is a common nature in all explanations. The formalists find this in the general logical or inferential structure of an explanation; the factualists, who take an ontological approach, find it in the particular objective content of explanation. It is doubtful that something like explanation has such essential features independently of the context.

Furthermore, it is easy to understand the historical reason for restricting the scope of the philosophers' theories of explanation to explanation in the natural sciences: the philosophers who developed these theories wanted to distinguish the sort of explanations found in the physical sciences (and possibly in the life sciences and social sciences as well, although that was not their major concern) from what they considered to be *pseudo-scientific* explanations given by religions and mythologies, astrology, phrenology, psychoanalysis, Marxist dialectical materialism, etc. From the time of Galileo and Descartes to that of Russell and Carnap, all were champions of the 'scientific' mode of explanation over a host of rival candidates sponsored by powerful interests of the time. (Popper, of course, is most famous for explicitly making this his career goal.) Their strategy, it seems to me, was to assume that the first priority was to distinguish the scientific explanation from the non- or pseudo-scientific claims to explanation. That end was best achieved by picking the clearest-cut example of what everyone agreed to be a genuinely scientific sort of explanation – that found in physics – and contrasting its form to the sort of explanations found in religions and pseudo-sciences. The question of where to put the teleological sorts of explanations found in the life sciences and the genetic, intentional, and historical sorts of explanations found in the social sciences was of secondary priority, though of course it was actively debated. Another factor in the twentieth century, of course, was the enthusiasm for symbolic logic, which lent itself rather nicely to a formal treatment of explanation in physics, but rather less well to the life sciences and the social sciences.

In contrast, I have defended the idea *that to understand explanation we must understand its rhetorical functions in providing understanding*: all serious requests for knowledge are information-seeking questions whose answers may provide explanations. We find such responses whenever the information given by an answer is put into a broader narrative by bringing in shared background knowledge. An explanatory answer is relevant and informative with respect to the context in which the question is posed and with respect to the background assumptions of the interlocutor and the respondent, and possibly their personal interests. Thus, the distinction between description and explanation is pragmatic rather than logical or semantic. I have also argued that explanatory force has *nothing* to do with truth, but a lot to do with making sense to the inquirer. The asymmetry between what explains and what has to be explained depends on the rules of discourse and the relevance of the available information, which is a function of our background knowledge or beliefs. Explanation is rhetorical in the sense that to be successful an explanation must make sense to the questioner. As we have seen, explanation is contingent on the concrete context of the communicative situation. In sum, I argued that information is relevant with respect to certain background knowledge if it fills out certain blanks in the questioner's knowledge, but is otherwise coherent with this knowledge, and if it provides him with the possibility of describing or taking action based on the content of the information. A good explanation expresses the explainer's understanding of the problem raised by an explanation-seeking question, while this answer hopefully provides the explainee with the same understanding.

Notes

1 Forms of Understanding

1. Newton-Smith (2000), pp. 130–131.
2. See, for instance, Salmon (1984), p. 10; Salmon (1998), 79ff.; Cushing (1994), p. 10; Schurz and Lambert (1994), p. 109; Weber (1996), p. 1; Faye (1999); Faye (2002), p. 29f.; de Regt and Dieks (2005); de Regt (2009); Dieks (2009).
3. Hempel (1965), p. 413.
4. Ibid., p. 327.
5. Trout (2002), p. 213.
6. Ibid., p. 214.
7. Ibid., p. 223.
8. However, I am not too happy about "genuine." Is it just a synonym for "true"? I see two problems: (a) As I shall emphasize later whether or not a discourse is considered explanatory depends on the audience; so there can be multiple true explanations of the same phenomenon expressed in the framework and vocabulary of different domains of science. Now "genuine," especially if we say "*the* genuine" one, suggests the real one amongst imposters. So the choice of words here suggests a "one and only genuine explanation" approach which I am arguing against. (b) A proposed explanation can be "true" in the sense that it makes no false assertions, and yet, because the explanans fails to have the right sort of connection to the explanandum (whatever that is); so "only true beliefs" does not necessarily give us a genuine explanation.
9. de Regt (2004), p. 107.
10. Ibid., p. 107.
11. de Regt and Dieks (2005).
12. Friedman (1974), p. 18.
13. Ibid., p. 15.
14. Ibid., p. 5.
15. In Note 33 of their seminal paper "Studies in the Logic of Explanation", see Hempel (1965) pp.247–290, Hempel and Oppenheim admit that they have not been able to establish clear-cut relevance criteria for explaining regularities in terms of more comprehensive regularities. The reason is that the purported explanation can be established as a conjunction of the law to be explained and any other law which is irrelevant for the explanation.
16. Friedman (1974), p. 15.
17. Expectation can be of two kinds: (a) expecting events based on general laws and conditions we believe are the case (this is Hempel's main concern); (b) expecting certain regularities to hold because we can deduce them from theoretical laws and limiting conditions. The first case (a) is by far the larger class for the simple reason that normally we don't start with the theoretical laws and only then deduce some empirical regularity and look at the world

and say "Ah, we expected that." In most real cases we start with the empirical generalizations and then build the model such that it enables us to deduce the generalizations, *which we already knew*. You can't be said to "expect" what you already know to be the case. Of course occasionally a hitherto unnoticed regularity is deduced from theoretical laws first (e.g. the Poisson-Fresnel bright spot in the shadow of a disk phenomenon); but this is the rare exception, it seems to me. This would seem to support Friedman's claim that Hempel's criterion of the satisfaction of expectation is not a necessary condition of understanding.

18. See, Lewis (1973), pp. 72–77; (1986), pp. xi–xvi.
19. The whole spirit of the unificationist approach is highly conditioned by the success of the parallel approach in logic and mathematics. Euclid was the "first unificationist"; he found nothing new, but unified a vast body of knowledge into a rudimentary deductive system with only 15 postulates whose truth was assumed. Thus it was natural to "feel" that geometric knowledge was better understood after Euclid's achievement. It was already implicit in Aristotle's notion of scientific knowledge that one should minimize the number of "basic truths," and I think it is fair to say the notion was already present in Plato's epistemology. That conception set in motion an intellectual challenge to reduce the 15 axioms and postulates to fewer and fewer, a tradition that ultimately produced Russell and Hilbert, logicism, and (at least some forms of) positivism. But this story suggests not that there is rivalry between "unificationists" and "deductivists" but rather that the unificationist (at least those who assume that unification is achieved by deduction) presupposes the basic outlook of the deductivist.
20. van Fraassen (1989), ch. 3.
21. This distinction is more or less taken over from Collin (1985), pp. 61f. But for obvious reasons I prefer to call his integration of ontology an integration of concepts.
22. It is not obvious that a conceptually more parsimonious description always marks an improvement in our understanding. At least in logic, it sometimes seems as if a more parsimonious logical vocabulary may make a proof more difficult to understand. Indeed, the point of introducing derived terms, defined in terms of the basic terms, is to aid pragmatically in how easily we can understand a language. Pure parsimony is no doubt a logician's virtue, (and I guess at least some ontologists' as well), but when we venture into the question of our understanding of phenomena it seems that it is possible to "overdo it" with respect to parsimony to the point where one can make oneself barely intelligible.
23. Kitcher (1981), p. 516.
24. Henry Folse has in a private communication made this objection.
25. Hempel (1965), p. 83.
26. Kitcher (1981), p. 508.
27. Einstein (1954), p. 228.
28. Einstein (1905/1923), p. 37.
29. Bohr (1998), p. 48 and p. 96.
30. Dieks (2009), p. 233.
31. Salmon (1984), p. 239.

32. Is the "goodness" of the reasons objective or subjective? Here is the problem as I see it: Aristotelians told causal stories in terms of entelechies and essences which they of course believed were real. (Let us suppose that these stories were without logical blemish.) No doubt this made them feel that they understood the relevant phenomena very nicely. If asked their reasons for this conviction of understanding, they would of course respond with a discourse about entelechies and essences. To them these reasons seemed very good reasons; to us (and everyone since the Renaissance) they seem preposterously bad. The ultimate arbiter of objective goodness of the reasons is that they have to appeal to things that are real; thus without a metaphysical foreknowledge of the true shape of reality, we can never know if our reasons are really good. But they do have to be really objectively good for our feeling that we understand the phenomena to be truly justified (and not just create the false conceit of holding a justificatory reason ("pseudo-understanding"), as was the case with the Aristotelians). Yet no one has such foreknowledge. So apparently we are never justified in feeling that we have understood a phenomenon.
33. See Salmon (1998); Dowe (2000).
34. For a further discussion of counterfactuals context-dependence one may consult Lowe (1995).
35. A detailed criticism of Salmon's attempt to provide a context-free description of causal processes can be found in Faye (1997a). Faced with this criticism, Salmon suggested that the notion of conjunctive forks could do the work of establishing causal priority. But this does not eliminate the problem as far as I see it. First 'conjunctive fork' is a statistically defined concept and therefore does not apply to singular causal processes; moreover, most physical interactions are not close to being conjunctive forks.
36. Davidson (1967/1975).
37. Faye (1991), pp. 120ff.
38. Cushing (1994), p. 10. See also Cushing (1991).
39. Ibid., p. 11.
40. Ibid., p. 20.
41. Bohr (1948/1998), p. 144.
42. The instrumentalist thinks what we have got with just the Born rule is enough; we don't need any metaphysical accompaniment for a 'scientific explanation.' One might very well complain that it provides no real "understanding" (as Cushing emphasizes when he introduces the distinction), but the instrumentalist just replies, "Well, that's all you're going to get." Similarly one might argue that the many-worlders are in the same boat; no one's understanding is enhanced by all this talk of "other worlds" because no one can really provide any "understanding" of this basic interpretive move. So I say that all this talk about other worlds where the system manifests some different state is nothing but an interpretive device, an instrument if you like, to get around interpretive difficulties engendered by a misunderstanding of the psi-function. It adds not one iota to one's "understanding," but it certainly greatly enhances the mysteriousness of the quantum description.
43. Salmon (1998), p. 77.
44. See de Regt and Dieks (2005); de Regt (2009); Dieks (2009).

2 Understanding As Organized Beliefs

1. Scriven (1962), pp. 224–225.
2. Achinstein (1983), p. 23.
3. Lipton (2004), p. 30.
4. See Trout (2002).
5. Grimm (2006) points to this striking fact.
6. In fact, I don't think they are that close since, as I shall argue, instincts constitute one form of understanding even though instincts do not make up beliefs.
7. John Locke's famous treaty about knowledge is called *An Essay Concerning Human Understanding*, but his use of "Understanding" in the title does not refer to knowledge of natural phenomena but to the function of the faculties of the mind.
8. Zagzebski (2001).
9. Grimm (2006), p. 518.
10. One might say that the question is whether the claim to understand X is ever justified by *internal* criteria alone. If by "internal criteria" we mean the coherence with the remainder of one's belief system, then it is obvious that to *feel* that one understands X, the beliefs one holds about X must to some degree be 'coherent' with the rest of my beliefs, but while necessary, such criteria hardly seem *sufficient*. But sufficient for what? Sufficient for having *understanding* or for having *correct* understanding? I would say sufficient for having understanding but insufficient for having correct understanding. In order for S to have understanding that is correct, there must be some relation with the "external world" as revealed in (future) experience of X (or other X-like things) that must be met. Suppose I feel that I understand, and that feeling is based on beliefs all of which cohere with the rest of my beliefs. In that case I am naturally led to have certain *expectations* about how X (and X-like things) will behave. If, all of a sudden, X starts behaving in ways very much at variance with my expectations about how it should behave, I say to myself, "I guess I was wrong; based on all my internal criteria, I thought I correctly understood X, but now the external world (future experience) has shown I didn't after all."
11. I have sympathy for the purpose of epistemic contextualism according to which our standards of knowledge ascription varies with the context. Nevertheless, I believe that Michael Williams (2001) is caught in a dilemma, which he does not address at all. On the one hand, as do most epistemologists, he agrees that truth is an essential part of our notion of knowledge; on the other hand, he argues that knowledge may be unstable, defeasible and fallible. Both claims cannot be true unless one thinks of truth in terms of provability, justification, etc., or one takes the connection between the belief's actual truth-value and its actual justification to be accidental. A deflationary theory of truth will not do because it does not allow truth to oscillate as the evidence accumulates.
12. What is this change of opinion? Is it (a) I used to think I understood X, but now I realize I never did; or (b) formerly I had one understanding of X based on the beliefs I then held, but now I have a new understanding of X based on new beliefs I now hold? Although we often speak like (a) it is (b) that expresses the change I have in mind.

13. What is the difference in meaning between saying that something is "true" and saying that it is "correct"? Sometimes these terms are used interchangeably as when a teacher makes the following response to a student's remark "What Jim just said is true" or "What Jim just said is correct." However, sometimes their meanings are different and they cannot replace one another. An action, a pronunciation, or a reproduction may be correct; here we will not say that any of these is true. An action is correct if, and only if, it is carried out according to certain rules of action, a pronunciation is correct if, and only if, it agrees with some phonetic rules, and a reproduction is correct if, and only if, there are structural resemblances between the reproduction and the reproduced. In all three cases what makes them correct is not that they have a particular content of meaning but that they exhibit a certain form similar to what the rules of action, pronunciation, or reproduction prescribe for them. Therefore, it makes sense to say that an action, a pronunciation, or a reproduction can be more or less 'precise', 'accurate', or 'exact', but not 'true'. In contrast a belief or a sentence may be true, and they are true because the particular content they express is in accordance with how things stand in the world.
14. See Schweder (2004), pp. 27ff.
15. Scriven (1975), p. 11.
16. See de Regt (2004); de Regt and Dieks (2005); de Regt (2009). There are some differences in the formulation of the central claims between the first and the third paper.
17. Trout (2002), p. 222.
18. de Regt (2009), p. 588.
19. de Regt (2004), p. 103. Apparently, "strength" is just the same as having the theoretical virtues listed in the quotation. Such selection criteria are very nice no doubt, but I would say "strength" has more to do with the degree and variety of confirmations. A theory might have all the virtues listed and still fail to be confirmed empirically, in which case I would not want to use the word "strength" to describe it. A young theory with only a few confirmations is still wobbly; once it has been around a long time and accumulated many confirmations, we are likely to call it "strong."
20. Ibid., p. 103.
21. de Regt (2009), p. 588.
22. It should be mentioned that actions could be carried out correctly or incorrectly. But this is not my point. Take the following example. Louis Pasteur demonstrated that it was possible to keep a nutritive substrate sterile to avoid infection caused by germs and bacteria. The result can be reached only by working aseptic. His colleague, Félix-Archimède Pouchet did not believe that Pasteur was right in assuming that spontaneous generation was due to infection because he could not reproduce Pasteur's experiments. What Pouchet lacked was sufficient experience in the form of practical knacks. He did not possess the skills of handling aseptic technics in the correct way. However, in my opinion, it makes sense to say that Pouchet did not understand the possible courses of infection in order to work aseptically and therefore did not have sufficient skills. However, you may ascribe to a certain person skills even though that person never exhibits these skills. An action can be correctly or incorrectly carried out, but the skills, which make it possible to carry it out, are not "correct" or "incorrect."

23. Polanyi (1958, 1966).
24. Scriven (1962), pp. 224–225.
25. Ibid.
26. Perkins (1986), p. 114.
27. Nickerson (1985), p. 234.
28. Ibid., p. 220.
29. Zagzebski (2001), p. 241.
30. Ibid., p. 244.
31. Ibid., p. 242. If I understand her correctly, it seems to me that here Zagzebski makes a mistake of conflating epistemology and ontology: she holds that understanding does not have a propositional form. Thus, I would have expected that she would have said, "understanding is the state of a non-propositional *comprehension* of structures of reality." It is not clear why there has to be a one-to-one correspondence between a propositional understanding and a propositional structure of reality, and between a non-propositional understanding and a non-propositional structure of reality. Furthermore, why should we suppose that reality is propositionally structured in the first place? A naturalist, like myself, might say that natural selection led us to have precisely those linguistic faculties that would lead to a language the propositional structure of which enabled it to express states of affairs in the world. It is no wonder that our language enables us to do this, for any language that did not would be useless for stating what is the case, which is after all a very useful thing to be able to do.
32. Ibid., p. 242.
33. Ryle published a paper "Knowing How and Knowing That" in 1945. It was later included in his *The Concept of Mind* (1949). His argument for the distinction goes like this: knowing that something is the case is the ability to contemplate a proposition, but being able to contemplate a proposition cannot be a case of knowing that something is the case, because one must be able to contemplate a proposition in order to know that something is the case. The infinite regress has to be stopped, and Ryle proposed that the correct way to do so is to posit that first one has to know how to contemplate a proposition before one can know that something is the case. 'Knowing-how' is then not only independent from 'knowing-that' but also prior to knowing-that (1949, pp. 30–31).
34. Dewey (1922), p. 178.
35. Searle (1969), pp. 19ff.
36. Polanyi (1966), p. 8.
37. Polanyi (1966), p. 7.
38. Ibid., p. 13. Polanyi's own emphasis. As far as I understand him, Polanyi's use of "term" may refer to either physical or linguistic entities.
39. Embodied cognition is regarded as the future of cognitive science by many of today's scientists in the field. Rather than having a full-fledged theory, the science of embodied cognition is a collection of theoretical and methodological claims about the causal or constitutive role of body and environment in cognitive processes. Its tenets are often built as negative claims about classical theories of cognition (Rupert 2009, p. 218) and are believed to constitute a radical break with traditional cognitivism according to which thinking can be characterized as manipulation of symbols in an algorithmic process

(see Markman and Dietrich 2000; Wilson 2002; Anderson 2003; Barsalou 2008, pp. 10–11; Calvo and Gomila 2008; Rowlands 2010; Shapiro 2011). However, such a view is probably an overreaction and rests on a mistaken overconfidence in the scope of embodied cognition as the new explanatory paradigm within the contemporary landscape of cognitive science. Because it originated as a reaction against the traditional explanations, the themes of embodied cognition are often seen as able to fully replace the classical theories, and to most researchers they appear to be the only sensible direction to follow in the future (cf. Davis and Markman 2012). Nonetheless, I think when it comes to abstract thinking and higher-order reasoning so far there is no evidence that rules out classical theories. Hence embodied cognition and classical theories may share a common subject matter and therefore should not be seen as mutually exclusive and competing explanatory paradigms. On the contrary, they may both contribute to the field of cognitive science and help us to reach a better understanding of human cognition.
40. Edelman and Tononi (2000), pp. 103ff.
41. See, for instance, Zuberbühler (2002); Ouattara *et al.* (2009).
42. In fact, there are also experiments within linguistics which strongly indicate that semantic (linguistic) representations are distinct from conceptual representations. It is not an easy task to separate the investigation into the conceptual content of our thoughts from the investigation into language itself, but apparently it can be done. See for instance Lucy and Gaskins (2001). A theoretical reflection on the relationship between semantic representations and conceptual representations, which is partly based on empirical investigation of the difference between thought and language, can be found in Levinson (1997). Levinson concludes that conceptual representations cannot be identical with semantic representations: the latter cannot be a subset of the former because there are essential distinctions between representations for linguistic meaning and representations for thoughts. Furthermore, he emphasizes that these two forms of representations are not only distinct kinds, but also they are not even isomorphic. Nonetheless, he also argues they are necessarily related, that the inner, private representations of thoughts cannot be totally independent from social, public ones.These findings go partly against the Sapir-Whorf thesis that language determines thought. A number of experiments support the universalists' claim that the categorical perception of colour is not constructed through language but that categorical perception is innate and universal and therefore that the categorical perception of colour is the same independent of cultures. Although the colour spectrum is continuous, there is strong evidence that this continuum is perceived in a number of discrete categories. Anna Franklin *et al.* (2005) have been able to demonstrate that the various colour categories do not change between pre-linguistic babies and post-linguistic children. Nor do they change across cultures. However, in a later study (2008) she and her co-workers report that semantic representation does affect our ability to discriminate between colours. Thus, Whorf was half right after all. Categorical perception of colour is lateralized to the right hemisphere in infants, but to the left hemisphere in adults; apparently because language has an impact on the brain so that the semantic categories become more dominant than the non-semantic categories in determining how the brain structures the visual world.

43. See Slobodchikoff *et al.* (2009).
44. Still, some may be reluctant to say that animals are operating with "concepts". It seems as if one could describe these observations without reference to "concepts" by saying simply that young prairie dogs acquire from the adults an ability to distinguish coyotes from hawks, etc. The ability to make distinctions must precede language in order for language learning to be possible. However, I would argue that an organism possess concepts if it is aware of something as a type instead of it as a particular, and this organism is aware of something as a type if it behaves similarly towards other instances of the same type. A concept is the ability to combine different perceptual organizations into a 'universal', and as long as the brains of prairie dogs manage to recognize common features of the coyotes, I would say that they possess a concept of a coyote. I see no reason to deny any organism the ability to have concepts if we, based on observation, can recognize from its behaviour that it sees different individuals as belonging to the same kind. What more would it take for prairie dogs to have a concept of 'coyote' and 'snake' other than being able to act functionally according to such a distinction? Furthermore, I take concepts to be vehicles of thoughts, so if science can demonstrate that thinking takes place in these animals, I think we have every reason to overcome the reluctance of ascribing concepts to animals.

    A possible objection against the assumption that animals possess concepts would be to suggest that most animals, even lower ones such a lizards, are able to identify their own species and to distinguish between 'male' and 'female' among the members of its own species. If this is true, it seems that these animals possess concepts. Indeed, one could argue that the possession of concepts only exists in connection with other species and not in relation to the animal's own species, but I think such a distinction would be arbitrary and evolutionary problematic.
45. Individual experiments have been reported in various research journals. Some of the most important observations are summarized in a survey article published by Heinrich and Bugnyar (2007) who are some of the leading scientists in this field. See some of newest findings in Marzluff and Angell (2012), which also focuses on the neurobiology behind the exceptional skills of crows and ravens.
46. Marzluff and Angell (2012) write about corvid language: "Most linguists, philosophers, and animal behaviourists would say that crows and other songbirds do not have a communication system organized like human language. Human language differs from other animal communication systems, even those where animals speak or sign their intentions, by the degree to which humans use rules to combine a few discrete, arbitrary sounds into a limitless and readily understood group of novel and complex words, phrases and sentence. Alex [an African grey parrot] taught us that birds are capable of such recombination, but this ability pales in comparison to those of a young child" (p. 61). But when they add that this skill is unknown in any wild animals, it seems not to be true as the investigations of Diana monkeys show. It seems safe to say that more and more experiments provide evidence that nothing exists in humans, which cannot be found in a rudimentary form in other higher animals.

47. Just think of the New Caledonian crows' use of tools to get to their food, and of monkeys' and apes' use of stone, sticks and branches to crack nuts, gather food, and defend themselves. The New Caledonian crow proves true instrumental insight in a three-part intelligence test. The crows had to retrieve a short stick to obtain a long stick before they could, by the help of the longer stick, access the food. See Marzluff and Angell (2012), pp. 99–100.
48. Salmon (1998), pp. 79–91.
49. Kant was probably the first to introduce the notion of schema into philosophy. He held that the pure concepts or categories of understanding have no empirical content and therefore couldn't have any similarity with anything empirical. For instance, a category of understanding like causation does not exist as something given in our sensation; it does not have a content, which represents what we see in contrast to the concept of a 'cat' that represents what we see a cat to be. Bridging the gulf between the categories and the sensation we need to introduce some third thing which is similar to both the pure concept and what is given to us in sensation. This third element is what Kant called the transcendental schema, which has to be time: because time as an a priori form of intuition is a condition of all experience. Only if a category is subsumed under the transcendental determination of time does it become a schema, and as a schema it can be used on what appears in the intuition. But a distinct difference between Kant's use and my use of "schema" is that a 'schema' to me does not have a priori origin, a schema is imposed on us by successful experience and is not a necessary condition for experience.
50. In a survey study of explanatory understanding in psychology, Keil (2006) concludes, "The processes of constructing and understanding explanations are intrinsic to our mental lives from an early age, with some sense of explanatory insight present before children are even able to speak" (p. 247). Thus, we see that empirical investigations seem to confirm the existence of innate, pre-linguistic schemata of understanding.
51. This is of course the naturalist's "standard model" of the evolution of human consciousness generally and of scientific understanding in particular, but how much is it confirmed by empirical evidence? There are evolutionary theorists who would dismiss this as way too much a "just so" story. One cannot deny that a certain amount of truth lies behind such contempt, but the more scientists get to know about the function of the brain and its genetics and can compare their observations with the function of brains of other primates and their genetics, indirect empirical evidence will eventually accumulate in support of the "standard model."
52. Causal schemata seem to have existed before the development of hominids. Some interesting observations of Diana monkeys indicate that these monkeys already have a causal understanding of their predators' behaviour. Zuberbühler (2000a) reports how leopards and chimpanzees prey upon them and how these predators use different hunting tactics to catch them. In response to their attackers the Diana monkeys have developed two distinct antipredator strategies: conspicuous alarm-calling behaviour in response to leopards and silent, cryptic behaviour in response to chimpanzees. Zuberbühler (2000b) also describes how Diana monkeys react to the alarm-calls of crested guinea fowls. These fowls give alarm calls of leopards, and

sometimes of human beings, but the monkeys respond always as if a leopard is present. Nonetheless, after having primed groups of Diana-monkeys to either leopards or human beings the play-back of the guinea fowl alarm calls suggests that the monkeys' response was not directly driven by the alarm calls themselves but by the calls' underlying cause, i.e. the predator most likely to have caused the calls.

53. One may ask whether 'explanatory misunderstanding' can be separated from 'explanatory understanding based upon false beliefs which cohered with my system' or they are the same. I think that the latter is part of the former. An explanatory misunderstanding may arise from the organization of acquired information using a wrong schema (not very likely, however), but it may also arise from the implementation of the right cognitive schema while also relating false beliefs.

3 On Interpretation

1. 'Data' or 'evidence' are neither external, mind-independent objects nor internal, mind-dependent phenomena but something in between. They are often designed. They have been designed with a purpose of providing relevant information about an external object from experiments or gathering of observations. I wouldn't say that the "data" are the phenomena themselves, but that in an experiment our data are records that describe or refer to the designed phenomena. Of course not all data are the result of designing phenomena to fit our purposes; field observations of animal behaviour, for example, provide scientific data by selecting what the observer finds important, but the behaviour that the data is about is not designed by the researchers.
2. Note that the interpreter and the questioner may be one and the same person but may also be two different persons. In the first situation the interpreter eventually answers his own interpretation-seeking question, in the second situation the questioner raises an interpretation-seeking question to another person in the hope of being informed by this person's answer.
3. See Faye (2012) for a discussion of interpretation in the humanities in the light of such a pragmatic-rhetorical theory.
4. Indeed, the situation is more complex since the positivists also assumed that one could in principle establish "correspondence rules" which give the abstract formalism of theory an empirical "interpretation." According to the positivist tradition, 'interpretation' goes from theory – which is in need of an interpretation – to the empirical world, thereby 'giving' the theory empirical significance. For the pragmatic tradition, you go the other way: you start with the empirical end – that is what is in need of an interpretation – and work your way towards the theoretical representation.
5. I use the word 'interpretee' for the recipient of an interpretation in the same way as an awardee is the recipient of an award.
6. IAU (2006).
7. It is well known that some social constructivists have had a hard time in distinguishing between objects and concepts, but some actually attempt to make such a distinction. David Bloor, for instance, says at one point that social

construction is concerned with concepts and not objects: "It is the concepts, not their real or imagined objects that are constructed. It is concepts that are institutions, not their presumed objects or the particulars that are currently said to fall under the concepts" (Bloor 1997, p. 17). According to Bloor, institutions contain an essential element of self-reference, which means that, "the referring practice is not directed at an object that exists independently of the practice of reference" (p. 12). However, it is not surprising for a realist that concepts are constructed, but a realist may claim two things in contrast to the social constructivists: (1) of all the objective similarities that exist between any two objects we usually choose those classificatory features that give us the most explanatory advances. I may be similar to other persons by being less than 2 meters tall, but in most situations it is not explanatorily relevant to classify me together with other things being less than 2 meters rather than together with other human beings. (2) Money may be part of a self-referential institution because money is whatever a community intentionally treats as money, but when I refer to a purely physical object, like a quark, I refer to whatever object that is understood as a quark. This is how chemists eventually discovered that there was no object corresponding to the concept of phlogiston.
8. Slipher (1913), p. 56. My emphasis.
9. Today the term "Nebulae" refers to the gas and matter that hurdles into space as a result of a supernova explosion. By the terminology of the 1920s all of them were nebulae, but what we call "nebulae" today, e.g. the Orion nebula, are in fact in the Milky Way. The problem of course was that these true nebulae were categorized with what we today call galaxies into one class. Only with Hubble was it really understood that these were two very different kinds of objects.
10. According to Pierre-Simon Laplace, the ardent French mathematician and physicist, nebulae were clouds of gases, which contract under their own weight, heating up in the process and creating new stars. The first observation of a supernova outside the Milky Way, the one observed in the Andromeda galaxy in 1885, was interpreted as a new-born star where the sudden flash on the sky was caused by its climatic ignition of the central gases. See Marschall (1988), pp. 94–95.
11. Slipher (1917), p. 409. The discussion of interpretations was not over yet. In April 1920 the National Academy of Science of America sponsored a debate between Harlow Shapley and Herber D. Curtis about the nature of nebulae. Shapley maintained that what we now call "galaxies," then called "nebulae," were placed inside the Milky Way and he suggested based on its luminosity in relation to Nova Persei, a common nova, discovered in 1901 and estimated to be 500 light years away, that the distance to S. Andromedae (the supernova discovered in the Andromeda Galaxy in 1885) was 10,000 light years. In opposition Curtis argued that these so-called nebulae were far beyond the outskirts of our galaxy, located at least several hundred thousand light years away. Thus, S. Andromedae could not be the same kind of object as Nova Persei. As Shapley pointed out, if S. Andromedae was as powerful as Curtis claimed, this would mean that its luminosity equals that of 100 million stars. However, Marschall (1988) summarizes their debate by concluding: "By 1920, there were clear indications from other quarters that there were indeed two

varieties of nova. Astronomers, scanning photographs of M31, had detected nearly a dozen novae in the nebula, all far fainter than the star of 1885. Yet depending on their preconceived notions about the distances of the nebulae, astronomers *interpreted* this in different ways" (p. 100, my emphasis). It was first by Erwin Hubble's work on Cepheid variable stars in the late 1920s that the astronomical community agreed about the great distance of galaxies like the Andromeda nebula.
12. Baade and Zwicky (1934a), p. 258.
13. Baade and Zwicky (1934b), p. 263.
14. See Hoyle and Fowler (1960).
15. Baade and Zwicky (1934a), p. 254.
16. It was true at the time. The Hubble space telescope and the Chandra X-ray Observatory have radically changed this situation. They have photographed a turbulent spherical cloud of debris at the position where Tycho observed his supernova, an observation that is interpreted as the supernova remnant.
17. Baade and Zwicky (1934a), p. 258.
18. Hempel (1952).
19. Quoted from Moyal (2001), p. 34.
20. Moyal (2001), pp. 49–50.
21. See Kragh (2012) for a detailed historical treatment of the developments in physics that lead to Bohr's 1913-model of atoms.
22. Rutherford (1911).
23. Bohr (1913).
24. Originally Peirce called abduction "*retroductive* inference."
25. As Henry Folse points out to me, there is a further option: quantum mechanics is incompatible with *classically* defined individual systems in isolation, but the metaphysical commitment to an underlying reality does not commit us to the notion that individual systems are defined by the classical properties.
26. See Dupré (1993), pp. 49–53 for a radical attitude to classification. Dupré calls his position "promiscuous realism" which covers the view that there are indefinite many ways of individuating and classifying the objects in the world. It depends on our cognitive interests which one we choose, and no one is more correct than any other. He also believes that all of them are equally referential in the sense that any taxonomy that forms part of a system of understanding refers to a set of kinds. Thus he holds that there is an indefinite number of sets of kinds. I don't think it is necessary to make such a sweeping move to reject reductionism or essentialism. Taxonomies and classifications are not only determined by our purposes. What is also important is that it is not every possible taxonomy that carves the world in its joints; some are more to the matter than others.
27. Hempel (1965), p. 179ff.
28. Hacking (1983), pp. 87–89.
29. See Heisenberg (1967), pp. 99–100: "In addition, during a short sick leave on Helgoland in June, I wrote the first draft of the quantum mechanics, which for me represented in a certain sense the quintessence of our discussions in Copenhagen – a mathematical formulation of Bohr's correspondence principle." Here he explicitly claims that following the correspondence principle was what led him to the matrix mechanics. In Bohr's own use of the word, 'the corresponding principle' (including other expressions for the same idea)

signifies an umbrella-concept. The precise content of Bohr's concrete formulations changes with the physical problem – which evidently is because the principle may be used heuristically. After the discovery of quantum mechanics Bohr promoted the principle in his endeavour to make use of a reinterpretation of the classical concepts to the greatest possible extent. But in 1923, before the discovery, Bohr attempted to argue that the principle of correspondence is a purely quantum theoretical hypothesis (see Faye 1991, p. 119). However, this was a mistake because the alleged quantum theoretical hypothesis (or law) is the very expression of a particular kind of reasoning which seems characteristic of Bohr's use of the principle.

4 Representations

1. I do not commit myself to any of the particular views regarding the nature and dynamics of scientific practices associated with these terms. In particular, I do not assume any kind of incommensurability thesis.
2. Models and their role in science are discussed in Giere (2004, 2006); van Fraassen (2008); Teller (2001); Frigg and Hartmann (2006); Faye (2002).
3. van Fraassen (2008), p. 7. Similarly Giere (2006) argues: "If we think of representing as a relationship, it would be a relationship with more than two components. One component should be the agents, the scientists who do the representing. Because scientists are intentional agents with goals and purposes, I propose explicitly to provide a space for purposes in my understanding of representational practices in science. So we are looking at a relationship with roughly the following form: S uses X to represent W for purposes P. Here S can be an individual scientist, a scientific group, or a larger scientific community. W is an aspect of the real world. So, more informally, the relationship to be investigated has the form. Scientists use X to represent some aspect of the world for specific purposes" (p.60).
4. Faye (1999, 2002), ch. 3, (2007a).
5. van Fraassen (2008), p. 23.
6. Indeed a footprint is not a 'representation' of a human being. A footprint is 'evidence' or perhaps a 'sign' of a human being, the only thing it "represents" is the shape of the sole of the maker's feet. If I measure my height at five foot nine inches, I would never say that's a "representation" of me; I might easily say "that is a measurement of my height." A measurement might help us to decide how to represent something, such as the height or weight or average rainfall of an area, but the measurement alone is not a representation, in the ordinary use of that word. We represent one stick figure of a human being as twice the height of another; does that show one is three foot and the other six foot? No, it shows one is twice as important as the other.
7. van Fraassen (2008), p. 24.
8. See Richard Rorty's Introduction to Sellars (1956/1997), p. 9. See also Brandon (1994), p. 8.
9. It is more complicated than I have just mentioned. Words like "horse," "hest," "pferd," and "cheval" are all different linguistic representans of a horse. These words of different languages express one and the same concept. This I take to indicate that the concept of horse is not a creation of a representation

302 Notes

but due to the human ability to make pre-reflective identification, separation, and categorization of whatever presents itself to the human senses. Hence the natural languages, perhaps in contrast to any other forms of representation, can be used to express pre-existing, non-linguistic concepts and to construct new linguistically based concepts.

10. Sellars (1956/1997), sec. 15.
11. See Dretske (1979, 1994); van Fraassen (1980), p. 15.
12. Sellars (1956/1997), sec. 43. See also Brandom's comments (p. 164).
13. Debs and Redhead (2007).
14. See note 3.
15. Debs and Redhead (2007), p. 158.
16. Ibid., p. 7.
17. Ibid., p. 24.
18. Ibid., p. 13.
19. Ibid., p. 158.
20. Ibid., p. 7.
21. Ibid., p. 18.
22. Ibid., p. 19.
23. Ibid., p. 21.
24. Giere (2006), p. 63.
25. Debs and Redhead (2007), p. 14.
26. Ibid., p. 15.
27. See Suárez (2003).
28. Suárez (2003), p. 8.
29. Debs and Redhead (2007), p. 22.
30. Ibid., p. 23.
31. Ibid., p. 24.
32. van Fraassen (2009).
33. Ibid., p. 84.
34. Suárez (2003), p. 226.
35. See Faye (2000b).
36. It is clear that, like many other philosophers, Cartwright does not always distinguish explicitly between 'laws of nature', i.e. what exist independently of our representation of it, and their expressions in our language or mathematical formulae, which are supposed to represent those laws of nature. But when she speaks about "laws," she takes it to be an abbreviation of "law expressions" and so do I, unless it the meaning is unclear from the context.
37. Cartwright (1983), p. 3.
38. Ibid., p. 4.
39. Ibid., p. 11.
40. Ibid., pp. 55, 69, 161–162.
41. Cartwright makes no distinction between different kinds of *ceteris paribus* clauses. However, a distinction can be made to distinguish between those clauses that require fulfilling of certain abstract and ideal circumstances, like being a point mass in the case of Newton's law of gravitation, and those that require the realization of certain factual and concrete circumstances such as the presence of oxygen in the case of combustion.
42. Giere (2006), p. 67.

43. Lars-Göran Johansson (2005) has made a similar observation, and I shall draw on some of his insights. Likewise, Ronald Giere holds a view that Newton's laws are definitions. Such laws, he says, are to be understood as definitions for providing adequate models (Giere, 1988, pp. 77–78, 84).
44. See Faye (2002), pp. 152ff.
45. Kuhn (1962/1969), p. 183.
46. Poincaré (1905), p. 100.
47. Ellis (1965), p. 52.
48. Hanson (1958), p. 112.
49. Let me highlight this point by quoting a telling passage from Johansson (2005): "Fundamental physics is a theoretical structure involving a lot of quantitative concepts such as mass, force, charge, electric field, magnetic field, energy, momentum, temperature, etc. By agreement among scientists some are treated as fundamental. These agreements are codified in the SI-system of quantities and units. The international research community has agreed on seven fundamental quantities: length, mass, time, electric current, thermodynamic temperature, amount of substance and luminous intensity. All other physical quantities are introduced by implicit definitions in the form of quantitative relations between quantities often called laws" (p. 157).
50. Ibid., p. 159.
51. Ellis (1963), p. 188.
52. Ellis (1965), p. 31.
53. See Sklar (1982).
54. Heelan (1983) expresses a similar view.
55. For a further discussion of models see Faye (2002), pp. 152–159.
56. Faye (2002), p. 162.
57. Giere (2004), p. 743. A critique of Giere's earlier, more semantic, and less pragmatic approach to models, in his (1988), can be found in Faye (2007b).

5 Scientific Explanation

1. Several philosophers like Hans Reichenbach and Wesley Salmon have argued along these lines. See, for example, Dowe (2000). In Faye (1989) I defended a position in which the causal processes can be defined in terms of positive energy transference.
2. See Faye (2005), p. 81ff.
3. Cartwright (1983) says: "Indeed not only are there no exceptionless laws, but in fact our best candidates are known to fail" (p. 46). However, she never explains why fundamental laws do not have a factual content.
4. Also Johansson (2005) argues that some fundamental laws, namely those I call theoretical laws, should be understood as definitions. At the same time he wishes to avoid conventionalism, as do I, and therefore takes such definitions to have a truth-value. Although I agree with much of what Johansson says, I disagree on this specific point. On my view theoretical laws are neither true nor false, but either adequate or inadequate in relation to specific purposes. They should be regarded as linguistic instructions that can be used to formulate empirical statements that are either true or false.
5. Keil (2005), pp. 194ff.

6. For a further discussion of nomological relevance, see Faye (1989), pp. 160–163.
7. Faye (2005), pp. 96–101.
8. Johansson (2005), pp. 164–165.
9. This does not exclude causal explanations of the relativistic effects a la Lorentz, as Dennis Dieks (2009) argues convincingly; the point here is merely to emphasize that Einstein's explanation by principles was a non-causal explanation.
10. Hempel (1965), pp. 352–354.
11. Russell (1912), p. 1.
12. Hempel (1965), p. 338.
13. See, for instance, Zeleňák (2009).
14. Hempel (1966), p. 48.
15. Ibid., p. 49.
16. According to Psillos (2002), p. 219, Hempel's concept of explanation is primarily epistemic because he thinks that nomic expectation is central to Hempel's tenets. I think this characterization is partly misguided. Hempel thought of explanation in terms of arguments and deductive relations between *explanans* and *explanandum* rather than in terms of beliefs and justification. He may be right so far that it might be reasonable to say that the hypothetico-deductive model of justification actually preceded Hempel's deductive-nomological view of explanation, and that in fact it was a prior commitment to a hypothetico-deductive model of justification that led Hempel to formulate his model of explanation this way. In this manner the two go together nicely.
17. van Fraassen (1980), pp. 132–134.
18. Culver (1974), p. 76
19. Hull (1992), p. 69.
20. Cartwright (1983), p. 45.
21. Faye (2005), p. 82.
22. Multiverse explanations attempt to account for the initial conditions of the actual universe in terms of statistical considerations based on quantum mechanics. However, this account by an appeal to multiverses is highly speculative, and I would not bet on it.
23. Cf. Hacking (1983), ch. 13.
24. While discussing the distinction between genuine laws and accidentally true generalization Stathis Psillos makes the following correct observation: "It is not the fact that a generalization has no exceptions that makes it a law. Rather it is the fact that a generalization *is* a law that deprives it of counterexamples, not just of actual exceptions, but also of *possible* exceptions to it" (Psillos 2002, p. 145). Causal generalizations may have exceptions, but they can be taken into consideration by incorporating them under the scope of one's generalization. Of course, the obvious danger is that such a specific generalization will have only one or a few instances. However, my claim is that in this case we do not come close to formulating causal laws because we cannot identify the nomologically relevant conditions that exclude all possible exceptions.
25. Salmon (1984) makes a distinction between causal laws and non-causal laws and associated causal laws with causal mechanism. Causal laws are regularities underpinned by causal processes. This implies that causal processes are

ontological fundamental and causal laws 'supervene' on causal processes. I too consider individual causal processes to be ontological fundamental and take 'causal laws' to be abstractions from singular causal processes. We do not need the notion of causal laws to identify causal processes. See the next chapter.
26. Hempel (1965), p. 359f.
27. Ibid., p. 369.

6 Causal Explanations

1. Of course sometimes we "gain causal beliefs" in the wrong circumstances, but these are unjustified beliefs; for example, they might commit the *post hoc* fallacy. We don't come forearmed with a criterion of when the "circumstances" are "right."
2. See Marzluff and Angell (2012), pp. 75–76. As they write in the opening of their book: 'Most people consider birds to be instinctual automatons acting out behaviors long ago scripted in their genes.' But particularly those birds in the corvid family 'are anything but mindless or robotic. These animals are exceptionally smart. Not only do they make tools, but they understand cause and effects. They use their wisdom to infer, discriminate, test, learn, remember, foresee, mourn, warn of impending doom, recognize people, seek revenge, lure or stampede other birds to their death, quaff coffee and beer, turn on lights to stay warm or expose danger, speak, steal, deceive, gift, windsurf, play with cats, and team up to satisfy their appetite for diverse foods whether soft cheese from a can or a meal of dead seal.' (p. 2). These attributions are the results of observations of crows and ravens living in the wild but also of many controlled experiments with corvids in captivity.
3. Presumably, lizards can learn causation in the Humean sense of expecting *B* when they experience *A*, because in the past they always experienced *B* type events following *A* type events. But most people would be very reluctant to say that lizards have a "concept of causality." They may say that (most) humans do, perhaps dogs and ravens too, but not mice. So when exactly do animals possess a concept of causation? I would say that they possess a concept of causation whenever they have the capacity of imposing successfully their innate schema of causation on relationships they have not experienced before. In those cases in which an animal has a capacity to distinguish between very different situations where various types of events are followed by various types of events and to act out its schema of causation in these different situations, I would hold that this animal possess a concept of causality. Thus, lizards can apply their innate schema only in identical situations, whereas ravens can apply it in a wide diversity of cases. Of course, it is an empirical question whether ravens, dogs, or mice have such a capacity. The naturalistic sort of account I am giving here is an account of how the organism comes to expect certain causal relations to hold between the items it senses in its environment. I suppose that this occurred at a pretty early level of evolutionary development. However, organisms with the ability to abstract from particular connections to a "notion of causality" must possess a level of consciousness

that we would tend to call a "mind." After all, a "notion" is just another name for a "concept" or an "idea." But I want to ascribe such a consciousness to higher animals. I think to embody a "notion of causation" is nothing other than "the capacity to make causal connections." Indeed there is a difference between being aware of having this capacity and not being aware of having this capacity. Thus, I distinguish between an embodied notion of causation (where you are not aware of such a capacity) and a reflective notion of causation, which I believe occurs only in highly developed conscious organisms. Even with young *Homo sapiens*, the infant knows that crying causes it to be nursed more or less instantly. But it's a long time, if ever, before it gets a reflective concept of causality.

4. Scientists, as well as ordinary people, use different kinds of causal explanation depending on both their epistemic stances and the domain of inquiry in which they are engaged. Whether or not there exist different sorts of causal schemata corresponding to different patterns of explanations is an empirical question. In his review of the results of psychological studies of explanations, Frank Keil (2006) groups explanations into four different types: (1) common cause; (2) common effect; (3) simple linear chains; and (4) causal homeostasis. The last listed seeks to explain how a set of properties arises from an interlocking set of causes and effects and forms an enduring entity. An example is an explanation of why feathers, hollow bones, nest building, flight, and a high metabolic rate might all reinforce their presence in birds.

5. Hempel (1965), p. 300.

6. Hempel (1965), pp. 415ff., 421ff.

7. My claim is that causal explanation is contrastive and/or contextual. Recently, several authors, Maslen (2004); Schaffer (2005); and Northcott (2008), go even further by claiming that already causation is a ternary or a quaternary relation. But I find contrastive causation ontologically suspicious. Thus, I limit the contrastive view to causal *explanation*, and I am not prepared to adopt the contrastive view of causation. Later, when I speak of context-dependent "causal statements," I mean merely statements used for explaining in certain contexts, not statements describing "contextual causation in reality."

8. Mackie (1974). Faye (1989, 2010) take a critical stand towards his inus-conditions.

9. Lewis (1986). An attempt to improve Lewis' view is found in Gjelsvik (2007). For a critical discussion consult Psillos (2002), pp. 236–239.

10. See Salmon (1984), pp. 139–144; (1998), p. 253. A critique of Salmon's view of causal processes is developed in Faye (1994, 1997a); and Dowe (1992, 2000). However, I think that any attempt to *define* "causation" in terms of the conservation of certain quantities is problematic because it is carried out on a background of a pre-fixed temporal order, and conserved quantities are invariant under normal time reversal. But if the temporal order is determined by the causal order we have a problem, especially because the causal order between the 'cause' and the 'effect' seems to be the same even if their temporal order is reversed. See Faye (1997b).

11. Faye (1989), pp. 146ff. Already in the Danish 1981-version I claimed that manipulation and intervention provide sufficient reason to give an 'ontological' separation of the cause and its effect: everything in the nomologically

relevant circumstances that is necessary for an event like the cause is also necessary for an event like the effect; while it is not the case that everything in the nomologically relevant circumstances that is necessary for an event similar to the effect is necessary for an event similar to the cause. See pp. 163–171. However, I don't believe that 'necessity' is an objective part of nature but a product of our mental abstraction and construction, based on our knowledge of what happens through intervention and manipulation under 'similar' relevant circumstances. (See also Faye 2010.) What is objective, of course, is the fact that an intervention, which changes the 'cause' under the "nomologically relevant" circumstances, also changes the 'effect' but not vice versa, and that the production of the 'cause' under the "nomologically relevant circumstances" brings about the 'effect'.

12. Cf. Mellor (1995), pp. 75–78.
13. Cf. Woodward (2003). In many respects Woodward's view on the physical possibility of intervention and manipulation for our understanding of causality is similar to the one I have defended since 1981. He also thinks that the physical possibility of an intervention is not constitutive of a causal connection between *c* and *e* but that it helps us to pick out the intrinsic feature that characterizes the relationship between *c* and *e*. Psillos (2007) has objected to Woodward's position that it is difficult to see whether the involved counterfactuals have truth conditions independently of their evidence conditions. Personally, I think they do have 'separated' truth conditions, but these are generalized from their evidence conditions. Ontologically speaking, the truth conditions of counterfactuals are abstracted objects whose existence depends on their evidence conditions. See Faye (2010).
14. Cf. Psillos (2002), p. 293. See also ch. 5 and his (2007), p. 102.
15. See Faye (1989, 1994, 1997b).
16. Cartwright (1999), p. 104.
17. Anscombe (1971), p. 68.
18. Cartwright (2002).
19. Dupré (2007).
20. Machamer, Darden, and Craver (2000), p. 3.
21. Glennan (1996).
22. Machamer, Darden, and Craver (2000), pp. 21–22.
23. Sterelny and Griffiths (1999).
24. See Rosen (2000); Wolkenhauer (2001).
25. Wolkenhauer (2001).
26. Emmeche *et al.* (2000).
27. See, for instance, Bird (1998), p. 66. In his dissertation Ylikoski (2001) makes a distinction between causal claims and causal explanatory claims, arguing that explanation requires more fine-grained relata. When we come up with an explanation, we want to have the cause described in the right way. But I cannot see that these grounds warrant his conclusion that "this makes it natural to think that the *relata* of an explanation are facts" (p. 20). Perhaps I need to add that facts, in my view, are considered to be actual states of affairs in the world and not pieces of information that exist in human minds. Therefore I would say that causal explanations consist of descriptions that refer to causal facts, and that causal explanations express the explainer's causal beliefs, and therefore that the relata of an explanation are beliefs.

28. Apparently, as Henry Folse points out to me, "there is a systematic ambiguity in English about how we use 'facts.' Sometimes we say 'facts' when speaking about actual states of affairs in the world; call these 'facts in the objective sense,' but at other times when we talk about 'facts' we really mean information about objective facts in the first sense; call these 'facts in the subjective sense.' These are the 'facts' that people know when they know a fact. The confusion between these two uses leads to no end of nonsense among writers who can't tell the difference between the world and information about the world. There are also 'facts' in the third sense of the objective state of affairs that people know certain facts (or hold certain beliefs) in the second sense. These are objective facts about subjective states of affairs. The important moral of this story is not whose usage is right but simply that there are these two senses and that they are often confused. For there to be true assertions about the world you need two things: assertions and a world."
29. Salmon (1984), p. 274. What Salmon says here seems to indicate that he has completely forgotten Hempel's assumption that the explanans and the explanandum are expressions that are able to participate in logical inferences.
30. This study shows that there is a time lag between temperature changes and changes in atmospheric CO_2 concentrations. Depending on whom you ask, the time lag is up to a thousand years. If this is true, then the two are causally connected, but the relationship is by no means linear. That is, a given amount of CO_2 increase as measured in the ice cores need not necessarily correspond with a certain amount of temperature increase. All one can say is that it is not as if the temperature increase has already ended when CO_2 starts to rise. Rather, they go very much hand in hand, with the temperature continuing to rise as the CO_2 goes up. In other words, CO_2 seems to act as an amplifier.
31. Mackie (1974), p. 35. In fact, Mackie tells us that he has borrowed it from John Anderson.
32. The accumulated effect of burning fossil fuels together with ocean currents, volcanic eruptions, etc. could be so well correlated with the changes in temperature and the increase in atmospheric CO_2 that it is impossible to distinguish the accumulated effect from an effect in which the isolated case of CO_2 is the only causal factor. So in the current greenhouse model the relative size of the forcing from CO_2 compared to the size of the other forcings is extremely large. It might not be rational to think so, but we cannot exclude it at this point. This can very easy be done by manipulating models. All that is required is to change a constant of motion within the model or an initial condition.
33. I want to thank Mark Tschaepe for stressing this point in our private correspondence.

7 Other Types of Explanations

1. See, for instance, Dennett (1985, 1991).
2. Cf. Salmon (1984, 1998).
3. Faye (2002), pp. 36–40; Faye (2012) pp. 65–71.

4. See Garfinkel (1981), p. 95.
5. Hempel (1967), p. 305.
6. Ibid., p. 305.
7. Ibid., p. 310.
8. Ibid., p. 310.
9. Salmon (1989), p. 30.
10. Ibid., p. 30.
11. Wright (1973/1999), p. 48.
12. Ibid., p. 45.
13. Boorse (1976), pp. 70–86.
14. Millikan (1984), pp. 23–31.
15. See Sober (1984), pp. 147–155.
16. See Boorse (2002), pp. 66–67.
17. In modern cosmology a new 'paradigm of explanation' has arisen within the last forty years which seems to involve a general appeal to a kind of functional explanation. I am thinking of the anthropic principle, which has been much debated. The anthropic principle is said to offer explanations of certain quantities and phenomena ranging from the neutron-proton mass difference to the age of the universe by a reference to our present. For instance, when Collins and Hawking (1973) suggest that the answer to a question like "Why is the universe isotropic?" is "Because we are here," the authors definitely do not intend that human beings cause the universe to be isotropic. Nor do they take such an explanation to be teleological by imagining that an isotropic universe exists for the sake of human life. Leslie (1989) has suggested that it should be understood in a logical sense like "saying that being a woman is a logical consequence of being a wife" (p. 137). However, I believe that this diagnosis is wrong. The anthropic explanation can hardly be considered to be "analytic," and if it is, then it tells us *nothing* about this particular world. There is nothing conceptual that links 'isotropic' and 'human beings'. What Collins and Hawking did (not unlike several other leading cosmologists) was that they appealed to the effect of an isotropic universe, namely the existence of human beings, and used this to explain one of the functions of an isotropic universe, which is to support biological evolution and human beings. So in general I will say that anthropic explanations are functional explanations, which thereby have been "reintroduced" into physics.

    But the use of "function" here is not without problems. I can say, for example, "The decrease in the vulture population was a function of the use of DDT." This means of course it was a causal factor, and we express it by saying there was a "functional" relationship between two variables, here successful vulture births and the extent of DDT usage. Hence this usage derives from the expression of a "mathematical function" as a way of talking about relationships between variables. A seemingly very different use is to say such and such is a thing's "function," is to say that is its purpose; that is what it was designed for (either by natural selection or by human intentions). The function of my umbrella is to keep me dry in the rain. In this latter sense, killing off vultures was certainly not the intention of the creators of DDT, but instead an undesirable (and unforeseen) side effect. I suspect the "explanatory" virtues of the anthropic argument for the value of the cosmological constant really trades on this equivocation. Yes, it is true that our being here is a function (in the sense of an effect) of the size of the cosmological constant or isotropic nature of space. But I don't doubt that 'explanations' derived from the anthropic principle are essentially teleological. So one might question whether they have really been introduced "into physics." When you ask why are the initial conditions what they are, if these are the "truly *initial* conditions" (i.e. no "conditions" are earlier) then, since there is no earlier time, the criteria of the cause and effect schema are not met, so you cannot give a causal explanation.

18. Hempel (1965), pp. 256–257.
19. A proponent of the Hempelian view would say that the pragmatic account does not explicate the concept of explanation as such, but only explanation-for-somebody-in-some-context. Another thing is that talk of explanatory feelings makes explanation a psychological or almost biological activity. For some authors this will be a problem – to link scientific explanation (which in their view should be something objective or at least intersubjective) to somebody's subjective feelings. But from the standpoint of pragmatic account this does not have to be a drawback.

8 The Pragmatics of Explanation

1. See Schweder (2004), p. 32.
2. Some recent studies in the pragmatics of explanation are to be found in Walton (2004); Tschaepe (2009); De Donato-Rodríguez and Zamora-Bonilla (2012); De Donato-Rodríguez and Zamora-Bonilla (2013).
3. Van Fraassen's earlier formalistic approach to explanation, according to which 'intention' plays no role, is not quite in agreement with his later position on representation where 'use' and 'intention' are at the centre.
4. Cartwright (1983), p. 78ff.
5. Hansson (1975).
6. Searle (1978).
7. Collin (1999) criticizes Searle for holding that the meaning of a sentence is a function of its truth conditions, but that a sentence only possesses truth conditions given a certain setting. I think, however, that Searle is consistent though not very explicit in his suggestion. As I read Searle, he imagines that the truth conditions of an indicative sentence consist of two parts: One is invariant from one setting to another; this constitutes the semantic content. Another varies from setting to setting. But both contribute to fixing the truth-value in a particular setting, i.e. a situation of utterance.
8. Danto (1985).
9. van Fraassen (1980), p. 125.
10. See Chapter 7, note 17.
11. See Bas van Fraassen (1980), p. 134.
12. Cf. ibid., p. 156.
13. Ibid., p. 4 and p. 88.
14. Cf. ibid., p. 137f.
15. Bengt Hansson gave his analysis in a manuscript that circulated in 1975 but was never published.
16. Cf. van Fraassen (1980), p. 129.

17. Ibid., p. 129.
18. Ibid., p. 91.
19. Cf. ibid., p. 130.
20. Ibid., p. 130.
21. Ibid., p. 143.
22. Ibid., p. 132. My emphasis.
23. Ibid., p. 144.
24. Ibid., p. 140.
25. Ibid., p. 141.
26. Scriven (1962), p. 2002.
27. van Fraassen (1980), pp. 132–133.
28. Kitcher and Salmon (1987), pp. 322ff. See also Salmon (1989), pp. 141–145.
29. van Fraassen (1980), pp. 146ff.
30. Kitcher and Salmon (1987), p. 322; Salmon (1989), pp. 142–143.
31. van Fraassen (1980), p. 126.
32. Achinstein (1984), pp. 277–279.

9 Not Just Why-questions

1. van Fraassen (1980), p. 134.
2. See Dray (1964).
3. Cf. Scriven (1962), pp. 173–174.
4. See Bromberger (1965).
5. See Achinstein (1983).
6. Cf. Salmon (1989), p. 138.
7. A reasonable understanding of Aristotle's Greek word 'aitia' is that it is better rendered in English as 'explanatory factor'; each of the four 'aitia' enters into the full "scientific knowledge" of the "ousia" or "being." They are not rival or alternative forms of explanation. And the Latin word 'causa', which was used to characterize the Aristotelian view in medieval times, is also poorly rendered as 'cause' in English, at least to contemporary ears. The change in the intention of the word 'cause' is of course at least partially a result of the development of modern science's very different understanding of what scientific knowledge aspires to provide.
8. See Kragh (2011) for a detailed historical treatment of the debate about the anthropic principle. The quotations due to Carter and Schramm are taken from him.
9. Ibid., p. 224.
10. Ibid., p. 232.
11. Salmon (1989), pp. 6–7. It seems Salmon draws on Hempel (1965), pp. 334–335.
12. It might be fair to point out that quantum theory has met with a lot of resistance precisely because of this fact. If we have a case where our worldview, whatever it might be, leads us to regard certain sorts of phenomena as objectively probabilistic, then probabilistic generalizations (which are truly 'laws' in this case) provide acceptable explanations. But with respect to phenomena which our worldview leads us to regard as in fact objectively determined, and the statistical nature of the correlations is regarded as reflecting only our partial ignorance,

then we are likely to regard such 'explanations' that appeal to these subjective probabilities as of an inferior sort, which we 'accept' only until some future time when we can get at some better knowledge of the real causes. Of course, if we have different worldviews that lead one of us to think a statistical correlation reflects an objective probability and the other to think it reflects only subjective ignorance, we are likely to disagree on the ultimate acceptability of the proffered explanation. The objectivist can happily accept it as in some sense a final explanation, but the person who sees the statistical correlation as merely a subjective matter will only accept such an explanation as a half-way house en route to the "real" (i.e. true) explanation. Cf. Bohr and Einstein on this.
13. Scriven (1962), pp. 174–176.
14. Hempel (1965), p. 246.
15. Scriven (1962), p. 175.
16. Ibid., p. 176.
17. Usually *intentional explanations* are considered as reason-giving explanations because this type of explanation cites the grounds (reasons) of actions, beliefs or other human phenomena. It is used in the human and social sciences but has no place in natural science (except in ethology). So within this context it seems to be a complete mistake to argue that responses to why-questions are reason-giving explanations since many of them are actually cause-giving explanations. Nevertheless, the present discussion focuses on whether or not a sentence stating a 'reason that' is a marker of an explanation as such, irrespective of whether this statement is causal, functional, or intentional. This means that we just consider an explanation to take part of a general public discourse and a candidate of explanation to be an intended response to a why-question.
18. See Bitzer (1968). His notion of the rhetorical situation and its application to explanation will be developed in the next Chapter.
19. See, for instance, Salmon (1984) p. 10. However, he later changed his mind; he then denied that all requests for scientific explanations can be formulated as why-questions. See his (1989), pp. 137–138.
20. van Fraassen (1980), p. 132.
21. Salmon (1989), pp. 137–138.

10 A Rhetorical Approach to Explanation

1. See Faye (2007a).
2. Gärdenfors (1990), p. 111.
3. Sintonen (1989). This and the following indent sentence are not quite Sintonen's own formulations. Instead I have borrowed the above formulation from my student, Thomas Basbøll, who, in his MA thesis, criticized that Sintonen writes "Why E," where E stands for a proposition mentioned, rather than "Why q," where q stands for a proposition used, i.e., a fact stated. A fact and a proposition are two different things: we express facts with propositions.
4. See Holton (1973).
5. Holton (2005), p. 140.
6. Ibid., p. 141.
7. See Faye (1991) for Bohr's motivation and thematic assumptions.

8. Achinstein (1983), p. 16.
9. Ibid., pp. 16–17.
10. Ibid., p. 42. The quotation expresses only a necessary condition, but later he takes it to express a sufficient condition as well (p. 57).
11. Salmon (1989), p. 148.
12. See John Searle (1969), p. 46 for his criticism of Paul Grice.
13. Cf. Austin (1962), p. 120ff.
14. See Faye (1999, 2007a); also Faye (2002), ch. 3.
15. Bitzer (1968/1999), p. 219.
16. Ibid., p. 220f.
17. Ibid., p. 221.
18. Bitzer takes a strong realist stand on the rhetorical situation by saying: "The exigence and the complex of persons, objects, events, and relations which generate rhetorical discourse are located in reality, are objective and publicly observable historic facts in the world of experience, are therefore available for scrutiny by an observer or critic who attends to them. To say the situation is objective, publicly observable, and historic means that is real or genuine – that our critical examination will certify its existence" (p. 223). This is at the same time a meta-rhetorical stand. The rhetorical situation itself may contain features belonging to the perspective of the persons involved in the discourse and which may not be publicly observable. These features exist nevertheless and may be revealed by other means.
19. See Kuhn (1977).
20. The formal-logical approach did not die with Hempel and his followers. Hållsten (2001) gives a modern defence of explanatory deductivism. The pragmatic-rhetorical view does not reject that deduction in the proper context is regarded as a condition for having explanation. But it claims that deduction is not a necessary condition for all contexts, not even for mathematical contexts where explanations in terms of informal proofs (high-level sketches) are rather the norm than the exception.
21. See Møller (1972), pp. 107, 219, and 232f. for details and further references. Møller himself used the old formulation in the first edition of this book that was published 1952. For instance, the Joule heat developed in the electric body per unit of time and volume with respect to a moving frame S would in the old formulation be expressed as $\varphi^0(1-u^2/c^2)^{1/2}$, whereas in the new formulation it becomes $\varphi^0/(1-u^2/c^2)^{1/2}$.
22. Møller (1972), p. 233.
23. Indeed, I have claimed that a false explanation is an explanation nonetheless. So Aristotelian explanations are really explanations even though we today believe that the questions Aristotelian physics was intended to answer are no longer our questions. What I would say is that they still have an explanatory capacity to anyone who can see the world and science as Aristotle saw them. To some extent historians do attempt to do this.

11 Pluralism and the Unity of Science

1. Though Giere (2006) argues for a pluralism and perspectivism, he also believes that human agency and the realm of intentions, norms, and values are not

really real because these things are not physically real. Atoms and neurophysiological processes exist objectively, but agency, intentionality, etc. are pure idealizations. In Searle's terminology Giere seems to hold that agency and intentionality are epistemologically objective but ontologically subjective.
2. See Collin (2011).
3. It might be a bit unfair to say it was "just" a symptom of such "trust," but I do think it is a worthwhile point that in Hempel's time both he and his audience were totally uncritical "believers" in the fact that scientific theories were "justified" and that (in their world) the question was never whether they were justified but always what was the "logic" behind that justification, i.e. how scientific knowledge rested on reason and the unquestionable direct deliverances of experience. That is why its concern was setting the normative ideal for genuine science, not describing the messy historical error-prone meanderings of real life scientists.
4. Apart from the discovery of the Higgs particle, the present lack of results from the Large Hadron Collider seems not to vindicate the existence of supersymmetry, and at least one influential particle physicist seems ready to draw the conclusion that the very idea of supersymmetry and superstrings cannot be substantiated and therefore should be put away. See Shifman's (2012) reflections upon the current situation in high energy physics.
5. See Carnap (1938), p. 58. See also Richardson (2006) who analyses the different views on the unity of science among some of the main subscribers.
6. See Faye (2002).
7. In the same way that the rules of methodology are abstracted and idealized from our common practice of conducing inquiry, I think that logical rules are abstracted and idealized from our actual reasoning practices. Some cognitive scientists, such as Johnson-Laird (1983), argue it is an empirical fact that people usually do not reason by following deductive rules, but by manipulating mental models representing various possibilities in their environment. Similar reasoning patterns may be used by higher animals; one can think that their (as well as our) innate cognitive schemata act as matrices for creating such individual mental models.
8. If all the empirical-type rules are discipline specific, then it would seem the only thing left to be "general" to all science is simply inductive logic. So how do scientific methodological rules (the general ones) differ from inductive logic? If one counts statistics as a branch of inductive logic, I think that inductive logic makes up a large part of the general methods that are used in data treatment. But so does deduction in terms of falsification. In addition to deductive and inductive logic we also use some more informal rules of reasoning such as abstraction, simplification, generalization, and identification. But rules of reasoning need to be supplemented with methodological prescriptions in order to have a "good" practice, as those stated by Alhazen, before we have a fully developed scientific methodology. For a discussion of this distinction between general and specific methods, see Faye (2002), ch. 6.
9. Putnam and Oppenheim (1958).
10. The two variants of reduction mentioned here correspond with two versions of emergence. One goes back to C. D. Broad, who argued that emergence is a synchronic, non-causal covariation of a novel property and its lower-level basis; the other goes even further back to J. S. Mill, who distinguished

between homopathic and heteropathic laws of causation while arguing that emergence was an effect of heteropathic causation.
11. Some may accuse me for not making a distinction between ontological reduction and epistemological reduction, but this is not my intent. I only want to claim that without any empirical evidence supporting a metaphysical principle it should be discarded as a fiction of the mind.
12. See Kim (2006), p. 295.

Literature

Achinstein, Peter (1983). *The Nature of Explanation*. New York and Oxford: Oxford University Press.
Achinstein, Peter (1984). "The Pragmatic Character of Explanation." In *PSA*, ii: 275–292.
Anderson, M. L. (2003). "Embodied Cognition: A Field Guide." In *Artificial Intelligence*, **149** (1): 91–130.
Anscombe, E. (1971). "Causality and Determination." Reprinted in E. Sosa (ed.), *Causation and Conditionals*. Oxford: Oxford University Press, 1975, 63–81.
Austin, John L. (1962). *How to Do Things with Words*. Oxford: Oxford University Press.
Baade, Wilhelm & Fritz Zwicky (1934a). "On Super-Novae." In *Proc. of Nat. Aca. Sci.*, **20**: 254–259.
Baade, Wilhelm & Fritz Zwicky (1934b). "Cosmic Rays from Supernovae." In *Proc. of Nat. Aca. Sci.*, **20**: 259–263.
Bailer-Jones, Daniela M. (2002). "Models, Metaphors and Analogies." In P. Machamer & M. Silberstein (eds), *The Blackwell Guide to the Philosophy of Science*. Malden: Blackwell.
Barsalou, L. (2008). "Grounded Cognition." In *Annual Review of Psychology*, **59**: 617–645.
Bird, Alexander (1998). *The Philosophy of Science*. London: UCL Press.
Bitzer, Lloyd (1968). "The Rhetorical Situation." In *Philosophy and Rhetoric*, **1**. Reprinted in J. L. Lucaites *et al.* (eds), *Contemporary Rhetorical Theory*. New York: The Guilford Press, 1999, 217–225.
Bloor, David (1997). "What Is a Social Construct?" In *Vest tidsskrift för ventenskapsstudier*, **10** (2): 9–21. Gothenburg University.
Bohr, Niels (1913). "On the Constitution of Atoms and Molecules." In *Philosophical Magazine*, **26** (1): 476.
Bohr, Niels (1998). *Causality and Complementarity: The Philosophical Writings of Niels Bohr*. Vol. 4. Eds, Jan Faye & Henry Folse, Woodbridge, Conn.: Ox Bow Press.
Boorse, Christopher (1976). "Wright on Functions." In *Philosophical Review*, **85**: 70–86.
Boorse, Christopher (2002). "A Rebuttal on Functions." In A. Ariew, R. Cummins & M. Perlman (eds), *Functions. New Essays in the Philosophy of Psychology and Biology*. Oxford: Oxford University Press.
Brandom, Robert (1994). *Making It Explicit: Reasoning, Representing, and Discursive Commitment*. Cambridge Mass.: Harvard University Press.
Bromberger, Silvain (1965). "An Approach to Explanation." In R. J. Butler (ed.), *Analytical Philosophy – Second Series*. Oxford: Basil Blackwell.
Brown, Michael E. (2006). "A World on the Edge." *NASA Solar System Exploration*. http://solarsystem.nasa.gov/scitech/display.cfm?ST_ID=105. Retrieved 2006-05-25.
Burbridge, E. M., G. R. Burbridge, W. A. Fowler & F. Hoyle (1957). "Synthesis of Elements in Stars." In *Reviews of Modern Physics*, **29**: 547–651.

Calvo, P. & Gomila, A. (eds) (2008). *Handbook of Cognitive Science: An Embodied Approach*. Amsterdam: Elsevier.
Carnap, Rudolph (1938). "Logical Foundations of the Unity of Science." In Otto Neurath, R. Carnap & C. Morris (eds), *The International Encyclopedia of Unified Science*. Chicago: University of Chicago Press, 42–62.
Cartwright, Nancy (1983). *How the Laws of Physics Lie*. Oxford: Clarendon Press.
Cartwright, Nancy (1999). *The Dappled World: A Study of the Boundaries of Science*. Oxford: Oxford University Press.
Cartwright, Nancy (2002). "Causation: One Word, Many Things." In J. Reiss (ed.), *Causality: Metaphysics and Methods*. Technical Report 07/03. London: Centre for Philosophy of Natural and Social Science.
Collin, Finn (1985). *Theory and Understanding: A Critique of Interpretive Social Science*. Oxford: Basil Blackwell.
Collin, Finn (1999). "Literal Meaning, Interpretation, and Objectivity." In Haapala & Naukkarinen (eds), *Interpretations and its Boundaries*. Helsinki: Helsinki University Press.
Collin, Finn (2011). *Science Studies as Naturalized Philosophy*. Series: *Synthese Library*, **348**. Dordrecht: Springer.
Collins, Barry & Hawking, Stephen W. (1973). "Why Is the Universe Isotropic?" In *Astrophysical Journal*, **180**: 317–334.
Culver, Roger B. (1974). *An Introduction to Experimental Astronomy*. San Francisco: W. H. Freeman and Company.
Cushing, James T. (1991). "Quantum Mechanics and Explanatory Discourse: Endgame for Understanding?" In *Philosophy of Science*, **58**: 337–358.
Cushing, James T. (1994). *Quantum Mechanics*. Chicago and London: The University of Chicago Press.
Danto, Arthur (1985). *Narration and Knowledge*. New York: Columbia University Press.
Davidson, Donald (1962/1980). "Actions, Reasons, and Causes." In his *Essays on Actions and Events*. Oxford: Clarendon Press.
Davidson, Donald (1967/1975). "Causal Relations." In *Journal of Philosophy*, **64**: 691–703. Reprinted in Ernst Sosa (ed.), *Causation and Conditionals*. Oxford: Oxford University Press, 82–95.
Davis, J. I. & Markman, A. B. (2012). "Embodied Cognition as a Practical Paradigm: Introduction to the Topic, the Future of Embodied Cognition." In *Topics in Cognitive Science*, **4** (4): 685–691.
Debs, Talal A. & Redhead, Michael L. G. (2007). *Objectivity, Invariance, and Convention: Symmetry in Physical Science*. London: Harvard University Press.
De Donato-Rodríguez, Xavier & Zamora-Bonilla, Jesús (2012). "Explanation and Modelization in a Comprehensive Inferential Account." In H. de Regt *et al.* (eds), *EPSA Philosophy of Science: Amsterdam 2009*. Amsterdam: Springer, 33–42.
De Donato-Rodríguez, Xavier & Zamora-Bonilla, Jesús (2013). "Scientific Explanation and Representation: An Inferentialist Viewpoint." In Evandro Aggazzi (ed.), *Representation and Explanation in the Sciences*. Milano: FrancoAngeli.
De Luca, Andrea *et al.* (2006). "A Long-period, Violently-variable X-ray Source in a Young SNR." In *Science Express*, **6**: July issue.
Dennett, Daniel C. (1985). "Intentional Systems." In *Brainstorms: Philosophical Essays on Mind and Psychology*. Brighton: Harvester Press.

Dennett, Daniel C. (1991). "Real Patterns." In *Journal of Philosophy*, **88**: 27–51.
De Regt, Henk W. (2004). "Discussion Note: Making Sense of Understanding." In *Philosophy of Science*, **71**: 98–109.
De Regt, Henk W. (2009). "The Epistemic Value of Understanding." In *Philosophy of Science*, **79**: 585–597.
De Regt, Henk W. & Dieks, Dennis (2005). "A Contextual Approach to Scientific Understanding." In *Synthese*, **144**: 137–170.
Dewey, John (1922). *Human Nature and Conduct. An Introduction to Social Psychology*. New York: Henry Holt and Company.
Dieks, Dennis (2009). "Bottom-Up and Top-Down: The Plurality of Explanation and Understanding in Science." In H. de Regt, K. Eigner & S. Leonelli (eds), *Scientific Understanding: Philosophical Perspectives*. Pittsburgh: University of Pittsburgh Press.
Donahue, W. H. (1988). "Kepler's Fabricated Figures: Covering Up the Mess in the New Astronomy." In *Journal of History of Astronomy*, **19** (4): 217–237.
Dorato, Mauro (2005). *The Software of the Universe*. Aldershot: Ashgate.
Dowe, Phil (1992). "Wesley Salmon's Process Theory of Causality and the Conserved Quantity Theory." In *Philosophy of Science*, **59**: 195–216.
Dowe, Phil (2000). *Physical Causation*. Cambridge: Cambridge University Press.
Dray, William H. (1964). *Laws and Explanation in History*. New York: Oxford University Press.
Dretske, Fred (1979). "Simple Seeing." In *Perception, Knowledge, and Belief*. Cambridge: Cambridge University Press, 2000, 97–112.
Dretske, Fred (1994). "Differences That Make No Difference." In *Perception, Knowledge, and Belief*. Cambridge: Cambridge University Press, 2000, 138–157.
Dupré, John (1993). *The Disorder of Things: Metaphysical Foundations of the Disunity of Science*. Cambridge, Mass.: Harvard University Press.
Dupré, John (2007). "Is Biology Reducible to the Laws of Physics?" In *American Scientist – LA English*, **95** (3): 274.
Edelman, Gerald M. & Giulio Tononi (2000). *Consciousness: How Matter Becomes Imagination*. Allen Lane: The Penguin Press.
Einstein, Albert (1905/1923). "On the Electrodynamics of Moving Bodies." In Lorentz, H. A., Einstein, A., Minkowski, H. & Weyl, H. (eds), *The Principle of Relativity*. New York: Dover Publication.
Einstein, Albert (1954). "What Is the Theory of Relativity?" In his *Ideas and Opinions*. New York: Crown Publishers.
Ellis, Brian (1963). "Universal and Differential Forces." In *The British Journal for the Philosophy of Science*, **14**: 177–194.
Ellis, Brian (1965). "The Origin and Nature of Newton's Laws of Motion." In R. C. Colodny (ed.), *Beyond the Edge of Certainty*. Englewood Cliffs: Prentice-Hall.
Emmeche, Claus, Simo Køppe & Frederik Stjernfeldt (2000). "Levels, Emergence, and Three Versions of Downward Causation." In P. B. Andersen, C. Emmeche, N. O. Finnemann & P. V. Christiansen (eds), *Downward Causation: Minds, Bodies and Matter*. Aarhus: Aarhus University Press, 13–33.
Faye, Jan (1989). *The Reality of the Future: An Essay on Time, Causation, and Backward Causation*. Odense: Odense University Press. This is an enlarged and translated version of a Danish book *Et naturfilosofisk essay om tid og kausalitet* (1981). København: Paludans Forlag.

Faye, Jan (1991). *Niels Bohr: His Heritage and Legacy: An Anti-Realist View of Quantum Mechanics*. Dordrecht: Kluwer Academic Publishers.
Faye, Jan (1994). "Causal Beliefs and Their Justification." In J. Faye, U. Scheffler & M. Urchs (eds), *Logic and Causal Reasoning*. Berlin: Akademie Verlag.
Faye, Jan (1997a). "Is the Mark Method Time Dependent?" In J. Faye, U. Scheffler & M. Urchs (eds), *Perspective on Time*. Series: *Boston Studies in the Philosophy of Science. Vol. 189*. Dordrecht: Kluwer Academic Publishers.
Faye, Jan (1997b). "Causation, Reversibility, and the Direction of Time." In J. Faye, U. Scheffler & M. Urchs (eds), *Perspective on Time*. Series: *Boston Studies in the Philosophy of Science. Vol. 189*. Dordrecht: Kluwer Academic Publishers.
Faye, Jan (1999). "Explanation Explained." In *Synthese*, **120**: 61–75.
Faye, Jan (2000a). *Athenes kammer*. København: Høst & Søn.
Faye, Jan (2000b). "Observing the Unobservable." In E. Agazzi & M. Pauri (eds), *The Reality of the Unobservable: Observability, Unobservability and Their Impact on the Issue of Scientific Realism*. Series: *Boston Studies in the Philosophy of Science. Vol. 215*. Dordrecht: Kluwer Academic Publishers, 165–177.
Faye, Jan (2002). *Rethinking Science: An Introduction to the Unity of Science*. Aldershot: Ashgate.
Faye, Jan (2005). "How Nature Makes Sense." In J. Faye, P. Needham, U. Scheffler & M. Urchs (eds), *Nature's Principles*. Series: *Logic, Epistemology, and the Unity of Science*, **4**: 77–102.
Faye, Jan (2007a). "The Pragmatic-Rhetorical Theory of Explanation." In Johannes Persson & Petri Ylikoski (eds), *Rethinking Explanation*. Series: *Boston Studies in the Philosophy of Science*, **252**: 43–68.
Faye, Jan (2007b). "Models, Theories, and Language." In Fabio Minazzi (ed.), *Filosofia, scienza e bioetica nel dibattito contemporaneo*. Presidenza del Consiglio dei Ministri, Rome: Poligrafico e Zecca dello Stato, 823–838.
Faye, Jan (2010). "Causality, Contiguity, and Construction." In *Organon F. Journal of Analytic Philosophy*, **17**: 443–460.
Faye, Jan (2012). *After Postmodernism: A Naturalistic Reconstruction of the Humanities*. London: Palgrave Macmillan.
Faye, Jan & Vincent Fella Hendricks (1999). "Abducting Explanation." In Lorenzo Magnani, Nancy J. Nersessian & Paul Thagard (eds), *Model-Based Reasoning in Scientific Discovery*. New York: Kluwer Academic/Plenum Press, 271–293.
Faye, Jan, Paul Needham, Uwe Scheffler & Max Urchs (eds) (2005). *Nature's Principles*. Series: *Logic, Epistemology, and the Unity of Science*, **4**. Dordrecht: Springer.
Franklin, Anna, et al. (2005). "Color Term Knowledge Does Not Affect Categorical Perception of Colors in Toddlers." In *Journal of Experimental Child Psychology*, **90** (2): 114–141.
Franklin, Anna, et al. (2008). "Categorical Perception of Color Is Lateralized to the Right Hemisphere in Infants, But to the Left Hemisphere in Adults." In *Proc of Nat Acad Sci U.S.A.*, **105**: 3221–3225.
Friedman, Michael (1974). "Explanation and Scientific Understanding." In *Journal of Philosophy*, **71**: 5–19.
Friedman, Michael (2001). *The Dynamics of Reason: Stanford Kant Lectures*. Stanford: CSLI Publications.

Frigg, Roman & Hartmann, Stephan (2006). "Models in Science." In *Stanford Encyclopaedia of Philosophy*. http://plato.stanford.edu/archives/entries/models-science/.
Garfinkel, Alan (1981). *The Forms of Explanation*. New Haven: Yale University Press.
Giere, Ronald N. (1988). *Explaining Science*. Chicago: The University of Chicago Press.
Giere, Ronald N. (2004). "How Models Are Used to Represent Reality." In *Philosophy of Science*, **71**: 742–752.
Giere, Ronald N. (2006). *Scientific Perspectivism*. Chicago: The University of Chicago Press.
Gjelsvik, Olav (2007). "Causal Explanation Provides Knowledge Why." In Johannes Persson & Petri Ylikoski (eds), *Rethinking Explanation*. Series: *Boston Studies in the Philosophy of Science*, **252**: 69–92.
Glennan, Stuart (1996). "Mechanisms and the Nature of Causation." In *Erkenntnis*, **44**: 49–71.
Grice, Paul (1975). "Logic and Conversation." In Peter Cole & Jerry L. Morgan (eds), *Syntax and Semantics. Vol. 3 Speech Acts*. New York: Academic Press, 41–58.
Grimm, Stephen R. (2006). "Is Understanding a Species of Knowledge?" In *British Journal for the Philosophy of Science*, **57**: 515–535.
Gärdenfors, Peter (1980). "A Pragmatic Approach to Explanations." In *Philosophy of Science*, **47**: 404–423.
Gärdenfors, Peter (1990). "An Epistemic Analysis of Explanations and Causal Beliefs." In *Topoi*, **9**: 109–124.
Hacking, Ian (1983). *Representing and Intervening*. Cambridge: Cambridge University Press.
Hanson, Norwood Russell (1958). *Patterns of Discovery*. Cambridge: Cambridge University Press.
Hansson, Bengt (1975). "Explanation – Of What?" (Unpublished manuscript).
Heelan, Patrick (1983). *Space-Perception and the Philosophy of Science*. Berkeley: University of California Press.
Heinrich, Bernd & Bugnyar, Thomas (2007). "Just How Smart Are Ravens?" In *Scientific American*, April 2007.
Heisenberg, Werner (1967). "Quantum Theory and Its Interpretation." In S. Rozental (ed.), *Niels Bohr: His Life and Work Seen by Friends and Colleagues*. Amsterdam: North Holland, 94–107.
Hempel, Carl G. (1952). "Fundamentals of concept formation in empirical science." In O. Neurath (ed.). *International Encyclopedia of United Science*. II, (7). Chicago: The University of Chicago.
Hempel, Carl G. (1965). *Aspects of Scientific Explanation and Other Essays in the Philosophy of Science*. New York: The Free Press.
Hempel, Carl G. (1966). *Philosophy of Natural Science*. Englewood Cliffs: Prentice-Hall.
Hesse, Mary (1961). *Forces and Fields: A Study of Action at a Distance in the History of Physics*. London: Thomas Nelson & Son.
Holton, Gerald (1973). *Thematic Origins of Scientific Thought: Kepler to Einstein*. Cambridge: Harvard University Press. Rev. Ed. 1988.

Holton, Gerald (2005). *Victory and Vexation in Science: Einstein, Bohr, Heisenberg and Others*. Cambridge, MA: Harvard University Press.
Hoyle, Fred & Fowler, William (1960). "Nucleosynthesis in Supernovae." In *Astrophysical Journal*, **132**: 565–590.
Hull, David L. (1992). "The Particular-Circumstance Model of Scientific Explanation." In M. H. Nitecki & D. V. Nitecki (eds), *History and Evolution*. Albany: State University of New York Press, 69–80.
Hållsten, Henrik (2001). *Explanation and Deduction: A Defence of Deductive Chauvinism*. Stockholm: Almqvist & Wiksell International.
Jammer, Max (1974). *The Philosophy of Quantum Mechanics: The Interpretation of Quantum Mechanics in Historical Perspective*. New York: John Wiley & Sons.
IAU (2006). "Definition of a Planet in the Solar System." http://www.iau.org/static/reso-lutions/-Resolution_GA26-5-6.pdf.
Johansson, Lars-Göran (2005). "The Nature of Natural Laws." In Jan Faye, Paul Needham, Uwe Scheffler & Max Urchs (eds), *Nature's Principles*. Series: *Logic, Epistemology, and the Unity of Science*, **4**: 151–166.
Johnson-Laird, Phillip N. (1983). *Mental Models: Towards a Cognitive Science of Language, Inference, and Consciousness*. Cambridge, MA: Harvard University Press.
Keil, Frank C. (2006). "Explanation and Understanding." In *Annual Review of Psychology*, **57**: 227–254.
Keil, Geert (2005). "How the Ceteris Paribus Laws of Physics Lie." In Jan Faye, Paul Needham, Uwe Scheffler & Max Urchs (eds), *Nature's Principles*. Series: *Logic, Epistemology, and the Unity of Science*, **4**: 167–200.
Kellert, Stephen H., Helena E. Longino & C. Kenneth Waters (eds) (2006). *Scientific Pluralism*. Minnesota Studies in the Philosophy of Science, XIX. Minneapolis: University of Minnesota Press.
Kim, Jaegwon (2006). *The Philosophy of Mind*. Cambridge, MA: Westview Press.
Kitcher, Philip (1981). "Explanatory Unification." In *Philosophy of Science*, **48**: 507–531.
Kitcher, Philip (1989). "Explanatory Unification and the Causal Structure of the World." In Philip Kitcher & Wesley C. Salmon (eds), *Scientific Explanation*. Series: *Minnesota Studies in the Philosophy of Science XIII*.
Kitcher, Philip & Wesley C. Salmon (1987). "van Fraassen on Explanation." In *Journal of Philosophy*, **84**: 315–330.
Kitcher, Philip & Wesley C. Salmon (eds) (1989). *Scientific Explanation*. Series: *Minnesota Studies in the Philosophy of Science XIII*. Minneapolis: Minnesota University Press.
Kragh, Helge S. (1990). *Dirac: A Scientific Biography*. Cambridge: Cambridge University Press.
Kragh, Helge S. (2011). *Higher Speculations: Grand Theories and Failed Revolutions in Physics and Cosmology*. Oxford: Oxford University Press.
Kragh, Helge (2012). *Niels Bohr and the Quantum Atom. The Bohr Model of Atomic Structure 1913–1925*. Oxford: Oxford University Press.
Kuhn, Thomas S. (1962/1969). *The Structure of Scientific Revolutions*. Chicago: The University of Chicago Press.
Kuhn, Thomas S. (1977). "Objectivity, Value Judgment, and Theory Choice." In his *The Essential Tension*. Chicago: The University of Chicago Press, 340–352.

Kuhn, Thomas S. (1991/2000). "The Natural and the Human Sciences." In D. R. Hiley, J. F. Bohman & R. Shusterman (eds), *The Interpretive Turn: Philosophy, Science, and Culture*. Ithaca, NY: Cornell University Press: 17–24. Reprinted in his *The Road Since Structure. Philosophical Essays 1970–1993, with an Autobiographical Interview*. Edited by J. Conant & J. Haugeland. Chicago and London: Chicago University Press, 216–223.

Leslie, John (1989). *Universes*. London: Routledge.

Levinson, Stephen C. (1997). "From Outer to Inner Space: Linguistic Categories and Non-linguistic Thinking." In J. Nuyts & E. Pederson (eds), *Language and Conceptualization*. Cambridge: Cambridge University Press, 13–45.

Lewis, David (1973). *Counterfactuals*. Cambridge, Mass.: Harvard University Press.

Lewis, David (1986). "Causal Explanation." In his *Philosophical Papers*. Oxford: Oxford University Press.

Lipton, Peter (2004). *Inference to the Best Explanation*. Second edition. New York: Routledge.

Lowe, Jonathan (1995). "The Truth About Counterfactuals." In *Philosophical Quarterly*, **45**: 41–59.

Lucy, John A. & Gaskins, Suzanne (2001). "Grammatical Categories and the Development of Classification Preferences: A Comparative Approach." In Melissa Bowerman & Stephen C. Levinson (eds), *Language Acquisition and Conceptual Development*. Cambridge: Cambridge University Press.

Machamer, P., Darden, L., & Craver, C. (2000). "Thinking About Mechanisms." In *Philosophy of Science*, **67**: 1–25.

Mackie, John L. (1974). *The Cement of the Universe*. Oxford: Clarendon Press.

Markman, A. B. & Dietrich, E. (2000). "Extending the Classical View of Representation." In *Trends in Cognitive Sciences*, **4** (12): 470–475.

Marschall, Laurence A. (1988). *The Supernova Story*. New York: Plenum Press.

Marzluff, John M. & Angell, Tony (2012). *Gifts of the Crow: How Perception, Emotion, and Thought Allow Smart Birds to Behave like Humans*. New York: Free Press.

Maslen, Cei (2004). "Causes, Contrasts, and the Nontransitivity of Causation." wam.umd.edu.

Mellor, D. Hugh (1995). *Facts of Causation*. London: Routledge.

Millikan, Ruth G. (1984). *Language, Thought, and Other Biological Categories*. Cambridge, Mass.: MIT Press.

Moyal, Ann (2001). *Platypus: The Extraordinary Story of How a Curious Creature Baffled the World*. Crows Nest: Allen & Unwin.

Møller, Christian (1972). *The Theory of Relativity*. Oxford: Clarendon Press.

Neurath, Otto, Carnap, Rudolf & Morris, C. (1938). *The International Encyclopedia of Unified Science*. Chicago: University of Chicago Press.

Newton, Isaac (1726). *Philosophiae Naturalis Principia Mathematica*, Third edition, I. Bernard Cohen & Anne Whitman's 1999 translation. Berkeley: University of California Press.

Newton-Smith, W. H. (2000). "Explanation." In W. H. Newton-Smith (ed.), *A Companion to the Philosophy of Science*. Oxford: Blackwell.

Nickerson, Raymond S. (1985). "Understanding Understanding." In *American Journal of Education*, **93**: 201–239.

Northcott, Robert (2008). "Causation and Contrast Classes." In *Philosophical Studies*, **139** (1): 111–123.

Ouattara, Karim, Lemasson, Alban & Zuberbühler, Klaus (2009). "Generating Meaning with Finite Means in Campbell's Monkeys." In *Proceedings of the National Academy of Sciences*, **106**: 48, 7 December 2009.
Perkins, David N. (1986). "Art as Understanding." In *Journal of Aesthetic Education*, **22**: 111–131.
Perlmutter, Saul (2003). "Supernova, Black Energy, and the Accelerating Universe." In *Physics Today*, **56** (4): 53–60.
Perlmutter, Saul *et al.* (1999). "Measurements of Ω and Λ from 42 High-redshift Supernovae." In *The Astrophysical Journal*, **517**: 565–586.
Persson, Johannes & Ylikoski, Petri (eds) (2007). *Rethinking Explanation*. Series: *Boston Studies in the Philosophy of Science*, **252**. Dordrecht: Springer..
Poincaré, Henri (1905/1952). *Science and Hypothesis*. New York: Dover Publication.
Polanyi, Michael (1958). *Personal Knowledge: Towards a Post-Critical Philosophy*. Chicago: The University of Chicago Press.
Polanyi, Michael (1966). *The Tacit Dimension*. London: Routledge.
Psillos, Stathis (2002). *Causation and Explanation*. Chesham: Acumen Press.
Psillos, Stathis (2007). "Causal Explanation and Manipulation." In Johannes Persson & Petri Ylikoski (eds), *Rethinking Explanation*. Series: *Boston Studies in the Philosophy of Science*, **252**: 93–107.
Putnam, Hilary & Oppenheim, Paul (1958). "Unity of Science as a Working Hypothesis." In H. Feigl, M. Scriven & G. Maxwell (eds), *Minnesota Studies in the Philosophy of Science, Vol. II*. Minneapolis: University of Minnesota Press, 3–36.
Richardson, Allan W. (2006). "The Many Unities of Science: Politics, Semantics, and Ontology." In Stephen H. Kellert, Helena E. Longino & C. Kenneth Waters (eds), *Scientific Pluralism*. Series: *Minnesota Studies in the Philosophy of Science*, **XIX**: 1–25.
Rosen, R. (2000). *Essays on Life Itself*. New York: Columbia University Press.
Rowlands, M. (2010). *The New Science of the Mind*. Cambridge, MA: MIT Press.
Ruben, David-Hillel (1990). *Explaining Explanation*. Oxford: Oxford University Press.
Rupert, R. D. (2009). *Cognitive System and the Extended Mind*. Oxford University Press.
Rutherford, Ernest (1911). "The Scattering of Alpha and Beta Particles by Matter and the Structure of the Atom." In *Philosophical Magazine*, **21**: 669–688.
Russell, Bertrand (1912). "On the Notion of Cause." In his *Mysticism and Logic and Other Essays*. London: G. Allen & Unwin, 1917.
Ryle, Gilbert (1949). *The Concept of Mind*. London: Hutchinson.
Salmon, Wesley (1984). *Scientific Explanation and the Causal Structure of the World*. Princeton: Princeton University Press.
Salmon, Wesley (1989). "Four Decades of Scientific Explanation." In Philip Kitcher & Wesley C. Salmon (eds), *Scientific Explanation. Minnesota Studies in the Philosophy of Science, vol. XIII*. Minneapolis: University of Minnesota Press.
Salmon, Wesley (1994). "Causality Without Counterfactuals." In *Philosophy of Science*, **61**: 297–312.
Salmon, Wesley (1998). *Causality and Explanation*. New York: Oxford University Press.

Schaffer, Jonathan (2005). "Contrastive Causation." In *Philosophical Review*, **114** (3): 327–358.
Schurz, Gerhard & Lambert, Karel (1994). "Outline of a Theory of Scientific Understanding." In *Synthese*, **101**: 65–120.
Schweder, Rebecca (2004). *A Unificationist Theory of Scientific Explanation*. Lund: University of Lund.
Scriven, Michael (1959). "Explanation and Prediction in Evolutionary Theory." In *Science*, **130**: 477–482.
Scriven, Michael (1962). "Explanation, Predictions and Laws." In H. Feigl & G. Maxwell (eds), *Scientific Explanation, Space and Time*. Series: *Minnesota Studies in the Philosophy of Science*, III.
Scriven, Michael (1975). "Causation as Explanation." In *Nous*, **9**: 3–16.
Searle, John R. (1969). *Speech Acts: An Essay in the Philosophy of Language*. Cambridge: Cambridge University Press.
Searle, John R. (1978). "Literal Meaning." In *Erkenntnis*, **13**: 207–224.
Sellars, Wilfrid (1956/1997). *Empiricism and the Philosophy of Mind*. Cambridge, Mass.: Harvard University Press. Study Guide by Robert Brandom. First printed in Herbert Feigl & Michael Scriven (eds), *Minnesota Studies in the Philosophy of Science, Volume I: The Foundations of Science and the Concepts of Psychology and Psychoanalysis*. Minneapolis: University of Minnesota Press, 253–329.
Shapiro, Lawrence (2011). *Embody Cognition*. New York: Routledge Press.
Shifman, Mikhail (2012). "Reflections and Impressionistic Portrait at the Conference. Frontiers Beyond the Standard Model." FTPI, October 2012. arXiv: 1211.0004v1. 31 October 2012.
Sintonen, Matti (1989). "Explanation: In Search of the Rationale." In Philip Kitcher & Wesley C. Salmon (eds), *Scientific Explanation. Minnesota Studies in the Philosophy of Science, Vol. XIII*. Minneapolis: University of Minnesota Press.
Sklar, Lawrence (1982). "Saving the Noumena." In *Philosophical Topics*, **13**: 49–72.
Slipher, Vesto (1913). "The Radial Velocity of the Andromeda Nebula." In *Lowell Observatory Bulletin*, **2** (8).
Slipher, Vesto (1917). "Nebulæ." In *Proceeding of American Philosophical Society*, **56**: 403–409.
Slurink, Pouwel (1996). "Back to Roy Wood Sellars: Why His Evolutionary Naturalism Is Still Worthwhile." In *Journal of History of Philosophy*, **34** (3): 425–449.
Slobodchikoff, C. N., Perla, B. S. & Verdolin J. L. (2009). *Prairie Dogs: Communication and Community in an Animal Society*. Cambridge, Mass.: Harvard University Press.
Sober, Elliot (1984). *The Nature of Selection*. Cambridge, Mass.: MIT Press.
Sterelny, K. & Griffiths, P. E. (1999). *Sex and Death: An Introduction to Philosophy of Biology*. Chicago: University of Chicago Press.
Suárez, Mauricio (2003). "Scientific Representation: Against Similarity and Isomorphism." In *International Studies in the Philosophy of Science*, **17** (3): 225–244.
Teller, Paul (2001). "Twilight of the Perfect Model." In *Erkenntnis*, **55**: 393–415.
Trout, J. D. (2002). "Scientific Explanation and the Sense of Understanding." In *Philosophy of Science*, **69**: 212–233.

Tschaepe, Mark D. (2009). "Pragmatics & Pragmatic Considerations in Explanation." In *Contemporary Pragmatism*, 6 (2): 25–44.
van Fraassen, Bas C. (1977). "The Pragmatics of Explanation." In *American Philosophical Quarterly*, 14: 143–150.
van Fraassen, Bas C. (1980). *The Scientific Image*. Oxford: Clarendon Press.
van Fraassen, Bas C. (1989). *Laws and Symmetry*. Oxford: Clarendon Press.
van Fraassen, Bas C. (2008). *Scientific Representation*. Oxford: Clarendon Press.
van Fraassen, Bas C. (2009). "Essay Review: Objectivity, Invariance, and Convention: Symmetry in Physical Science." In *Studies in History and Philosophy of Modern Physics*, 40: 84–87.
van Fraassen, Bas C. & Sigman, Jill (1993). "Interpretation in Science and in the Arts." In George Levine (ed.), *Realism and Representation*. Madison: The University of Wisconsin Press, 73–99.
Walton, D. (2004). "A New Dialectical Theory of Explanation." In *Philosophical Explorations*, 7: 71–89.
Weber, Erik (1996). "Explanation, Understanding, and Scientific Theories." In *Erkenntnis*, 44: 1–23.
Williams, Michael (2001). *The Problems of Knowledge*. Oxford: Oxford University Press.
Wilson, M. (2002). "Six Views of Embodied Cognition." In *Psychonomic Bulletin and Review*, 9 (4): 625–636.
Wolkenhauer, O. (2001). "Systems Biology: The Reincarnation of Systems Theory Applied in Biology." In *Briefings in Bioinformatics*, 2 (3): 258–270.
Woodward, James (2003). *Making Things Happen: A Theory of Causal Explanation*. New York: Oxford University Press.
Wright, Larry (1973/1999). "Functions." In *Philosophical Review*, 82: 139–168. Reprinted in D. J. Buller (ed.), *Function, Selection and Design*. New York: SUNY Press.
Ylikoski, Petri (2001). *Understanding Interests and Causal Explanation*. Helsinki: University of Helsinki.
Zagzebski, Linda (2001). "Recovering Understanding." In Matthias Steup (ed.), *Knowledge, Truth, and Duty: Essays on Epistemic Justification, Responsibility, and Virtue*. New York: Oxford University Press, 235–251.
Zeleňák, Eugen (2009). "On Explanatory Relata in Singular Causal Explanation." In *Theoria*, 75 (2009): 179–195.
Zuberbühler, Klaus (2000a). "Causal Knowledge of Predators' Behaviour in the Wild Diana Monkeys." In *Animal Behaviour*, 59: 209–220.
Zuberbühler, Klaus (2000b). "Causal Cognition in a Non-human Primate: Field Playback Experiments with Diana Monkeys." In *Cognition*, 75: 195–207.
Zuberbühler, Klaus (2002). "A Syntactic Rule in Forest Monkey Communication." In *Animal Behaviour*, 63: 293–299.

Index

abduction, 69, 75, 76–7, 79–80, 83, 277, 278, 300
abstraction, 14, 19, 51, 80, 81, 89, 97, 110, 132, 139, 140, 141, 162, 190, 277, 305, 307, 314
Achinstein, Peter, ix, 26, 203, 207–9, 211, 248–50, 292, 311, 313
Anderson, John, 308
Anderson, M.L., 295
Angell, Tony, 296, 297, 305
Anscombe, Elisabeth, 149, 307
Aristotle, ix, 39, 71, 82, 108, 123, 211, 253, 290, 311, 313
Arzeliès, H., 266
Austin, John L., 154, 155, 248, 250–1, 313

Baade, Wilhelm, 67, 68, 69–71, 300
Barsalou, L., 295
Basbøll, Thomas, 312
Becquerel, Henri, 72
Bennett, Christopher, 128
Bird, Alexander, 307
Bitzer, Lloyd, xii, 230, 256–7, 312, 313
Bloor, David, 298–9
Bohm, David, 16–17, 18
Bohr, Niels, 3, 11, 15, 16, 17, 18, 19–21, 72, 74, 77, 78, 82, 83, 89, 247, 260, 290, 291, 300–1, 312
Boorse, Christopher, 176–7, 309
Brahe, Tycho, 67, 70
Brandom, Robert, 302
Bridgman, Percy W., 190
Broad, C.D., 314
Bromberger, Sylvain, 126, 211, 311
Brown, Michael E., 65
Bugnyar, Thomas, 296

Calvo, P., 295
Carnap, Rudolph, 125, 287, 314
Carter, Brandon, 212

Cartwright, Nancy, 99–100, 103, 104, 119, 128–9, 148–9, 186, 273, 302, 303, 304, 307, 310
causal statement, *passim*
  truth conditions of, 157–9
causation,
  in biology, 149–54
  and counterfactuals, 14, 142, 146, 149, 157, 191–2, 291, 307
  without counterfactuals, 14, 184
  downward, 152–3
  and manipulation, 140, 141, 306, 307
  mechanism, 4, 12, 13, 33, 53, 110, 11, 117, 132, 133, 137, 146, 147, 150, 155, 166, 171, 192, 214, 250, 304
  notion of, 136–42, 184, 306
  in physics, 147–9
  schema of, 56, 58, 137, 139, 156, 306
cause *vs.* causal process, 148, 151
Chadwick, James, 67
classification, 10, 61, 62, 65, 71, 72, 75, 78–9, 80, 81, 300
cognition,
  adapted, 46–7, 48, 56, 57
  distinct from comprehension, 45
  embedded, 41–6, 90
  embodied, 41–6, 48, 90, 91, 294, 295
  knowledge and understanding, 45
  reflective, 48, 51, 145
cognitive schemata, 26, 55, 56–7, 136, 137, 138, 139, 141, 298, 314
Collin, Finn, 290, 310, 314
Collins, Barry, 309
comprehension, 1, 39, 40, 43, 45, 46, 47, 62, 71, 78, 122, 141, 145, 195, 294
  see also understanding

concept, *passim*
  definition of, 47, 88, 139, 296
  and representation, 88–9
Craver, Carl F., 150, 307
Culver, Roger B., 304
Curtis, Herbert D., 299
Cushing, James T., 4, 17–18, 289, 291

Danto, Arthur, 190, 310
Darden, Lindley, 150, 307
Darwin, Charles, vii, 264
Davidson, Donald, 14–15, 155, 291
Davis, J.I., 295
Debs, Talal A., 92–8, 302
De Donato-Rodríguez, Xavier, 310
definition, *passim*
  implicit, 105, 107, 119, 303
  nominal, 70–1
  real, 70–1
  semantic, 76, 80–3
Dennett, Daniel, 162, 163, 308
De Regt, Henk W., 3, 4, 23, 34–6, 289, 291, 293
Descartes, Réne, vii, 88, 287
Dewey, John, 23, 41, 294
Dewitt, Bryce, 21
Dieks, Dennis, 4, 12, 23, 289, 290, 291, 293, 304
Dietrich, E., 295
Dowe, Phil, 14, 291, 303, 306
Dray, William H., 210, 311
Dretske, Fred, 90, 302
Dupré, John, 150, 151, 273, 300, 307

Edelman, Gerald, 47, 295
Einstein, Albert, 4, 10–12, 16, 21, 29, 41, 54, 74, 83, 86, 105, 108, 121, 247, 260, 266, 274, 290, 304, 312
Einstein's mass-energy relation, 67, 68, 70
Ellis, Brian, 104, 106, 303
Emmeche, Claus, 307
Everett, Hugh, 21
explanation, *passim*
  actual, 264–5
  analytic, 122
  asperlocutionary act, 250–1, 264
  asymmetrical, 126, 143, 201, 202–3, 237–8, 254–5, 287
  causal, 13–15, 18, 57, 11, 113, 115–17, 118, 120, 121, 123, 127, 130, 132, 133, 136–61, 164–7, 176, 178, 179, 183, 186, 187, 190–2, 194, 206, 211, 237–8, 243, 250, 261–2, 304, 305, 306, 307, 310
  complete, 143, 198, 200–1, 208–9, 235, 249
  covering law model of, 3, 4, 5, 9, 10, 26, 50, 100, 112, 117, 118, 126–34, 146, 172, 186, 187, 199, 214, 217, 267
  credibility of, 133, 243, 285–6
  deductive-nomological model of, 12, 64, 117, 124, 126, 127, 130, 133, 145, 271, 309
  see also the covering law model
  description-giving, 225
  epistemic theory of, 186, 241–4
  evolutionary, 171 ff., 178
  formal-logical theory of, 10, 12, 186, 187, 188, 190, 193, 203, 242, 244, 263, 266, 286, 313
  functional, 13, 15, 153, 161, 162, 164–6, 171–81, 194, 195, 309
  genre of, 261
  as illocutionary act, 248–51
  inductive-statistical model of, 117–18
  inference to the best, 75, 76, 278
  and intention, 192, 194, 208–9, 228, 233, 247–50, 255–6
  intentional, 13, 15, 16, 154, 162, 163, 165, 167, 190, 312
  and interest, 154, 184–5, 193, 197–8, 244–8, *passim*
  and interpretation, *see* interpretation
  interpretive 187
  level of, 192–3
  nomic, 50, 117–22, 123
  ontic theory of, 13–15, 183, 217, 266
  possible, 123, 264–5
  pragmatic theory of, 33, 160, 181, 183–209, 241

and prediction, 124
reason-giving, 225
reduction-to-answer, 185
reduction-to-intention, 185
rhetorical theory of, 241–69
standards of, 254
structural, 15, 161, 164, 168–71
successful, 244, 250–2, 255, 258, 263, 267–9, 286, 288
synthetic, 122
teleological, 164–5, 168, 171, 176, 181, 287, 309–10
and truth, 249, 252, 253–4, 263–4, *passim*
and understanding, 33, 34, *passim*
explanation-seeking question, *passim*
how-question, 67, 210, 211, 212, 219, 220, 227, 233–4, 236, 238
what-question, 67, 210, 218, 220, 222, 225, 226–7, 229–30, 231–3, 237, 238, 234, 237, 238, 239
when-question, 218, 220, 221–2
where-question, 218, 220
which-question, 210, 234
why-question, 67, 116, 122, 124, 143, 154, 163, 166–8, 187–8, 193–5, 196–7, 198, 199–201, 203–4, 209, 210–40
explanatory force, 3, 100, 125, 214, 234, 253, **262–9**, 286, 287
explanatory relevance, 125, 144, 151, 203–5, 237–8, 250, **254**, 255, 259–62, 263, 266, 267, 268, 269, 285, 299
explanatory situation, 12, 132, 160, 228, 230, 235, 256, 258, 259, 260, 261, 262, 269

Folse, Henry, 290, 300, 308
Fowler, William, 68, 300
Franklin, Anna, 295
Friedman, Michael, 4–6, 289, 290
Frigg, Roman, 301

Galileo, Galilei, 3, 287
Gärdenfors, Peter, 242–3, 244, 248, 249, 312
Garfinkel, Alan, 309
Gaskins, Suzanne, 295

Geiger, Hans, 73, 74
Ghirardi, GianCarlo, 18
Giere, Ronald N., 92, 96, 101, 102, 111, 273, 301, 302, 303, 313–14
Gjelsvik, Olav, 306
Glennan, Stuart, 150, 307
Gornila, A., 295
Grice, Paul, 313
Griffiths, P. E., 307
Grimm, Stephen R., 29, 30–1, 33, 292

Hacking, Ian, 82, 130, 300, 304
Hållsten, Henrik, 313
Hanson, Norwood Russell, 104, 105, 190, 191, 303
Hansson, Bengt, 187, 188, 197, 310
Hartmann, Stephan, 301
Hawking, Stephen, W., 309
Heelan, Patrick, 303
Hegel, Friedrich, 88
Heinrich, Bernd, 296
Heisenberg, Werner, 11, 16, 17, 18, 77, 83, 89, 300
Hempel, ix, Carl G., 2, 3, 5, 9, 10, 13, 33, 64, 70, 71, 76, 80, 112, 114, 117, **122–7**, 130, 131, 133, 134, 143, **172–4**, 175, 181, 190, 195, 198, 199, 201, 204, 207–9, 210, 216, 217, 228, 234, 239, 271, 286, 289, 290, 300, 304, 305, 306, 308, 309, 310, 311, 312, 313, 314
Hilbert, David, 290
Holton, Gerald, 245–7, 312
Hoyle, Fred, 68, 300
Hull, David L., 127–8, 304
Hume, David, 14, 122, 125, 137, 138
hypothesis, *passim*
  discovery of, 75–9
  scientific, 75–84
  semantic, 80–4

Ibn al-Haytham, 277–8
induction, 69, 71, 75, 76, 80, 81, 140, 214, 277, 278
interpretation, *passim*
  as construction of meaning, 60 ff.
  determinative, 62–3, 66, 67
  and experience, 90–1

interpretation, *passim* – *continued*
  as explanation of meaning, 60 ff.
    67, 69
  and intention, 63
  investigative, 62–3, 65, 67, 75–6
  and understanding, 60–1

Johansson, Lars-Göran, 105, 120,
  303, 304
Johnson-Laird, Phillip N., 214

Kant, Immanuel, 29, 88, 287
Keil, Frank C., 297, 306
Keil, Geert, 119, 303
Kim, Jaegwon, 315
Kitcher, Philip, 4, 9–10, 203–7,
  290, 311
knowledge, *passim*
  empirical, 54, 155, 257, 259, 276
  practical, 41–3
  propositional, 27, 39, 41–2
  scientific, 32, 48, 50, 137, 196, 212,
    257, 290, 311, 314
  tacit, 2, 35, 43–4, 59, 89
Kragh, Helge, 300, 311
Kuhn, Thomas S., 15, 54, 83, 85, 89,
  104–5, 260, 303, 313

Lakatos, Imre, 89
Lamarck, Jean-Baptiste, 72, 264, 285
Lambert, Karel, 289
Laplace, Pierre-Simon, 299
Laudan, Larry, 85, 89, 248
law, *passim*
  causal, 13, 101, 102, 103, 117, 118,
    119, 123, 124, 128–32, 143,
    149, 159–60, 275–6, 304–5
  ceteris paribus, 100–2, 103, 109,
    110, 119–20, 128–9, 132, 143,
    145, 151, 157–61, 186, 190,
    275, 302
  of co-existence, 123
  conservation, *see* principle
  counterfactuals, 107–8
  empirical, 5, **99**, 102, 103, 105, 105,
    121, 122
  functional, 102
  fundamental, 5, **99–100**, 102, 103,
    118–20, 186, 232, 263, 276, 303

phenomenological, **99–100**, 102,
  104, 109
structural, 102
of succession, 123
theoretical 6, 7, **99**, 102, 103, 105,
  107–8, **119**, 122, 128, 129, 289,
  290, 303
Leslie, John, 309
Levinson, Stephen C., 295
Lewis, David, 6–7, 146, 191, 290, 306
Lipton, Peter, 26, 292
Lister, Joseph, 202
Locke, John, 88, 292
Longino, Helena E., 273
Lorentz, H.A., 12, 81, 121
Lowe, Jonathan, 291
Lucy, John A., 295

Machamer, Peter, 150, 307
Mackie, John L., 146, 157, 190,
  306, 308
Markman, A.B., 295
Marschall, Laurence A., 299–300
Marsden, Ernest, 73, 74
Marzluff, John M., 138, 296, 297, 305
Maslen, Cei, 306
meaning,
  construction of, 61
  explanation of, 61
Mellor, D. Hugh, 146, 307
Mill, John, S., 6, 314
Millikan, Ruth G., 176–8, 179, 309
Minkowski, Hermann, 20
Minkowski, Rudolph, 67
model, *passim*
  vs. hypothesis, 86
  inconsistent, 112, 170, 193
  and purpose, 111–13
  and representation, 92–8, 108–13
  vs. theory, 103, 108–9
Moyal, Ann, 300
Møller, Christian, 266–7, 313

Nagel, Ernst, 210, 239
Nagel, Thomas, 40
Newton, Isaac, 3, 4, 8, 29, 82
Newton's laws, *passim*
  status of, 104–8
Newton-Smith, W. H., xi, 1, 289

Nickerson, Raymond S., 37, 38, 294
Northcott, Robert, 306

Oersted (Ørsted), Hans C., 130
Oppenheim, Paul, 5, 123, 124, 134, 181, 216–17, 279, 283, 289, 314
Ott, H., 266
Ouattara, Karim, 295

Pasteur, Louis, 202, 293
Peirce, Charles S., 23, 76, 300
Penzias, Arno, 129
Perkins, David N., 37–8, 294
Planck, Max, 74, 81, 266
Plato, 166, 290
pluralism
  epistemic, 274
  account, **274**
Podolsky, Boris, 21
Poincaré, Henri, 104, 303
Polanyi, Michael, 35, 43–4, 294
Popper, Karl R., vii, 10, 64, 69, 76, 277, 287
Pouchet, Felix A., 293
practice
  explanatory, 134–5, 143, 170, 181, 231, 252, 255, 286
  scientific, 3, 22, 42, 45, 53, 64, 70, 85, 86, 89, 90, 91, 93, **103**, 127, 150 220, 244, 270, 277, 284, 286, 301
principle
  anthropic, 145, 211–12, 309–10. 311
  causal, 16, 145
  conservation, 77, 102, 118–20, 147, 164, 306
  as constitutive rule, 8, 102, 121
  correspondence, 17, 300–1
  of emergence, 283
  of evolution, 279
  of ontogenesis, 279
  of reduction, 280
Psillos, Stathis, 146, 304, 306, 307
Putnam, Hilary, 279, 283, 314

quantum mechanics, 11, 16–22, 72–8, 83, 89, 148, 186, 193, 247, 260, 274
Quine, Willard V.O, 270

Ramsay, Frank P., 6
Ray, John, 71–2
Redhead, Michael L.G., 92–8, 302
Reichenbach, Hans, 64, 151, 303
representation, *passim*
  conceptual, 54, 61, 62, 295
  as intentional activity, 85–9, 92, 97–9
  isomorphic, 93, 95–8
  vs. presentation, 89
  semantic, 25, 295
  similarity, 92, 95, 96–9
  structure, 92–8
  and understanding, 91–8
Rescher, Nicolas, 23
rhetoric, 93, 192
rhetorical situation, 255–9, 266, 312, 313
Richardson, Allan W., 314
Rimini, Alberto, 18
Rorty, Richard, 301
Rosen, Nathan, 21
Rosen, Robert, 152, 307
Rowlands, M., 295
Royds, Thomas, 73
Rupert, R.D., 294
Russell, Bertrand, 123, 125, 287, 290, 304
Rutherford, Ernest, 18, 72–4, 115, 300
Ryle, Gilbert, 41, 294

Salmon, Wesley, 4, 13, 14, 22, 52, 112, 122, 146, 155, 163, 174, 184, 203–207, 210, 211, 229, 236, 249, 250, 289, 290, 291, 297, 303, 304, 306, 308, 309, 311, 312, 313
Schaffer, Jonathan, 306
Schramm, David, 212
Schurz, Gerhard, 289
Schweder, Rebecca, 293
scientific pluralism,
  epistemic, 272–6
  and methodology, 276–9
  vs. ontological monism, 279–84
  pragmatic, 272, 284
Scriven, Michael, 23, 26, 33, 37, 127, 133, 183, 201, 210, 216, 217–18, 292, 283, 294, 311, 312

Searle, John R., 43, 189, 198, 294, 310, 313, 314
Sellars, Roy W., xiv
Sellars, Wilfrid, 90, 91, 301, 302
Semmelweis, Ignaz, 202
Shifman, Mikhail, 314
Sintonen, Matti, 242, 243–4, 248, 312
skills, 2, 4, 21, 33, 34–7, 41, 43, 44, 49, 60, 88, 89, 90, 259, 293, 296
Sklar, Lawrence, 107, 303
Slipher, Vesto, 66, 299
Slobodchikoff, C.N., 296
Slurink, Pouwel, xiv
Sober, Elliot, 309
Stalnaker, Robert, 191
Sterelny, K., 307
Strawson, Peter, 154
Suárez, Mauricio, 96, 97, 98, 302

Teller, Paul, 301
theory, *passim*
  constructive, 10–12
  linguistic approach, 103–8
  principle-, 10–12
  semantic view, 109
Thomson, J.J., 72, 73
Tombaugh, Clyde, 65
Tononi, Giulio, 295
Trout, J.D., 3–4, 27, 34, 289, 292, 293
Tschaepe, Mark D., 308, 310

understanding, *passim*
  ability view of, 35–7
  causal, 15, 122, 136, 142–7, 154, 156, 163, 194, 297
  as comprehension, 45
  content view of, **26–34**, 35, 38
  embedded, 41 ff, **44**, 45, 59, 85
  embodied, **44–6**, 48, 49, 57, 59, 90, 91, 141
  expectation, 2, 5, 6, 27, 133–4, 137, 140, 289–90, 304
  experiential, 48–9
  explanatory, 49–50, 12, 15, 17, 22, 25, 29, 49–50, 51, 55, 57–8, 179, 224, 297, 298
  imaginative, 49
  instrumental, 49
  interpretive, 50, 91
  and intelligibility, 1, 4, 5, 15, 17, 25, 34–7, 51, 53, 56, 150
  and knowledge, 26–34, 37–8, 41, 42, 45, 201, 250
  level of, 41, 48–51
  linguistic, 49
  metaphysical 50,
  norms of, 4, 15, 51–6
  and objectivity, 3–4
  ontological view of, 12–15
  organization view of, **26**, 32, **37–41**, 45, 55, 139, 296, 298
  as part of cognition, 45–7
  reflective, 30, 45, 46, 49, **54–5**, 57, 58, 142
  schemata, 56–9, 139, 141, 287
  sense of, 3, 27, 181
  and skills, 34–7
  and subjectivity, 3–4, 28, 181
  and truth, 3–4, 26–8, 31, 32, 37, 42, 58
  as unification, 4–10, 12, 33, 34, 50, 52, 53, 11, 241, 247, 290
  as visualization, 14–17, 19
unity of science, 272–3, 276

van Fraassen, Bas, ix, xii, 7, 23, 87–8, 90, 97, 126, 184, 185, 186, 190, 191, **195–209**, 210, 211, 212, 222, 232, 233, 234, 235, 241, 290, 301, 302, 304, 310, 311, 312
Villard, Paul, 72

Walton D., 204
Weber, Erik, 289
Weber, Tulio, 18
Whorf, Benjamin, L., 295
Williams, Michael, 292
Wilson, M., 205

Wilson, Robert W., 129
Wittgenstein, Ludwig, 98, 105, 252
Wolkenhauer, O., 307
Woodward, James, 146, 307
Wright, Larry, 165, 176–7, 179, 309

Ylikoski, Petri, 307
Yukawa, Hideki, 82

Zagzebski, Linda, vii, 29, 30, 38–40, 41, 292, 294
Zamora-Bonilla, Jesús, 310
Zeleňák, Eugen, 304
Zuberbühler, Klaus, 295, 297–8
Zwicky, Fritz, 67, 68, 69–71, 300

CPSIA information can be obtained at www.ICGtesting.com
Printed in the USA
LVOW03*2126090215

426377LV00009B/81/P